BASIC
STATISTICAL
TABLES

EDITOR

WILLIAM H. BEYER, Ph.D.

*Professor of Mathematics and
Head of Department of Mathematics,
University of Akron,
Akron, Ohio*

Published by

THE CHEMICAL RUBBER CO.

18901 Cranwood Parkway, Cleveland, Ohio 44128

Preface

Statistics is the key technology of the present day.
It is an important component of scientific reasoning as well as an integral part of academic curricula, business, and technology. Statisticians and scientists working in diverse fields need an authoritative reference handbook of up-dated statistical tables to "aid" in the investigation and solution of present day problems involving space and undersea exploration, ecology, etc., where the importance of data collection and statistical analysis is basic.

The purpose of this publication is to present a format of Basic Statistical Tables that are logically arranged, documented, and readily accessible for use, as well as being moderately priced to the user. Special attention has been given to include those tables which are most commonly employed in all phases of human endeavor rather than highly specialized tables which are rarely used. For those who desire a more extensive collection of statistical tables, The Chemical Rubber Co. provides the Professional Edition of the Handbook of Tables for Probability and Statistics, now in its second edition (1968).

The Editorial Staff invites and welcomes suggestions and comments from users of this edition to assist in the continuous improvements of the contents.

WILLIAM H. BEYER

January 1971

Acknowledgments

Acknowledgment is made to the following authors, editors, and publishers whose material has been used in the Handbook of Tables for Probability and Statistics, and for which permission has been received.

AMERICAN SOCIETY FOR TESTING MATERIALS STP-15C;
  ASTM Manual on Quality Control of Materials (1951)
  XI.I—Factors for Computing Control Limits

AMERICAN STATISTICAL ASSOCIATION, JOURNAL OF
  Vol. 32 (1937) 349–386, W. E. Ricker
  III.6—Confidence Limits for the Expected Value of a Poisson Distribution
  Vol. 41 (1946) 557–566, W. J. Dixón and A. M. Mood
  X.I—Critical Values for the Sign Test
  Vol. 46 (1951) 68–78, F. J. Massey, Jr.
  Vol. 47 (1952) 425–441, Z. W. Birnbaum
  X.6—Critical Values for the Kolmogorov-Smirnov One-Sample Statistic
  Vol. 47 (1952) 583–621, W. H. Kruskal and W. A. Wallis
  X.8—Kruskal-Wallis One-Way Analysis of Variance by Ranks

BIOMETRIKA TRUSTEES; E. S. PEARSON AND H. O. HARTLEY,
  Cambridge University Press
  Biometrika
  Vol. 38 (1951) 112–130, E. S. Pearson and H. O. Hartley
  VI.2—Power Functions of the Analysis-of-Variance Tests
  Vol. 48 (1961) 151–165, H. L. Harter
  VII.1—Expected Values of Order Statistics from a Standard Normal
       Population
  Vol. 32 (1942) 301–310, E. S. Pearson and H. O. Hartley
  VII.6—Simple Estimates in Small Samples
  Vol. 34 (1947) 41–67, E. Lord
  VIII.4—Substitute t-Ratios
  Biometrika Tables for Statisticians

Vol. 1 (1962) 204–205	III.5 —Confidence Limits for Proportions
Vol. 1 (1962) 234–235	III.7 —Various Functions of p and q = 1 − p
Vol. 1 (1962) 135	IV.2 —Power Function of the t-Test
Vol. 1 (1962) 130–131	V.1 —Percentage Points, Chi-Square Distribution
Vol. 1 (1962) 157–163	VI.1 —Percentage Points, F-Distribution
Vol. 1 (1962) 166–171	VIII.1 —Probability Integral of the Range
Vol. 1 (1962) 165	VIII.2 —Percentage Points, Distribution of the Range
Vol. 1 (1962) 176–177	VIII.3 —Percentage Points, Studentized Range
Vol. 1 (1962) 138	IX.1 —Percentage Points, Distribution of the Correlation Coefficient When $\rho = 0$

Vol. 1 (1962) 140–141   IX.2 —Confidence Limits for the Population Correlation Coefficient

Vol. 1 (1962) 139   IX.3 —The Transformation $Z = \tanh^{-1} r$ for the Correlation Coefficient

Vol. 1 (1962) 211   X.10—Critical Values of Kendall's Rank Correlation Coefficient

Vol. 1 (1962) 165   XI.2 —Percentage Points of the Distribution of the Mean Deviation

INSTITUTE OF EDUCATION RESEARCH,
Indiana University, Bloomington, Indiana
Vol. 1, No. 2 (1953), D. Auble
   X.4—Critical Values of U in the Wilcoxon (Mann-Whitney) Two Sample Statistic

LEDERLE LABORATORIES
Some Rapid Approximate Statistical Procedures (1964) 28, F. Wilcoxon and R. A. Wilcox
   X.2—Critical Values of T in the Wilcoxon Matched-Pairs Signed-Ranks Test

MATHEMATICAL STATISTICS, ANNALS OF, D. L. BURKHOLDER, EDITOR
Vol. 27 (1956) 427–451, A. E. Sarhan and B. G. Greenberg
   VII.2—Variances and Covariances of Order Statistics
Vol. 17 (1946) 377–408, F. Mosteller
   VII.5—Percentile Estimates in Large Samples
Vol. 31 (1960) 1122–1147, H. L. Harter
   VIII.2—Percentage Points, Distribution of the Range
   VIII.3—Percentage Points, Studentized Range
Vol. 20 (1949a) 257–267, J. E. Walsh
   VIII.4—Substitute t-Ratios
Vol. 21 (1950) 112–116, R. F. Link
   VIII.5—Substitute F-Ratio
Vol. 18 (1947) 50–60, H. B. Mann and D. R. Whitney
   X.3—Probabilities for the Wilcoxon (Mann-Whitney) Two-Sample Statistic
Vol. 14 (1943) 66–67, C. Eisenhart and F. C. Swed
   X.5—Distribution of the Total Number-of-Runs Test
Vol. 23 (1952) 435–441, F. J. Massey, Jr.
   X.7—Critical Values for the Kolmogorov-Smirnov Two Sample Statistic
Vol. 9 (1938) 133–148, E. G. Olds
   X.9—Critical Values of Spearman's Rank Correlation Coefficient

McGRAW-HILL BOOK COMPANY
Selected Techniques of Statistical Analysis, C. Eisenhart, M. W. Hastay, W. A. Wallis
   (1947) 102–107   II.2—Tolerance Factors for Normal Distributions
   (1947) 284–309   V.3—Number of Observations for the Comparison of a Population Variance With a Standard Value Using

the Chi-Square Test
VI.3—Number of Observations Required for the Comparison of Two Population Variances Using the F-Test
Introduction to Statistical Analysis, 2nd Edition, W. J. Dixon, F. J. Massey, Jr.
(1957) 405–407   VII.6—Simple Estimates in Small Samples

OLIVER AND BOYD, LTD., EDINBURGH, SCOTLAND
Design and Analysis of Industrial Experiments, O. L. Davies
Research Vol. 1 (1948) 520–525
(1956) 606–607   IV.3—Number of Observations for t-Test of Mean
(1956) 609–610   IV.4—Number of Observations for t-Test of Difference Between Two Means
(1956) 613–614— V.3—Number of Observations for the Comparison of a Population Variance With a Standard Value Using the Chi-Square Test
Statistical Tables for Biological, Agricultural and Medical Research
R. A. Fisher, F. Yates
(1938) 46   IV.1—Percentage Points, Student's t-Distribution
(1938) 98–103   XII.7—Orthogonal Polynomials

THE ROYAL SOCIETY, LONDON, ENGLAND
Royal Society Mathematical Tables
Vol. 3 (1954) 2   XII.2—Number of Combinations

SPRINGER-VERLAG NEW YORK, INC.
Fünfstellige Funktionentafeln (1930), Hayashi, K.
XII.1—Number of Permutations

STANFORD UNIVERSITY PRESS
Tables of the Hypergeometric Probability Distribution, G. J. Lieberman and D. B. Owen (1961)
III.8—Hypergeometric Distribution

VIRGINIA POLYTECHNIC INSTITUTE, R. E. BARGMANN AND J. E. WHITE
"Some Contributions to the Evaluation of Pearsonian Distribution Functions" Research Sponsored by the National Institutes of Health, Epidemiology, and Biometry, Technical Report No. 1 (1960)
XII.8—Percentage Points of Pearson Curves

JOHN WILEY & SONS, INC.
Statistical Tables and Formulas, A. Hald
(1952) 44–45   V.2—Percentage Points, Chi-Square Over Degrees of Freedom Distribution Statistics and Experimental Design
Statistics and Experimental Design, N. L. Johnson, F. C. Leone
Vol. 1 (1964) 412   X.9—Critical Values of Spearman's Rank Correlation Coefficient

Table of Contents

PART II—NORMAL DISTRIBUTION

PART III—BINOMIAL, POISSON, HYPERGEOMETRIC, AND NEGATIVE BINOMIAL DISTRIBUTIONS

PART IV—STUDENT'S t-DISTRIBUTION

GREEK ALPHABET

Greek letter	Greek name	English equivalent	Greek letter	Greek name	English equivalent
A α	Alpha	a	N ν	Nu	n
B β	Beta	b	Ξ ξ	Xi	x
Γ γ	Gamma	g	O o	Omicron	ŏ
Δ δ	Delta	d	Π π	Pi	p
E ε	Epsilon	ĕ	P ρ	Rho	r
Z ζ	Zeta	z	Σ σ s	Sigma	s
H η	Eta	ē	T τ	Tau	t
Θ θ ϑ	Theta	th	Υ υ	Upsilon	u
I ι	Iota	i	Φ φ φ	Phi	ph
K κ	Kappa	k	X χ	Chi	ch
Λ λ	Lambda	l	Ψ ψ	Psi	ps
M μ	Mu	m	Ω ω	Omega	ō

I. Probability and Statistics

CENTRAL MEASURES

Here the range of i is from 1 to n. With each value x_i is associated a weighting factor $f_i \geq 0$ (such as the frequency, the probability, the mass, the reliability, or other multiplier).

N, the **total weight,** $= \Sigma f_i$.

$\bar{x}$, the **arithmetic mean,** $= \Sigma f_i x_i / N = \Sigma f_i x_i / \Sigma f_i$.

GM, the **geometric mean** (available when each x_i is positive), $= \sqrt[N]{\Pi x_i^{f_i}}$. Log $GM =$ $\Sigma f_i \log x_i / N$.

Mo, the **mode,** $=$ value among $(x_1, \ldots, x_n)$ having maximum associated f_i (usually obtained by interpolating after the date are graduated). For unweighted items, x_i, a mode is a value about which the values of x_i cluster most densely.

RMS, the **root-mean-square,** $= \sqrt{\Sigma f_i x_i^2 / N}$.

Md, the **median** (see below). For unweighted items, the median is the value, equaled or exceeded by exactly half of the values x_i in the given list. In case of a central pair, the median is usually taken as the arithmetic mean of this pair.

Mm, the **mid-mean** (see below). For unweighted items, the mid-mean is the arithmetic mean of the half-list obtained upon dropping out the highest quarter and lowest quarter of the items.

$\text{Cum} f \big|_X$, the value of "cumulative f" at X, $= \displaystyle\sum_{x_i < X} f_i$ (interpolation being used for X if necessary).

THE M-TILES

For **ungrouped data,** X is called the rth **m-tile** (or rth **m-tile mark**) ($r = 0$, 1, $\ldots$, m) if simultaneously, $\displaystyle\sum_{x_i < X} f_i / N \leq r/m$, and $\displaystyle\sum_{x_i > X} f_i / N \leq (m - r)/m$. In particular the zeroth m-tile is **min,** the minimal value among the list $(x_1, \ldots, x_n)$, and the mth m-tile is **max,** the maximal value among the list.

For **grouped data,** the rth m-tile mark, X, is such that

$$\text{Cum} f \big|_X = Nr/m, \quad (r = 0, 1, 2, \ldots, m).$$

$$\text{Cum} f \big|_{\min} = 0, \qquad \text{Cum} f \big|_{\max} = N.$$

In particular, certain intermediate ($0 < r < m$) m-tile marks are named as follows:

m	$r = 1$	2	3	$\cdots$
2	Md (median)			
3	T_1 (lower tertile)	T_2 (upper tertile)		
4	Q_1 (lower quartile)	Md	Q_3 (upper quartile)	
10	D_1 (first decile)	D_2	D_3	etc.
100	PC_1 (first percentile)	PC_2	PC_3	etc.

The term "rth **m-tile**" ($r = 1, \ldots, m$) is also used to denote the class interval extending from the $(r - 1)$st to rth m-tile mark as defined above.

$$Mm, \text{ the } \textbf{mid-mean,} = 2 \sum_{Q_1 \leq x_i \leq Q_3} f_i x_i / N = \sum_{Q_1 \leq x_i \leq Q_3} f_i x_i / \sum_{Q_1 \leq x_i \leq Q_3} f_i .$$

When each x_i is positive, and not all are equal, one always has $0 < \min < GM < \bar{x} < RMS < \max$.

For moderately-skewed distributions, one has approximately $Mo - \bar{x} = 3(Md - \bar{x})$, or $3Md = Mo + 2\bar{x}$.

MEASURES OF DISPERSION AND SKEWNESS

Here A is an arbitrary reference value, usually a convenient integral measure near $\bar{x}$.

ν_k', kth **moment about** A, $= \Sigma f_i(x_i - A)^k/N$, $(k = 0, 1, \ldots)$.

$\nu_0' = 1$, $\nu_1' = \bar{x} - A$. ν_2' as function of A is minimum for $A = \bar{x}$.

ν_k, kth moment about the origin, $= \Sigma f_i x_i^k/N$, $(k = 0, 1, 2, \ldots)$ $\nu_0 = 1$, $\nu_1 = \bar{x}$.

μ_k, kth **moment about** $\bar{x}$, $= \Sigma f_i(x_i - \bar{x})^k/N$, $(k = 0, 1, \ldots)$.

$$\mu_0 = 1,$$
$$\mu_1 = 0,$$
$$\mu_2 = \nu_2 - \nu_1^2 \ (\mu_2 = \textbf{variance}),$$
$$\mu_3 = \nu_3 - 3\nu_1\nu_2 + 2\nu_1^3,$$
$$\mu_4 = \nu_4 - 4\nu_1\nu_3 + 6\nu_1^2\nu_2 - 3\nu_1^4.$$
$$\beta_1 = \mu_3^2/\mu_2^3, \quad \beta_2 = \mu_4/\mu_2^2.$$

σ, **standard deviation**, $= \sqrt{\mu_2}$.

$\alpha_3/2$, **momental skewness**; $\alpha_3 = \sqrt{\beta_1} = \mu_3/\sigma^3$.

$(\alpha_4 - 3)/2$, **kurtosis**; $\alpha_4 = \beta_2$.

MD, **mean deviation** (from the mean), $= \Sigma f_i|x_i - \bar{x}|/N = 2\left[\bar{x}\sum_{x_i<\bar{x}} f_i - \sum_{x_i<\bar{x}} f_i x_i\right]/N$.

(This latter form is convenient for computation.)

s, **quartile deviation**, $= |Q_3 - Q_1|/2$.

$P.E.$, **probable error**, $= 0.6745\sigma$.

V, **coefficient of variation**, $= 100\sigma/\bar{x}\%$.

Pearson's measure of skewness $= (\bar{x} - Mo)/\sigma$. (Usually approximately $\alpha_3/2$.)

Bowley's measure of skewness $= (Q_3 - 2Md + Q_1)/(2s)$.

(Bowley's measure of skewness lies between -1 and $+1$.)

THE CLASS INTERVAL

$$\Delta x_i = x_{i+1} - x_i.$$

For equi-spaced arguments, $\Delta x_i = h$, the **length of the class interval**, x_i is the **mid-value** or **class mark**. The interval from $x_i - (h/2)$ to $x_i + (h/2)$ is the **class interval** with these as given **initial** and **terminal end values**.

$$u_i = (x_i - A)/h.$$
$$\bar{u} = \Sigma f_i u_i/N, \quad \bar{x} = h\bar{u} + A.$$
$$(\mu_k)_x = h^k(\mu_k)_u, \ (k = 0, 1, \ldots).$$
$$\sigma_u^2 = [\Sigma f_i u_i^2/N] - \bar{u}^2, \quad \sigma_x = h\sigma_u.$$
$$(\beta_1)_x = (\beta_1)_u, \quad (\beta_2)_x = (\beta_2)_u.$$

Sheppard's corrections (to correct approximately for the error due to treating all elements in a given class interval of length h as though concentrated at the class mark).

For μ_0, μ_1, μ_3, no corrections.

In x-units,

corrected $(\mu_2)_x$ = uncorrected $(\mu_2)_x - h^2/12$,

corrected $(\mu_4)_x$ = uncorrected $(\mu_4)_x - h^2$ uncorrected $(\mu_2)_x/2 + 7h^4/240$.

In u-units, replace h by 1 in the formulae given above.

LEAST SQUARES

The **normal equations** for finding coefficients, $a_0, a_1, \ldots, a_m$, in fitting a curve of the form $y = a_0 + a_1 x + \ldots + a_m x^m$ to data (X_i, Y_i), $i = 1, \ldots, n$, $(n > m)$, are $m + 1$ in numbers as follows:

$$\Sigma Y_i = a_0 n + a_1 \Sigma X_i + a_2 \Sigma X_i{}^2 + \ldots + a_m \Sigma X_i{}^m,$$
$$\Sigma X_i Y_i = a_0 \Sigma X_i + a_1 \Sigma X_i{}^2 + a_2 \Sigma X_i{}^3 + \ldots + a_m \Sigma X_i{}^{m+1},$$
$$\cdots \cdots \cdots \cdots \cdots \cdots \cdots \cdots \cdots \cdots \cdots \cdots \cdots$$
$$\Sigma X_i{}^m Y_i = a_0 \Sigma X_i{}^m + a_1 \Sigma X_i{}^{m+1} + a_2 \Sigma X_i{}^{m+2} + \ldots + a_m \Sigma X_i{}^{2m}.$$

Deviation from fitted curve,

$$d_i = Y_i - (a_0 + a_1 X_i + \cdots + a_m X_i{}^m).$$
$$\Sigma d_i{}^2 = \Sigma Y_i{}^2 - (a_0 \Sigma Y_i + a_1 \Sigma X_i Y_i + \cdots + a_m \Sigma X_i{}^m Y_i).$$

For $z = ab^x$, use $y = \log z$, $a_0 = \log a$, $a_1 = \log b$.
For $z = at^p$, use $y = \log z$, $a_0 = \log a$, $a_1 = p$, $x = \log t$.

S_y, **standard error of estimate,** = root-mean-square of the y-deviations about a fitted curve = $\sqrt{\Sigma d_i{}^2/n} = \sigma_y \sqrt{1 - r^2}$.

SIMPLE CORRELATION

PRODUCT MOMENT METHOD

Given n equi-spaced measurements X_i, $i = 1, 2, \ldots, n$, with $h = X_{i+1} - X_i$, $x_i = X_i - \bar{X}$; and m equi-spaced measurements Y_j, $j = 1, 2, \ldots, m$, with $k = Y_{j+1} - Y_j$, $y_j = Y_j - \bar{Y}$; and a weight (frequency, probability, etc.) e_{ij} (≥ 0), associated with (X_i, Y_j). Here e_{ij} is an entry in the table.

$$f_i = \sum_j e_{ij}, \; g_j = \sum_i e_{ij}.$$

$$N = \sum_{ij} e_{ij} = \sum_i f_i = \sum_j g_j. \; \text{(Check)}$$

$$\bar{x} = \sum_{ij} e_{ij} X_i / N = \sum_i f_i X_i / N; \; \bar{y} = \sum_{ij} e_{ij} Y_j / N = \sum_j g_j Y_j / N.$$

Let A and B be arbitrary reference values, usually convenient integral measures near $\bar{X}$ and $\bar{Y}$, respectively.

$$u_i = (X_i - A)/h, \; v_j = (Y_j - B)/k;$$
$$\bar{u} = \Sigma f_i u_i / N, \; \bar{X} = h\bar{u} + A, \; \bar{v} = \Sigma g_j v_j / N, \; \bar{Y} = k\bar{v} + B.$$
$$\sigma_u{}^2 = (\mu_2)_u = (\Sigma f_i u_i{}^2 / N) - \bar{u}^2, \; \sigma_x = h\sigma_u. \left.\vphantom{\begin{matrix}a\\a\end{matrix}}\right\} \text{ Apply Sheppard's corrections.}$$
$$\sigma_v{}^2 = (\mu_2)_v = (\Sigma g_j v_j{}^2 / N) - \bar{v}^2, \; \sigma_y = k\sigma_v.$$
$$U_j = \sum_i e_{ij} u_i, \; V_i = \sum_j e_{ij} v_j, \; P = \sum u_i V_i = \sum v_j U_j \; . \text{(Check)}$$
$$p_{uv} = \sum_{ij} e_{ij}(u_i - \bar{u})(v_j - \bar{v})/N$$
$$= (P/N) - \overline{uv} \; .$$
$$p_{xy} = hk p_{uv} \; .$$

$r = p_{uv}/(\sigma_u\sigma_v) = p_{xy}/(\sigma_x\sigma_y)$ (product-moment) coefficient of correlation. In every case $-1 \le r \le 1$.

$Y - \bar{Y} = r\dfrac{\sigma_y}{\sigma_x}(X - \bar{X})$, or $y = r\dfrac{\sigma_y}{\sigma_x}x$, regression line of y on x.

$X - \bar{X} = r\dfrac{\sigma_x}{\sigma_y}(Y - \bar{Y})$, or $x = r\dfrac{\sigma_x}{\sigma_y}y$, regression line of x on y.

EXAMPLE OF COMPUTATION FOR PRODUCT-MOMENT COEFFICIENT OF CORRELATION

v_j \ u_i / x_i	y_j	-3 / 12	-2 / 16	-1 / 20	0 / 24	1 / 28	2 / 32	g_j	g_jv_j	$g_jv_j{}^2$	$\left(U_j = \sum_i e_{ij}u_i\right)$	v_jU_j
2	21			1	5	7	1	14	28	56	8	16
1	18		1	3	7	5	2	18	18	18	4	4
0	15			2	3	4	1	10	0	0		0
-1	12			3	1	1		5	-5	5	-7	7
-2	9	2	1					3	-6	12	-8	16
f_i		2	7	8	17	13	3	50	35	91		43
f_iu_i		-6	-14	-8	0	13	6	-9				
$f_iu_i{}^2$		18	28	8	0	13	12	79				
$V_i,\left(=\sum_j e_{ij}v_j\right)$		-4	-4	4		19	4					
u_iV_i		12	8	-4	0	19	8	43				

$A = 24, B = 15,$
$h = 4, k = 3,$
$N = \Sigma f_i = \Sigma g_j = 50,$
$\Sigma f_iu_i = -9, \Sigma g_jv_j = 35,$
$\Sigma f_iu_i{}^2 = 79, \Sigma g_jv_j{}^2 = 91,$
$P = \Sigma u_iV_i = \Sigma v_jU_j = 43.$

$\bar{u} = -\tfrac{9}{50} = -.18$ $\bar{v} = \tfrac{35}{50} = .70$
$\sigma_u{}^2 = (\tfrac{79}{50}) - (-.18)^2 - .083 = 1.46,$ $\sigma_u = 1.21$
$\sigma_v{}^2 = (\tfrac{91}{50}) - (.70)^2 - .083 = 1.247,$ $\sigma_v = 1.117$
$p_{uv} = (\tfrac{43}{50}) - (-.18)(.70) = +0.986$
$r = +0.986/(1.21 \times 1.117) = +0.730$ Ans. $r = +0.730$

RANK DIFFERENCE METHOD

Given n corresponding pairs of measured items (X_i, Y_i), $(i = 1, \ldots, n)$. Let (u_i, v_i) be the corresponding rank numbers. Here $u_i = 1$ for the largest X_i, 2 for the next largest X_i, etc.; and similarly $v_i = 1$ for the largest Y_i, 2 for the next largest Y_i, etc.

$\rho = 1 - \dfrac{6\Sigma(u_i - v_i)^2}{n(n^2 - 1)}$, (rank difference) coefficient of correlation.

In every case $-1 \le \rho \le 1$. Check: $\Sigma(u_i - v_i) = 0$.

EXAMPLE OF COMPUTATION FOR RANK-DIFFERENCE COEFFICIENT OF CORRELATION

X_i	Y_i	u_i	v_i	$u_i - v_i$	$(u_i - v_i)^2$	
76	52	3	1	$+2$	4	Check: $\Sigma(u_i - v_i) = 0$.
66	34	8	9	-1	1	
63	32	10	10	0	0	$\rho = 1 - \dfrac{6 \times 62}{10(10^2 - 1)}$
74	45	4	4	0	0	
79	50	1	2	-1	1	$= +0.63$
69	37	7	7	0	0	
77	35	2	8	-6	36	Ans. $\rho = +0.63$
65	42	9	5	$+4$	16	
71	40	6	6	0	0	
73	48	5	3	$+2$	4	
$N = 10$				0	62	

Normal curve (x measured in σ-units from the mean, and with area $= 1$):

$$y = \frac{1}{\sqrt{2\pi}}\, e^{-x^2/2} = 0.3989 e^{-x^2/2}.$$

MD (mean deviation from the mean) $= \sigma\sqrt{2/\pi} = 0.7979\sigma$.
s (quartile deviation from the mean) $= 0.6745\sigma = 0.845 MD$.
Percentage areas, under normal curve, for successive class intervals measured from the mean:

Multiples of σ: 34%, 14%, 2%.
Multiples of s: 25%, 16%, 7%, 2%.

Normal surface (x measured in σ_x-units, y in σ_y-units, from their means),

$$z = \frac{1}{2\pi\sqrt{1 - r^2}}\, e^{-(x^2 - 2rxy + y^2)/[2(1 - r^2)]}.$$

* Goodness of Fit. For a universe of objects falling into n mutually exclusive classes with class marks, $x_i (i = 1, 2, \ldots, n)$, let p_i be the probability for the ith class. Given a sample of N items, with f_i items in the ith class ($\Sigma f_i = N$), the probability that a random sample of N items gives no better fit, expressed in terms of n and χ^2 ("Chi square"), $= \Sigma (f_i - Np_i)^2/(Np_i)$, is given by a table, portions of which are as follows:

PROBABILITY THAT A RANDOM SAMPLE GIVES NO BETTER FIT

n \ χ^2	1	2	3	4	6	8	10	15	20
3	.607	.368	.223	.135	.050	.018	.007	.001	.000
4	.801	.572	.392	.261	.112	.046	.019	.002	.000
5	.910	.736	.558	.406	.199	.092	.040	.005	.000
6	.963	.849	.700	.549	.306	.156	.075	.010	.001
7	.986	.920	.809	.677	.423	.238	.125	.020	.003
8	.995	.960	.885	.780	.540	.333	.189	.036	.006
9	.998	.981	.934	.857	.647	.433	.265	.059	.010
10	.999	.991	.964	.911	.740	.534	.350	.091	.018
11	1.000	.996	.981	.947	.815	.629	.440	.132	.029
12	1.000	.998	.991	.970	.873	.713	.530	.182	.045

n \ χ^2	8	10	12	14	16	18	20	25	30
10	.534	.350	.213	.122	.067	.035	.018	.003	.000
11	.629	.440	.285	.173	.100	.055	.029	.005	.001
12	.713	.530	.363	.233	.141	.082	.045	.009	.002
13	.785	.616	.446	.301	.191	.116	.067	.015	.003
14	.844	.694	.528	.374	.249	.158	.095	.023	.005
15	.889	.762	.606	.450	.313	.207	.130	.035	.008
16	.924	.820	.679	.526	.382	.263	.172	.050	.012
17	.949	.867	.744	.599	.453	.324	.220	.070	.018
18	.967	.904	.800	.667	.524	.389	.274	.095	.026
19	.979	.932	.847	.729	.593	.456	.333	.125	.037
20	.987	.953	.886	.784	.657	.522	.395	.161	.052

BASIC CONCEPTS FOR ALGEBRA OF SETS

I. *Algebra of Sets*

1. *A* set is a collection or an aggregate of objects, called "the elements of the set". If a is an element of set A, we write $a \, \varepsilon \, A$. If not then $a \, \cancel{\varepsilon} \, A$. If a set contains only the element a, we denote it by $\{a\}$.

* See index for t-Test and F-Test tables.

2. The *null set*, denoted by ϕ is the set which has no elements.

3. Two sets A and B are called "equal",* written $A = B$
   if (1) every element of A is an element of B
   and (2) every element of B is an element of A.

4. If every element of set A is an element of set B, we call set A a "subset" of set B,
   written $A \subset B$ (or $B \supset A$).
   By convention $\phi \subset A$ for every set A.

5. If $A \subset B$ and if $B \subset A$, then A is called an *improper* subset of B—also $A = B$ by (3).
   If $A \subset B$ and if B includes at least one element which is not an element of A, then
   A is a *proper* subset of B.
   The symbol $\subset$ is sometimes used to mean *proper* inclusion, with $\subseteq$ meaning inclusion as defined above.
   $\not\subset$ is sometimes also used for proper inclusion.

6. If all of the elements under consideration are elements of a universal set I, then for
   all sets A, $A \subset I$.

7. The set A', called the *complementary* set of A (relative to I) is the set which contains
   all the elements of I which are not elements of A.

8. Two binary operations on sets are $\cup$ and $\cap$.
   $A \cup B$, called the *union* (sometimes called the *join*) of sets A and B is the set of all
   elements which are elements of A or of B or of both.
   $A \cap B$, called the *intersection* (sometimes called the *meet*) of sets A and B is the set
   of all elements which are elements of *both A and B*.

9. Some properties of sets involving these relations:
   For all sets A, B, C in a universal set I. Only in those rules explicitly involving I,
   or those using complementation (which is defined in terms of I) is it necessary to
   assume that $A, B, C, \ldots$, lie in a universal set.

A. (Closure)
   A_1: There is a unique set $A \cup B$
   A_2: There is a unique set $A \cap B$

B. (Commutative Laws)
   B_1: $A \cup B = B \cup A$
   B_2: $A \cap B = B \cap A$

C. (Associative Laws)
   C_1: $(A \cup B) \cup C = A \cup (B \cup C)$
   C_2: $(A \cap B) \cap C = A \cap (B \cap C)$

D. (Distributive Laws)
   D_1: $A \cup (B \cap C) = (A \cup B) \cap (A \cup C)$
   D_2: $A \cap (B \cup C) = (A \cap B) \cup (A \cap C)$

E. (Idempotent Laws)
   E_1: $A \cup A = A$
   E_2: $A \cap A = A$

F. Properties of I and ϕ
   F_1: $A \cap I = A$
   F_2: $A \cup \phi = A$
   F_3: $A \cap \phi = \phi$
   F_4: $A \cup I = I$

---

* This equality, as well as ordinary equality between numbers, is a special case of an *equivalence relation*. In general, an equivalence relation is any relation with the following three properties:
 1. Reflexive Law: $A = A$.
 2. Symmetric Law: If $A = B$, then $B = A$.
 3. Transitive Law: If $A = B$ and $B = C$, then $A = C$.

G. Properties of $\subset$.

G_1: $A \subset (A \cup B)$

G_2: $(A \cap B) \subset A$

G_3: $A \subset I$

G_4: $\phi \subset A$

G_5: If $A \subset B$, then $A \cup B = B$

   If $B \subset A$, then $A \cap B = B$

H. Properties of $'$.

H_1: For every set A, there is a unique set A'

H_2: $A \cup A' = I$

H_3: $A \cap A' = \phi$

H_4: $(A \cup B)' = A' \cap B'$

H_5: $(A \cap B)' = A' \cup B'$

I. *Duality*

If we interchange $\left\{ \begin{array}{c} \cup \text{ and } \cap \\ \phi \text{ and } I \\ \subset \text{ and } \supset \end{array} \right\}$ in any correct formula we obtain another correct formula.

J. The above Algebra of Sets is a representation of a *Boolean* Algebra, which may be defined:

Undefined concepts: Set H of elements a, b, c, . . .

   2 binary operations $\oplus$, $\otimes$

Postulates for all a, b, c, of H

P_1: $a \oplus b \, \varepsilon \, H$; P_1': $a \otimes b \, \varepsilon \, H$

P_2: $a \oplus b = b \oplus a$; P_2': $a \otimes b = b \otimes a$

P_3: $(a \oplus b) \oplus c = a \oplus (b \oplus c)$;

P_3': $(a \otimes b) \otimes c = a \otimes (b \otimes c)$

P_4: $a \oplus (b \otimes c) = (a \oplus b) \otimes (a \oplus c)$

P_4': $a \otimes (b \oplus c) = (a \otimes b) \oplus (a \otimes c)$

P_5: There exists an element Z in H, such that for every element a of H, $a \oplus Z = a$

P_5': There exists an element U in H, such that for every element a of H, $a \otimes U = a$

P_6: For every element a of H, there exists an element a' such that $a \oplus a' = U$ and $a \otimes a' = Z$

PROBABILITY

Definitions

A sample space S associated with an experiment is a set S of elements such that any outcome of the experiment corresponds to one and only one element of the set. An event E is a subset of a sample space S. An element in a sample space is called a sample point or a simple event (Unit subset of S).

Definition of Probability

If an experiment can occur in n mutually exclusive and equally likely ways, and if exactly m of these ways correspond to an event E, then the probability of E is given by

$$P(E) = \frac{m}{n}.$$

If E is a subset of S, and if to each unit subset of S, a non-negative number, called its probability, is assigned, and if E is the union of two or more different simple events, then the probability of E, denoted by $P(E)$, is the sum of the probabilities of those simple events whose union is E.

Marginal and Conditional Probability

Suppose a sample space S is partioned into rs disjoint subsets where the general subset is denoted by $E_i \cap F_j$. Then the marginal probability of E_i is defined as

$$P(E_i) = \sum_{j=1}^{s} P(E_i \cap F_j)$$

and the marginal probability of F_j is defined as

$$P(F_j) = \sum_{i=1}^{r} P(E_i \cap F_j) .$$

The conditional probability of E_i, given that F_j has occurred, is defined as

$$P(E_i/F_j) = \frac{P(E_i \cap F_j)}{P(F_j)} , \qquad P(F_j) \neq 0$$

and that of F_j, given that E_i has occurred, is defined as

$$P(F_j/E_i) = \frac{P(E_i \cap F_j)}{P(E_i)} , \qquad P(E_i) \neq 0 .$$

Probability Theorems

1. If ϕ is the null set, $P(\phi) = 0$.
2. If S is the sample space, $P(S) = 1$.
3. If E and F are two events

$$P(E \cup F) = P(E) + P(F) - P(E \cap F).$$

4. If E and F are mutually exclusive events,

$$P(E \cup F) = P(E) + P(F).$$

5. If E and E' are complementary events,

$$P(E) = 1 - P(E').$$

6. The conditional probability of an event E, given an event F, is denoted by $P(E/F)$ and is defined as

$$P(E/F) = \frac{P(E \cap F)}{P(F)},$$

where $P(F) \neq 0$.
7. Two events E and F are said to be independent if and only if

$$P(E \cap F) = P(E) \cdot P(F).$$

E is said to be statistically independent of F if $P(E/F) = P(E)$ and $P(F/E) = P(F)$.
8. The events $E_1, E_2, \ldots, E_n$ are called mutually independent for all combinations if and only if every combination of these events taken any number at a time is independent.

9. *Bayes Theorem.*

If $E_1, E_2, \ldots, E_n$ are n mutually exclusive events whose union is the sample space S, and E is any arbitrary event of S such that $P(E) \neq 0$, then

$$P(E_k/E) = \frac{P(E_k) \cdot P(E/E_k)}{\sum\limits_{j=1}^{n} [P(E_j) \cdot P(E/E_j)]}$$

EXAMPLE: Two unbiased six sided dice, one red and one green, are tossed and the number of dots appearing on their upper faces is observed.

The sample space S consists of 36 elements, i.e.

$$S = \{(1, 1), (1, 2), \ldots, (6, 5), (6, 6)\} \ .$$

1. What is the probability of throwing a seven, denoted by the event A?

$$A = \{(1, 6)\} \cup \{(2, 5)\} \cup \{(3, 4)\} \cup \{(4, 3)\} \cup \{(5, 2)\} \cup \{(6, 1)\}$$
$$P(A) = \tfrac{1}{36} + \tfrac{1}{36} + \tfrac{1}{36} + \tfrac{1}{36} + \tfrac{1}{36} + \tfrac{1}{36} = \tfrac{1}{6}$$

2. What is the probability of throwing a seven or a ten? Denote by A the event "throwing a 7" and by B the event "throwing a ten" .

$$P(A) = \tfrac{1}{6} \qquad P(A \cup B) = P(A) + P(B)$$
$$P(B) = \tfrac{1}{12} \qquad\qquad = \tfrac{1}{6} + \tfrac{1}{12}$$
$$\qquad\qquad\qquad = \tfrac{1}{4}$$

3. What is the probability that the red die shows a number less than or equal to three and the green die shows a number greater than or equal to five? Denote by C the event "red die shows number ≤ 3" and by D the event "green die shows number ≥ 5" .

$$C \cap D = \{(1, 5), (2, 5), (3, 5), (1, 6), (2, 6), (3, 6)\}$$
$$P(C \cap D) = \tfrac{6}{36} = \tfrac{1}{6}$$

Note:
$$P(C) = \tfrac{18}{36} = \tfrac{1}{2}$$
$$P(D) = \tfrac{12}{36} = \tfrac{1}{3}$$
$$P(C) \cdot P(D) = \tfrac{1}{2} \cdot \tfrac{1}{3} = \tfrac{1}{6}$$

Thus $P(C \cap D) = P(C) \cdot P(D)$ and events C and D are independent.

4. What is the probability that the green die shows a one, given that the sum of the number on the two dice is less than four? Denote by E the event "green die shows 1" and by F the event "sum of numbers on dice <4" .

$$E = \{(1, 1), (1, 2), (1, 3), (1, 4), (1, 5), (1, 6)\}$$
$$F = \{(1, 1), (1, 2), (2, 1)\}$$
$$E \cap F = \{(1, 1), (1, 2)\}$$
$$P(E/F) = \frac{P(E \cap F)}{P(F)} = \frac{\tfrac{2}{36}}{\tfrac{3}{36}} = \tfrac{2}{3}$$

5. What is the probability that the sum of the numbers on the two dice is not seven?

$$P(A) = \tfrac{1}{6}$$
$$P(A') = 1 - \tfrac{1}{6} = \tfrac{5}{6}$$

Random Variable

A function whose domain is a sample space S and whose range is some set of real numbers is called a random variable, denoted by **X**. The function **X** transforms sample points of S into points on the x-axis. **X** will be called a discrete random variable if it is a random variable that assumes only a finite or denumerable number of values on the x-axis.

X will be called a continuous random variable if it assumes a continuum of values on the x-axis.

Probability Function (Discrete Case)

The random variable **X** will be called a discrete random variable if there exists a function f such that $f(x_i) \geq 0$ and $\sum_i f(x_i) = 1$ for $i = 1, 2, 3, \ldots$ and such that for any event E,

$$P(E) = P[\mathbf{X} \text{ is in } E] = \sum_E f(x)$$

where $\sum_E$ means sum $f(x)$ over those values x_i that are in E and where $f(x) = P[\mathbf{X} = x]$. The probability that the value of **X** is some real number x, is given by $f(x) = P[\mathbf{X} = x]$, where f is called the probability function of the random variable **X**.

Cumulative Distribution Function (Discrete Case)

The probability that the value of a random variable **X** is less than or equal to some real number x is defined as

$$F(x) = P(\mathbf{X} \leq x)$$
$$= \Sigma f(x_i), \qquad -\infty < x < \infty,$$

where the summation extends over those values of i such that $x_i \leq x$.

Probability Density (Continuous Case)

The random variable **X** will be called a continuous random variable if there exists a function f such that $f(x) \geq 0$ and $\int_{-\infty}^{\infty} f(x)\, dx = 1$ for all x in interval $-\infty < x < \infty$ and such that for any event E

$$P(E) = P(\mathbf{X} \text{ is in } E) = \int_E f(x)\, dx.$$

$f(x)$ is called the probability density of the random variable **X**. The probability that **X** assumes any given value of x is equal to zero and the probability that it assumes a value on the interval from a to b, including or excluding either end point, is equal to

$$\int_a^b f(x)\, dx.$$

Cumulative Distribution Function (Continuous Case)

The probability that the value of a random variable **X** is less than or equal to some real number x is defined as

$$F(x) = P(\mathbf{X} \leq x), \qquad -\infty < x < \infty$$
$$= \int_{-\infty}^{x} f(x)\, dx.$$

From the cumulative distribution, the density, if it exists, can be found from

$$f(x) = \frac{dF(x)}{dx}.$$

From the cumulative distribution

$$P(a \leq \mathbf{X} \leq b) = P(\mathbf{X} \leq b) - P(\mathbf{X} \leq a)$$
$$= F(b) - F(a)$$

Mathematical Expectation

A. EXPECTED VALUE

Let X be a random variable with density $f(x)$. Then the expected value of X, $E(X)$, is defined to be

$$E(X) = \sum_x xf(x)$$

if X is discrete and

$$E(X) = \int_{-\infty}^{\infty} xf(x)\, dx$$

if X is continuous. The expected value of a function g of a random variable X is defined as

$$E[g(X)] = \sum_x g(x) \cdot f(x)$$

if X is discrete and

$$E[g(X)] = \int_{-\infty}^{\infty} g(x) \cdot f(x)\, dx$$

if X is continuous.

Theorems

1. $E[aX + bY] = aE(X) + bE(Y)$
2. $E[X \cdot Y] = E(X) \cdot E(Y)$ if X and Y are statistically independent.

B. MOMENTS

a. Moments About the Origin. The moments about the origin of a probability distribution are the expected values of the random variable which has the given distribution. The rth moment of X, usually denoted by ν_r, is defined as

$$\nu_r = E[X^r] = \sum_x x^r f(x)$$

if X is discrete and

$$\nu_r = E[X^r] = \int_{-\infty}^{\infty} x^r f(x)\, dx$$

if X is continuous.

The first moment, ν_1, is called the mean of the random variable X and is usually denoted by μ.

b. Moments About the Mean. The rth moment about the mean, usually denoted by μ_r, is defined as

$$\mu_r = E[(X - \mu)^r] = \sum_x (x - \mu)^r f(x)$$

if X is discrete and

$$\mu_r = E[(X - \mu)^r] = \int_{-\infty}^{\infty} (x - \mu)^r f(x)\, dx$$

if X is continuous.

The second moment about the mean, μ_2, is given by

$$\mu_2 = E[(X - \mu)^2] = \nu_2 - \nu_1^2$$

and is called the variance of the random variable X, and is denoted by σ^2. The square root of the variance, σ, is called the standard deviation.

Theorems

1. $\sigma^2_{cX} = c^2\sigma^2_X$
2. $\sigma^2_{c+X} = \sigma^2_X$
3. $\sigma^2_{aX+b} = a^2\sigma^2_X$

c. Factorial Moments. The rth factorial moment of a probability distribution is defined as

$$E[X^{[r]}] = \sum_x x^{[r]} f(x)$$

if X is discrete and

$$E[X^{[r]}] = \int_{-\infty}^{\infty} x^{[r]} f(x) \, dx$$

if X is continuous, where the symbol $x^{[r]}$ denotes the factorial

$$x^{[r]} = x(x-1)(x-2) \cdots (x-r+1), r = 1, 2, 3, \ldots$$

C. GENERATING FUNCTIONS

a. Moment Generating Functions. The moment generating function (m.g.f.) of the random variable X is defined as

$$m_x(t) = E(e^{tX}) = \sum_x e^{tx} f(x)$$

if X is discrete and

$$m_x(t) = E(e^{tX}) = \int_{-\infty}^{\infty} e^{tx} f(x) \, dx$$

if X is continuous.
$E(e^{tX})$ is the expected value of e^{tX}. If $m_x(t)$ and its derivatives exist, $|t| < h^2$, the rth moment about the origin is

$$\nu_r = m_x^{(r)}(0), \qquad r = 0, 1, 2, \ldots$$

where $m_x^{(r)}(0)$ is the rth derivative of $m_x(t)$ with respect to t, evaluated at $t = 0$. For

$$m_x(t) = E(e^{tX})$$
$$= E\left[1 + Xt + \frac{(Xt)^2}{2!} + \cdots \right]$$
$$= 1 + \nu_1 t + \nu_2 \frac{t^2}{2} + \cdots \qquad .$$

Thus, the moments ν_r appear as coefficients of $\dfrac{t^r}{r!}$, and $m_x(t)$ may be regarded as generating the moments ν_r. The moments μ_r may be generated by the generating function

$$M_x(t) = E[e^{t(X-\mu)}] = e^{-\mu t} E(e^{tX}) = e^{-\mu t} m_x(t) \ .$$

b. Factorial Moment Generating Function. The factorial moment generating function is defined as

$$E(t^X) = \sum_x t^x f(x) \qquad \text{(probability generating function)}$$

if X is discrete and

$$E(t^X) = \int_{-\infty}^{\infty} t^x f(x) \, dx$$

if X is continuous.
The rth factorial moment is obtained from the factorial moment generating function by differentiating it r times with respect to t and then evaluating the result when $t = 1$.

Theorems

1. If c is a constant, the m.g.f. of $c + X$ is $e^{ct} m_x(t)$.
2. If c is a constant, the m.g.f. of cX is $m_x(ct)$.

3. If $\mathbf{Y} = \sum_{i=1}^{n} \mathbf{X}_i$, and $m_x(t)$ is the m.g.f. of $\mathbf{X}_i$, where $\mathbf{X}_1, \ldots, \mathbf{X}_n$ is a random sample from $f(x)$, then the m.g.f. of $\mathbf{Y}$ is $[m_x(t)]^n$.

D. CUMULANT GENERATING FUNCTION

Let $m_x(t)$ be a m.g.f. If $\ln m_x(t)$ can be expanded in the form

$$c(t) = \ln m_x(t) = \kappa_1 t + \kappa_2 \frac{t^2}{2!} + \kappa_3 \frac{t^3}{3!} + \cdots + \kappa_r \frac{t^r}{r!} + \cdots,$$

then $c(t)$ is called the cumulant generating function (semi-invariant generating function) and κ_r are called the cumulants (semi-invariants) of a distribution.

$$\kappa_r = c^{(r)}(0)$$

where $c^{(r)}(0)$ is the rth derivative of $c(t)$ with respect to t evaluated at $t = 0$.

E. CHARACTERISTIC FUNCTIONS

The characteristic function of a distribution is defined as

$$\phi(t) = E(e^{it\mathbf{X}}) = \sum_{x} e^{itx} \cdot f(x)$$

if $\mathbf{X}$ is discrete and

$$\phi(t) = E(e^{it\mathbf{X}}) = \int_{-\infty}^{\infty} e^{itx} \cdot f(x) \, dx$$

if $\mathbf{X}$ is continuous.

Here t is a real number, $i^2 = -1$, and $e^{it\mathbf{X}} = \cos(t\mathbf{X}) + i \sin(t\mathbf{X})$. The characteristic function also generates moments, if they exist for

$$i^r \nu_r = \phi^{(r)}(0)$$

where $\phi^{(r)}(0)$ is the rth derivative of $\phi(t)$ with respect to t evaluated at $t = 0$.

Multivariate Distributions

A. DISCRETE CASE

The k-dimensional random variable $(\mathbf{X}_1, \mathbf{X}_2, \ldots, \mathbf{X}_k)$ is a k-dimensional discrete random variable if it assumes values only at a finite or denumerable number of points $(x_1, x_2, \ldots, x_k)$. Define

$$P[\mathbf{X}_1 = x_1, \mathbf{X}_2 = x_2, \ldots, \mathbf{X}_k = x_k] = f(x_1, x_2, \ldots, x_k)$$

for every value that the random variable can assume. $f(x_1, x_2, \ldots, x_k)$ is called the joint density of the k-dimensional random variable. If E is any subset of the set of values that the random variable can assume, then

$$P(E) = P[(\mathbf{X}_1, \mathbf{X}_2, \ldots, \mathbf{X}_k) \text{ is in } E] = \sum_{E} f(x_1, x_2, \ldots, x_k)$$

where the sum is over all those points in E. The cumulative distribution is defined as

$$F(x_1, x_2, \ldots, x_k) = \sum_{x_1} \sum_{x_2} \cdots \sum_{x_k} f(x_1, x_2, \ldots, x_k).$$

B. CONTINUOUS CASE

The k random variables $\mathbf{X}_1, \mathbf{X}_2, \ldots, \mathbf{X}_k$ are said to be jointly distributed if there exists a function f such that $f(x_1, x_2, \ldots, x_k) \geq 0$ for all $-\infty < x_i < \infty$,

$i = 1, 2, \ldots, k$ and such that for any event E

$$P(E) = P[(\mathbf{X}_1, \mathbf{X}_2, \ldots, \mathbf{X}_k) \text{ is in } E]$$
$$= \int_E f(x_1, x_2, \ldots, x_k)\, dx_1\, dx_2 \cdots dx_k.$$

$f(x_1, x_2, \ldots, x_k)$ is called the joint density of the random variables $\mathbf{X}_1, \mathbf{X}_2, \ldots, \mathbf{X}_k$. The cumulative distribution is defined as

$$F(x_1, x_2, \ldots, x_k) = \int_{-\infty}^{x_1} \int_{-\infty}^{x_2} \cdots \int_{-\infty}^{x_k} f(x_1, x_2, \ldots, x_k)\, dx_k \cdots dx_2\, dx_1 .$$

Given the cumulative distribution, the density may be found by

$$f(x_1, x_2, \ldots, x_k) = \frac{\partial}{\partial x_1} \cdot \frac{\partial}{\partial x_2} \cdots \frac{\partial}{\partial x_k} F(x_1, x_2, \ldots, x_k) .$$

Moments

The rth moment of $\mathbf{X}_i$, say, is defined as

$$E(\mathbf{X}_i{}^r) = \sum_{x_1} \sum_{x_2} \cdots \sum_{x_k} x_i{}^r f(x_1, x_2, \ldots, x_k)$$

if the $\mathbf{X}_i$ are discrete and

$$E(\mathbf{X}_i{}^r) = \int_{-\infty}^{\infty} \int_{-\infty}^{\infty} \cdots \int_{-\infty}^{\infty} x_i{}^r f(x_1, x_2, \ldots, x_k)\, dx_k \cdots dx_2\, dx_1$$

if the $\mathbf{X}_i$ are continuous.
Joint moments about the origin are defined as

$$E(\mathbf{X}_1{}^{r_1} \mathbf{X}_2{}^{r_2} \cdots \mathbf{X}_k{}^{r_k})$$

where $r_1 + r_2 + \cdots + r_k$ is the order of the moment.
Joint moments about the mean are defined as

$$E[(\mathbf{X}_1 - \mu_1)^{r_1}(\mathbf{X}_2 - \mu_2)^{r_2} \cdots (\mathbf{X}_k - \mu_k)^{r_k}].$$

Marginal and Conditional Distributions

If the random variables $\mathbf{X}_1, \mathbf{X}_2, \ldots, \mathbf{X}_k$ have the joint density function $f(x_1, x_2, \ldots, x_k)$, then the marginal distribution of the subset of the random variables, say, $\mathbf{X}_1, \mathbf{X}_2, \ldots, \mathbf{X}_p$ $(p < k)$, is given by

$$g(x_1, x_2, \ldots, x_p) = \sum_{x_{p+1}} \sum_{x_{p+2}} \cdots \sum_{x_k} f(x_1, x_2, \ldots, x_k)$$

if the $\mathbf{X}$'s are discrete, and

$$g(x_1, x_2, \ldots, x_p) = \int_{-\infty}^{\infty} \int_{-\infty}^{\infty} \cdots \int_{-\infty}^{\infty} f(x_1, x_2, \ldots, x_k)\, dx_{p+1} \cdots dx_{k-1}\, dx_k$$

if the $\mathbf{X}$'s are continuous.
The conditional distribution of a certain subset of the random variables is the joint distribution of this subset under the condition that the remaining variables are given certain values. The conditional distribution of $\mathbf{X}_1, \mathbf{X}_2, \ldots, \mathbf{X}_p$ given $\mathbf{X}_{p+1}, \mathbf{X}_{p+2}, \ldots, \mathbf{X}_k$ is

$$h(x_1, x_2, \ldots, x_p | x_{p+1}, x_{p+2}, \ldots, x_k) = \frac{f(x_1, x_2, \ldots, x_k)}{g(x_{p+1}, x_{p+2}, \ldots, x_k)}$$

if $g(x_{p+1}, x_{p+2}, \ldots, x_k) \neq 0.$

The variance σ_{ii} of $\mathbf{X}_i$ and the covariance σ_{ij} of $\mathbf{X}_i$ and $\mathbf{X}_j$ are given by

$$\sigma_{ii} = \sigma_i{}^2 = E[(\mathbf{X}_i - \mu_i)^2]$$

and

$$\sigma_{ij} = \rho_{ij}\sigma_i\sigma_j = E[(\mathbf{X}_i - \mu_i)(\mathbf{X}_j - \mu_j)]$$

where ρ_{ij} is the correlation coefficient and σ_i and σ_j are the standard deviations of $\mathbf{X}_i$ and $\mathbf{X}_j$. A joint m.g.f. is defined as

$$m(t_1, t_2, \ldots, t_k) = E[e^{t_1\mathbf{X}_1 + t_2\mathbf{X}_2 + \cdots + t_k\mathbf{X}_k}]$$

if it exists for all values of t_i such that $|t_i| < h^2$.

The rth moment of $\mathbf{X}_i$ may be obtained by differentiating the m.g.f. r times with respect to t_i and then evaluating the result when all t's are set equal to zero. Similarly, a joint moment would be found by differentiating the m.g.f. r_1 times with respect to t_1, $\ldots$, r_k times with respect to t_k, and then evaluating the result when all t's are set equal to zero.

Probability Distributions

A. DISCRETE CASE

1. *Binomial Distribution.* If the random variable $\mathbf{X}$ has a probability function given by

$$P(\mathbf{X} = x) = f(x) = \binom{n}{x} p^x q^{n-x}, \qquad x = 0, 1, 2, \ldots, n$$

where

$$p + q = 1 \text{ and } \binom{n}{x} = \frac{n!}{x!(n-x)!},$$

then the variable $\mathbf{X}$ is said to possess a binomial distribution. $f(x)$ is the general term of the expansion of $(q + p)^n$.

Properties

$$\text{Mean} = \mu = np$$
$$\text{Variance} = \sigma^2 = npq = np(1 - p)$$
$$\text{Standard Deviation} = \sigma = \sqrt{npq}$$
$$\text{Moment Generating Function} = m_x(t) = (pe^t + q)^n$$

2. *Multinomial Distribution.* If a set of random variables $\mathbf{X}_1, \mathbf{X}_2, \ldots, \mathbf{X}_n$ has a probability function given by

$$P(\mathbf{X}_1 = x_1, \mathbf{X}_2 = x_2, \ldots, \mathbf{X}_n = x_n) = f(x_1, x_2, \ldots, x_n) = \frac{N!}{\prod\limits_{i=1}^{n} x_i!} \prod_{i=1}^{n} p_i{}^{x_i}$$

where x_i are positive integers and each $p_i > 0$ for $i = 1, 2, \ldots, n$ and

$$\sum_{i=1}^{n} p_i = 1, \qquad \sum_{i=1}^{n} x_i = N,$$

then the joint distribution of $\mathbf{X}_1, \mathbf{X}_2, \ldots, \mathbf{X}_n$ is called the multinomial distribution. $f(x_1, x_2, \ldots, x_n)$ is the general term of the expansion of $(p_1 + p_2 + \cdots + p_n)^N$.

Properties

$$\text{Mean of } \mathbf{X}_i = \mu_i = Np_i$$
$$\text{Variance of } \mathbf{X}_i = \sigma_i{}^2 = Np_i(1 - p_i)$$
$$\text{Covariance of } \mathbf{X}_i \text{ and } \mathbf{X}_j = \sigma_{ij}{}^2 = -Np_ip_j$$
$$\text{Joint Moment Generating Function} = (p_1e^{t_1} + \cdots + p_ne^{t_n})^N$$

3. *Poisson Distribution.* If the random variable $\mathbf{X}$ has a probability function given by

$$P(\mathbf{X} = x) = f(x) = \frac{e^{-m}m^x}{x!}, \qquad m > 0, x = 0, 1, \ldots,$$

then the variable $\mathbf{X}$ is said to possess a Poisson distribution.

Properties

$$\text{Mean} = \mu = m$$
$$\text{Variance} = \sigma^2 = m$$
$$\text{Standard Deviation} = \sigma = \sqrt{m}$$
$$\text{Moment Generating Function} = m_\mathbf{x}(t) = e^{m(e^t - 1)}$$

4. *Hypergeometric Distribution.* If the random variable $\mathbf{X}$ has a probability function given by

$$P(\mathbf{X} = x) = f(x) = \frac{\binom{k}{x}\binom{N-k}{n-x}}{\binom{N}{n}}, \qquad x = 0, 1, 2, \ldots, [n, k],$$

where $[n, k]$ means the smaller of the two numbers n, k, then the variable $\mathbf{X}$ is said to possess a hypergeometric distribution.

Properties

$$\text{Mean} = \mu = \frac{kn}{N}$$
$$\text{Variance} = \sigma^2 = \frac{k(N-k)n(N-n)}{N^2(N-1)}$$
$$\text{Standard Deviation} = \sigma = \sqrt{\frac{k(N-k)n(N-n)}{N^2(N-1)}}$$

5. *Negative Binomial Distribution.* If the random variable $\mathbf{X}$ has a probability function given by

$$P(\mathbf{X} = x) = f(x) = \binom{x+r-1}{r-1} p^r q^x, \qquad x = 0, 1, 2, \ldots; p + q = 1,$$

then the variable $\mathbf{X}$ is said to possess a negative binomial distribution, known also as the Pascal or Pólya distribution.

Properties

$$\text{Mean} = \mu = \frac{rq}{p}$$
$$\text{Variance} = \sigma^2 = \frac{r}{p}\left(\frac{1}{p} - 1\right) = \frac{rq}{p^2}$$
$$\text{Standard Deviation} = \sigma = \sqrt{\frac{r}{p}\left(\frac{1}{p} - 1\right)} = \sqrt{\frac{rq}{p^2}}$$
$$\text{Moment Generating Function} = m_\mathbf{x}(t) = p^r(1 - qe^t)^{-r}$$

B. Continuous Case

1. *Normal Distribution.* The random variable **X** is said to be normally distributed if its density function is given by

$$f(x) = \frac{1}{\sqrt{2\pi}\,\sigma} e^{-(x-\mu)^2/2\sigma^2} , \qquad -\infty < x < \infty$$

where μ and σ are parameters, called the mean and the standard deviation of the random variable **X**, respectively.

Properties

$$\text{Mean} = \mu$$
$$\text{Variance} = \sigma^2$$
$$\text{Standard Deviation} = \sigma$$
$$\text{Moment Generating Function} = m_{\mathbf{x}}(t) = e^{t\mu + \frac{\sigma^2 t^2}{2}}$$

Cumulative Distribution

$$F(x) = \int_{-\infty}^{x} \frac{1}{\sqrt{2\pi}\,\sigma} e^{-(x-\mu)^2/2\sigma^2} \, dx$$

Set $y = \dfrac{x - \mu}{\sigma}$ to obtain the cumulative standard normal.

2. *Gamma Distribution.* A random variable **X** is said to be distributed as the Gamma Distribution if the density function is given by

$$f(x) = \frac{1}{\Gamma(\alpha + 1)\beta^{\alpha+1}} x^\alpha e^{-x/\beta} , \qquad 0 < x < \infty$$

where α and β are parameters with $\alpha > -1$ and $\beta > 0$.

Properties

$$\text{Mean} = \mu = \beta(\alpha + 1)$$
$$\text{Variance} = \sigma^2 = \beta^2(\alpha + 1)$$
$$\text{Standard Deviation} = \sigma = \beta\sqrt{\alpha + 1}$$
$$\text{Moment Generating Function} = m_{\mathbf{x}}(t) = (1 - \beta t)^{-(\alpha+1)} , \qquad t < \frac{1}{\beta} .$$

3. *Beta Distribution.* A random variable **X** is said to be distributed as the Beta Distribution if the density function is given by

$$f(x) = \frac{\Gamma(\alpha + \beta + 2)}{\Gamma(\alpha + 1)\Gamma(\beta + 1)} x^\alpha (1 - x)^\beta , \qquad 0 < x < 1$$

where α and β are parameters with $\alpha > -1$ and $\beta > -1$.

Properties

$$\text{Mean} = \mu = \frac{\alpha + 1}{\alpha + \beta + 2}$$
$$\text{Variance} = \sigma^2 = \frac{(\alpha + 1)(\beta + 1)}{(\alpha + \beta + 2)^2(\alpha + \beta + 3)}$$
$$r\text{th moment about the origin} = \nu_r = \frac{\Gamma(\alpha + \beta + 2)\Gamma(\alpha + r + 1)}{\Gamma(\alpha + \beta + r + 2)\Gamma(\alpha + 1)} .$$

ampling Distributions

Population—A finite or infinite set of elements of a random variable X.
Random Sample—If the random variables X_1, X_2, . . . , X_n have a joint density,

$$g(x_1, x_2, \ldots, x_n) = f(x_1)f(x_2) \cdots f(x_n)$$

here the density of each X_i is $f(x)$, then X_1, X_2, . . . , X_n is said to be a random sample size n from the population with density $f(x)$.

ampling Distributions

A random sample is selected from a population in which the form of the probability nction is known, and from the joint density of the random variables a distribution, lled the sampling distribution, of a function of the random variables is derived.

1. *Chi-Square Distribution.* If Y_1, Y_2, . . . , Y_n are normally and independently istributed with mean 0 and variance 1, then

$$\chi^2 = \sum_{i=1}^{n} Y_i^2$$

distributed as Chi-Square (χ^2) with n degrees of freedom. The density function is given

$$f(\chi^2) = \frac{(\chi^2)^{\frac{1}{2}(n-2)}}{2^{\frac{n}{2}}\Gamma\left(\frac{n}{2}\right)} e^{-\chi^2/2}, \quad 0 < \chi^2 < \infty .$$

roperties

$$\text{Mean} = \mu = n$$
$$\text{Variance} = \sigma^2 = 2n$$

eproductive Property of χ^2 - Distribution

If χ_1^2, χ_2^2, . . . , χ_k^2 are independently distributed according to χ^2 - distributions ith n_1, n_2, . . . , n_k degrees of freedom, respectively, then $\sum_{j=1}^{k} \chi_j^2$ is distributed according

a χ^2 - distribution with $n = \sum_{j=1}^{k} n_j$ degrees of freedom.

2. *Snedecor's F-Distribution.* If a random variable X is distributed as χ^2 with m egrees of freedom (χ_m^2) and a random variable Y is distributed as χ^2 with n degrees of eedom (χ_n^2) and if X and Y are independent, then $F = \dfrac{X/m}{Y/n}$ is distributed as Snedecor's with m and n degrees of freedom, denoted by $F(m, n)$. The density function of the -distribution is given by

$$f(F) = \frac{\Gamma\left(\frac{m+n}{2}\right)\left(\frac{m}{n}\right)^{m/2} F^{(m-2)/2}}{\Gamma\left(\frac{m}{2}\right)\Gamma\left(\frac{n}{2}\right)\left(1 + \frac{m}{n}F\right)^{(m+n)/2}}, \quad 0 < F < \infty .$$

Properties

$$\text{Mean} = \mu = \frac{n}{n-2}, \qquad n > 2$$

$$\text{Variance} = \sigma^2 = \frac{2n^2(m+n-2)}{m(n-2)^2(n-4)}, \qquad n > 4 .$$

The transformation $w = \dfrac{mF/n}{1 + \dfrac{mF}{n}}$ transforms the F-density into a Beta density.

3. *Student's t-Distribution.* If a random variable X is normally distributed with mean 0 and variance σ^2, and if Y^2/σ^2 is distributed as χ^2 with n degrees of freedom and if X and Y are independent, then

$$t = \frac{X\sqrt{n}}{Y}$$

is distributed as Student's t with n degrees of freedom. The density function is given by

$$f(t) = \frac{\Gamma\left(\dfrac{n+1}{2}\right)}{\sqrt{n\pi}\,\Gamma\left(\dfrac{n}{2}\right)\left(1 + \dfrac{t^2}{n}\right)^{\frac{1}{2}(n+1)}} , \qquad -\infty < t < \infty .$$

Properties

$$\text{Mean} = \mu = 0$$

$$\text{Variance} = \sigma^2 = \frac{n}{n-2}, \qquad n > 2 .$$

II. Normal Distribution

II.1 THE NORMAL PROBABILITY FUNCTION AND RELATED FUNCTIONS

This table gives values of:

a) $f(x)$ = the probability density of a standardized random variable

$$= \frac{1}{\sqrt{2\pi}} e^{-\frac{1}{2}x^2}$$

For negative values of x, one uses the fact that $f(-x) = f(x)$.

b) $F(x)$ = the cumulative distribution function of a standardized normal random variable

$$= \int_{-\infty}^{x} \frac{1}{\sqrt{2\pi}} e^{-\frac{1}{2}t^2} dt$$

For negative values of x, one uses the relationship $F(-x) = 1 - F(x)$. Values of x corresponding to a few special values of $F(x)$ are given in a separate table following the main table. (See page 124.)

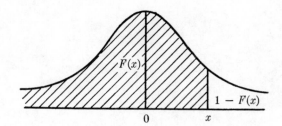

c) $f'(x)$ = the first derivative of $f(x)$ with respect to x

$$= -\frac{x}{\sqrt{2\pi}} e^{-\frac{1}{2}x^2} = -xf(x)$$

d) $f''(x)$ = the second derivative of $f(x)$ with respect to x

$$= \frac{(x^2 - 1)}{\sqrt{2\pi}} e^{-\frac{1}{2}x^2} = (x^2 - 1)f(x)$$

e) $f'''(x)$ = the third derivative of $f(x)$ with respect to x

$$= \frac{3x - x^3}{\sqrt{2\pi}} e^{-\frac{1}{2}x^2} = (3x - x^3)f(x)$$

f) $f^{iv}(x)$ = the fourth derivative of $f(x)$ with respect to x

$$= \frac{x^4 - 6x^2 + 3}{\sqrt{2\pi}} e^{-\frac{1}{2}x^2} = (x^4 - 6x^2 + 3)f(x)$$

21

It should be noted that other probability integrals can be evaluated by the use of these tables. For example,

$$\int_0^x f(t)dt = \tfrac{1}{2} \operatorname{erf}\left(\frac{x}{\sqrt{2}}\right),$$

where $\operatorname{erf}\left(\dfrac{x}{\sqrt{2}}\right)$ represents the error function associated with the normal curve.

To evaluate erf (2.3) one proceeds as follows: Since $\dfrac{x}{\sqrt{2}} = 2.3$, one finds $x = (2.3)(\sqrt{2}) = 3.25$. In the entry opposite $x = 3.25$, the value 0.9994 is given. Subtracting 0.5000 from the tabular value, one finds the value 0.4994. Thus erf (2.3) $= 2(0.4994) = 0.9988$.

NORMAL DISTRIBUTION AND RELATED FUNCTIONS

x	$F(x)$	$1 - F(x)$	$f(x)$	$f'(x)$	$f'''(x)$	$f''''(x)$	$f^{\mathrm{v}}(x)$
.00	.5000	.5000	.3989	− .0000	− .3989	.0000	1.1968
.01	.5040	.4960	.3989	− .0040	− .3989	.0120	1.1965
.02	.5080	.4920	.3989	− .0080	− .3987	.0239	1.1956
.03	.5120	.4880	.3988	− .0120	− .3984	.0359	1.1941
.04	.5160	.4840	.3986	− .0159	− .3980	.0478	1.1920
.05	.5199	.4801	.3984	− .0199	− .3975	.0597	1.1894
.06	.5239	.4761	.3982	− .0239	− .3968	.0716	1.1861
.07	.5279	.4721	.3980	− .0279	− .3960	.0834	1.1822
.08	.5319	.4681	.3977	− .0318	− .3951	.0952	1.1778
.09	.5359	.4641	.3973	− .0358	− .3941	.1070	1.1727
.10	.5398	.4602	.3970	− .0397	− .3930	.1187	1.1671
.11	.5438	.4562	.3965	− .0436	− .3917	.1303	1.1609
.12	.5478	.4522	.3961	− .0475	− .3904	.1419	1.1541
.13	.5517	.4483	.3956	− .0514	− .3889	.1534	1.1468
.14	.5557	.4443	.3951	− .0553	− .3873	.1648	1.1389
.15	.5596	.4404	.3945	− .0592	− .3856	.1762	1.1304
.16	.5636	.4364	.3939	− .0630	− .3838	.1874	1.1214
.17	.5675	.4325	.3932	− .0668	− .3819	.1986	1.1118
.18	.5714	.4286	.3925	− .0707	− .3798	.2097	1.1017
.19	.5753	.4247	.3918	− .0744	− .3777	.2206	1.0911
.20	.5793	.4207	.3910	− .0782	− .3754	.2315	1.0799
.21	.5832	.4168	.3902	− .0820	− .3730	.2422	1.0682
.22	.5871	.4129	.3894	− .0857	− .3706	.2529	1.0560
.23	.5910	.4090	.3885	− .0894	− .3680	.2634	1.0434
.24	.5948	.4052	.3876	− .0930	− .3653	.2737	1.0302
.25	.5987	.4013	.3867	− .0967	− .3625	.2840	1.0165
.26	.6026	.3974	.3857	− .1003	− .3596	.2941	1.0024
.27	.6064	.3936	.3847	− .1039	− .3566	.3040	0.9878
.28	.6103	.3897	.3836	− .1074	− .3535	.3138	0.9727
.29	.6141	.3859	.3825	− .1109	− .3504	.3235	0.9572
.30	.6179	.3821	.3814	− .1144	− .3471	.3330	0.9413
.31	.6217	.3783	.3802	− .1179	− .3437	.3423	0.9250
.32	.6255	.3745	.3790	− .1213	− .3402	.3515	0.9082
.33	.6293	.3707	.3778	− .1247	− .3367	.3605	0.8910
.34	.6331	.3669	.3765	− .1280	− .3330	.3693	0.8735
.35	.6368	.3632	.3752	− .1313	− .3293	.3779	0.8556
.36	.6406	.3594	.3739	− .1346	− .3255	.3864	0.8373
.37	.6443	.3557	.3725	− .1378	− .3216	.3947	0.8186
.38	.6480	.3520	.3712	− .1410	− .3176	.4028	0.7996
.39	.6517	.3483	.3697	− .1442	− .3135	.4107	0.7803
.40	.6554	.3446	.3683	− .1473	− .3094	.4184	0.7607
.41	.6591	.3409	.3668	− .1504	− .3051	.4259	0.7408
.42	.6628	.3372	.3653	− .1534	− .3008	.4332	0.7206
.43	.6664	.3336	.3637	− .1564	− .2965	.4403	0.7001
.44	.6700	.3300	.3621	− .1593	− .2920	.4472	0.6793
.45	.6736	.3264	.3605	− .1622	− .2875	.4539	0.6583
.46	.6772	.3228	.3589	− .1651	− .2830	.4603	0.6371
.47	.6808	.3192	.3572	− .1679	− .2783	.4666	0.6156
.48	.6844	.3156	.3555	− .1707	− .2736	.4727	0.5940
.49	.6879	.3121	.3538	− .1734	− .2689	.4785	0.5721
.50	.6915	.3085	.3521	− .1760	− .2641	.4841	0.5501

Normal Distribution

NORMAL DISTRIBUTION AND RELATED FUNCTIONS

x	$F(x)$	$1 - F(x)$	$f(x)$	$f'(x)$	$f''(x)$	$f'''(x)$	$f^{\mathrm{iv}}(x)$
.50	.6915	.3085	.3521	−.1760	−.2641	.4841	.5501
.51	.6950	.3050	.3503	−.1787	−.2592	.4895	.5279
.52	.6985	.3015	.3485	−.1812	−.2543	.4947	.5056
.53	.7019	.2981	.3467	−.1837	−.2493	.4996	.4831
.54	.7054	.2946	.3448	−.1862	−.2443	.5043	.4605
.55	.7088	.2912	.3429	−.1886	−.2392	.5088	.4378
.56	.7123	.2877	.3410	−.1920	−.2341	.5131	.4150
.57	.7157	.2843	.3391	−.1933	−.2289	.5171	.3921
.58	.7190	.2810	.3372	−.1956	−.2238	.5209	.3691
.59	.7224	.2776	.3352	−.1978	−.2185	.5245	.3461
.60	.7257	.2743	.3332	−.1999	−.2133	.5278	.3231
.61	.7291	.2709	.3312	−.2020	−.2080	.5309	.3000
.62	.7324	.2676	.3292	−.2041	−.2027	.5338	.2770
.63	.7357	.2643	.3271	−.2061	−.1973	.5365	.2539
.64	.7389	.2611	.3251	−.2080	−.1919	.5389	.2309
.65	.7422	.2578	.3230	−.2099	−.1865	.5411	.2078
.66	.7454	.2546	.3209	−.2118	−.1811	.5431	.1849
.67	.7486	.2514	.3187	−.2136	−.1757	.5448	.1620
.68	.7517	.2483	.3166	−.2153	−.1702	.5463	.1391
.69	.7549	.2451	.3144	−.2170	−.1647	.5476	.1164
.70	.7580	.2420	.3123	−.2186	−.1593	.5486	.0937
.71	.7611	.2389	.3101	−.2201	−.1538	.5495	.0712
.72	.7642	.2358	.3079	−.2217	−.1483	.5501	.0487
.73	.7673	.2327	.3056	−.2231	−.1428	.5504	.0265
.74	.7704	.2296	.3034	−.2245	−.1373	.5506	.0043
.75	.7734	.2266	.3011	−.2259	−.1318	.5505	−.0176
.76	.7764	.2236	.2989	−.2271	−.1262	.5502	−.0394
.77	.7794	.2206	.2966	−.2284	−.1207	.5497	−.0611
.78	.7823	.2177	.2943	−.2296	−.1153	.5490	−.0825
.79	.7852	.2148	.2920	−.2307	−.1098	.5481	−.1037
.80	.7881	.2119	.2897	−.2318	−.1043	.5469	−.1247
.81	.7910	.2090	.2874	−.2328	−.0988	.5456	−.1455
.82	.7939	.2061	.2850	−.2337	−.0934	.5440	−.1660
.83	.7967	.2033	.2827	−.2346	−.0880	.5423	−.1862
.84	.7995	.2005	.2803	−.2355	−.0825	.5403	−.2063
.85	.8023	.1977	.2780	−.2363	−.0771	.5381	−.2260
.86	.8051	.1949	.2756	−.2370	−.0718	.5358	−.2455
.87	.8078	.1922	.2732	−.2377	−.0664	.5332	−.2646
.88	.8106	.1894	.2709	−.2384	−.0611	.5305	−.2835
.89	.8133	.1867	.2685	−.2389	−.0558	.5276	−.3021
.90	.8159	.1841	.2661	−.2395	−.0506	.5245	−.3203
.91	.8186	.1814	.2637	−.2400	−.0453	.5212	−.3383
.92	.8212	.1788	.2613	−.2404	−.0401	.5177	−.3559
.93	.8238	.1762	.2589	−.2408	−.0350	.5140	−.3731
.94	.8264	.1736	.2565	−.2411	−.0299	.5102	−.3901
.95	.8289	.1711	.2541	−.2414	−.0248	.5062	−.4066
.96	.8315	.1685	.2516	−.2416	−.0197	.5021	−.4228
.97	.8340	.1660	.2492	−.2417	−.0147	.4978	−.4387
.98	.8365	.1635	.2468	−.2419	−.0098	.4933	−.4541
.99	.8389	.1611	.2444	−.2420	−.0049	.4887	−.4692
1.00	.8413	.1587	.2420	−.2420	.0000	.4839	−.4839

NORMAL DISTRIBUTION AND RELATED FUNCTIONS

x	$F(x)$	$1 - F(x)$	$f(x)$	$f'(x)$	$f''(x)$	$f'''(x)$	$f^{iv}(x)$
1.00	.8413	.1587	.2420	−.2420	.0000	.4839	−.4839
1.01	.8438	.1562	.2396	−.2420	.0048	.4790	−.4983
1.02	.8461	.1539	.2371	−.2419	.0096	.4740	−.5122
1.03	.8485	.1515	.2347	−.2418	.0143	.4688	−.5257
1.04	.8508	.1492	.2323	−.2416	.0190	.4635	−.5389
1.05	.8531	.1469	.2299	−.2414	.0236	.4580	−.5516
1.06	.8554	.1446	.2275	−.2411	.0281	.4524	−.5639
1.07	.8577	.1423	.2251	−.2408	.0326	.4467	−.5758
1.08	.8599	.1401	.2227	−.2405	.0371	.4409	−.5873
1.09	.8621	.1379	.2203	−.2401	.0414	.4350	−.5984
1.10	.8643	.1357	.2179	−.2396	.0458	.4290	−.6091
1.11	.8665	.1335	.2155	−.2392	.0500	.4228	−.6193
1.12	.8686	.1314	.2131	−.2386	.0542	.4166	−.6292
1.13	.8708	.1292	.2107	−.2381	.0583	.4102	−.6386
1.14	.8729	.1271	.2083	−.2375	.0624	.4038	−.6476
1.15	.8749	.1251	.2059	−.2368	.0664	.3973	−.6561
1.16	.8770	.1230	.2036	−.2361	.0704	.3907	−.6643
1.17	.8790	.1210	.2012	−.2354	.0742	.3840	−.6720
1.18	.8810	.1190	.1989	−.2347	.0780	.3772	−.6792
1.19	.8830	.1170	.1965	−.2339	.0818	.3704	−.6861
1.20	.8849	.1151	.1942	−.2330	.0854	.3635	−.6926
1.21	.8869	.1131	.1919	−.2322	.0890	.3566	−.6986
1.22	.8888	.1112	.1895	−.2312	.0926	.3496	−.7042
1.23	.8907	.1093	.1872	−.2303	.0960	.3425	−.7094
1.24	.8925	.1075	.1849	−.2293	.0994	.3354	−.7141
1.25	.8944	.1056	.1826	−.2283	.1027	.3282	−.7185
1.26	.8962	.1038	.1804	−.2273	.1060	.3210	−.7224
1.27	.8980	.1020	.1781	−.2262	.1092	.3138	−.7259
1.28	.8997	.1003	.1758	−.2251	.1123	.3065	−.7291
1.29	.9015	.0985	.1736	−.2240	.1153	.2992	−.7318
1.30	.9032	.0968	.1714	−.2228	.1182	.2918	−.7341
1.31	.9049	.0951	.1691	−.2216	.1211	.2845	−.7361
1.32	.9066	.0934	.1669	−.2204	.1239	.2771	−.7376
1.33	.9082	.0918	.1647	−.2191	.1267	.2697	−.7388
1.34	.9099	.0901	.1626	−.2178	.1293	.2624	−.7395
1.35	.9115	.0885	.1604	−.2165	.1319	.2550	−.7399
1.36	.9131	.0869	.1582	−.2152	.1344	.2476	−.7400
1.37	.9147	.0853	.1561	−.2138	.1369	.2402	−.7396
1.38	.9162	.0838	.1539	−.2125	.1392	.2328	−.7389
1.39	.9177	.0823	.1518	−.2110	.1415	.2254	−.7378
1.40	.9192	.0808	.1497	−.2096	.1437	.2180	−.7364
1.41	.9207	.0793	.1476	−.2082	.1459	.2107	−.7347
1.42	.9222	.0778	.1456	−.2067	.1480	.2033	−.7326
1.43	.9236	.0764	.1435	−.2052	.1500	.1960	−.7301
1.44	.9251	.0749	.1415	−.2037	.1519	.1887	−.7274
1.45	.9265	.0735	.1394	−.2022	.1537	.1815	−.7243
1.46	.9279	.0721	.1374	−.2006	.1555	.1742	−.7209
1.47	.9292	.0708	.1354	−.1991	.1572	.1670	−.7172
1.48	.9306	.0694	.1334	−.1975	.1588	.1599	−.7132
1.49	.9319	.0681	.1315	−.1959	.1604	.1528	−.7089
1.50	.9332	.0668	.1295	−.1943	.1619	.1457	−.7043

Normal Distribution

NORMAL DISTRIBUTION AND RELATED FUNCTIONS

x	$F(x)$	$1 - F(x)$	$f(x)$	$f'(x)$	$f''(x)$	$f'''(x)$	$f^{\mathrm{iv}}(x)$
1.50	.9332	.0668	.1295	−.1943	.1619	.1457	−.7043
1.51	.9345	.0655	.1276	−.1927	.1633	.1387	−.6994
1.52	.9357	.0643	.1257	−.1910	.1647	.1317	−.6942
1.53	.9370	.0630	.1238	−.1894	.1660	.1248	−.6888
1.54	.9382	.0618	.1219	−.1877	.1672	.1180	−.6831
1.55	.9394	.0606	.1200	−.1860	.1683	.1111	−.6772
1.56	.9406	.0594	.1182	−.1843	.1694	.1044	−.6710
1.57	.9418	.0582	.1163	−.1826	.1704	.0977	−.6646
1.58	.9429	.0571	.1145	−.1809	.1714	.0911	−.6580
1.59	.9441	.0559	.1127	−.1792	.1722	.0846	−.6511
1.60	.9452	.0548	.1109	−.1775	.1730	.0781	−.6441
1.61	.9463	.0537	.1092	−.1757	.1738	.0717	−.6368
1.62	.9474	.0526	.1074	−.1740	.1745	.0654	−.6293
1.63	.9484	.0516	.1057	−.1723	.1751	.0591	−.6216
1.64	.9495	.0505	.1040	−.1705	.1757	.0529	−.6138
1.65	.9505	.0495	.1023	−.1687	.1762	.0468	−.6057
1.66	.9515	.0485	.1006	−.1670	.1766	.0408	−.5975
1.67	.9525	.0475	.0989	−.1652	.1770	.0349	−.5891
1.68	.9535	.0465	.0973	−.1634	.1773	.0290	−.5806
1.69	.9545	.0455	.0957	−.1617	.1776	.0233	−.5720
1.70	.9554	.0446	.0940	−.1599	.1778	.0176	−.5632
1.71	.9564	.0436	.0925	−.1581	.1779	.0120	−.5542
1.72	.9573	.0427	.0909	−.1563	.1780	.0065	−.5452
1.73	.9582	.0418	.0893	−.1546	.1780	.0011	−.5360
1.74	.9591	.0409	.0878	−.1528	.1780	−.0042	−.5267
1.75	.9599	.0401	.0863	−.1510	.1780	−.0094	−.5173
1.76	.9608	.0392	.0848	−.1492	.1778	−.0146	−.5079
1.77	.9616	.0384	.0833	−.1474	.1777	−.0196	−.4983
1.78	.9625	.0375	.0818	−.1457	.1774	−.0245	−.4887
1.79	.9633	.0367	.0804	−.1439	.1772	−.0294	−.4789
1.80	.9641	.0359	.0790	−.1421	.1769	−.0341	−.4692
1.81	.9649	.0351	.0775	−.1403	.1765	−.0388	−.4593
1.82	.9656	.0344	.0761	−.1386	.1761	−.0433	−.4494
1.83	.9664	.0336	.0748	−.1368	.1756	−.0477	−.4395
1.84	.9671	.0329	.0734	−.1351	.1751	−.0521	−.4295
1.85	.9678	.0322	.0721	−.1333	.1746	−.0563	−.4195
1.86	.9686	.0314	.0707	−.1316	.1740	−.0605	−.4095
1.87	.9693	.0307	.0694	−.1298	.1734	−.0645	−.3995
1.88	.9699	.0301	.0681	−.1281	.1727	−.0685	−.3894
1.89	.9706	.0294	.0669	−.1264	.1720	−.0723	−.3793
1.90	.9713	.0287	.0656	−.1247	.1713	−.0761	−.3693
1.91	.9719	.0281	.0644	−.1230	.1705	−.0797	−.3592
1.92	.9726	.0274	.0632	−.1213	.1697	−.0832	−.3492
1.93	.9732	.0268	.0620	−.1196	.1688	−.0867	−.3392
1.94	.9738	.0262	.0608	−.1179	.1679	−.0900	−.3292
1.95	.9744	.0256	.0596	−.1162	.1670	−.0933	−.3192
1.96	.9750	.0250	.0584	−.1145	.1661	−.0964	−.3093
1.97	.9756	.0244	.0573	−.1129	.1651	−.0994	−.2994
1.98	.9761	.0239	.0562	−.1112	.1641	−.1024	−.2895
1.99	.9767	.0233	.0551	−.1096	.1630	−.1052	−.2797
2.00	.9772	.0228	.0540	−.1080	.1620	−.1080	−.2700

NORMAL DISTRIBUTION AND RELATED FUNCTIONS

x	$F(x)$	$1 - F(x)$	$f(x)$	$f'(x)$	$f''(x)$	$f'''(x)$	$f^{iv}(x)$
2.00	.9773	.0227	.0540	−.1080	.1620	−.1080	−.2700
2.01	.9778	.0222	.0529	−.1064	.1609	−.1106	−.2603
2.02	.9783	.0217	.0519	−.1048	.1598	−.1132	−.2506
2.03	.9788	.0212	.0508	−.1032	.1586	−.1157	−.2411
2.04	.9793	.0207	.0498	−.1016	.1575	−.1180	−.2316
2.05	.9798	.0202	.0488	−.1000	.1563	−.1203	−.2222
2.06	.9803	.0197	.0478	−.0985	.1550	−.1225	−.2129
2.07	.9808	.0192	.0468	−.0969	.1538	−.1245	−.2036
2.08	.9812	.0188	.0459	−.0954	.1526	−.1265	−.1945
2.09	.9817	.0183	.0449	−.0939	.1513	−.1284	−.1854
2.10	.9821	.0179	.0440	−.0924	.1500	−.1302	−.1765
2.11	.9826	.0174	.0431	−.0909	.1487	−.1320	−.1676
2.12	.9830	.0170	.0422	−.0894	.1474	−.1336	−.1588
2.13	.9834	.0166	.0413	−.0879	.1460	−.1351	−.1502
2.14	.9838	.0162	.0404	−.0865	.1446	−.1366	−.1416
2.15	.9842	.0158	.0396	−.0850	.1433	−.1380	−.1332
2.16	.9846	.0154	.0387	−.0836	.1419	−.1393	−.1249
2.17	.9850	.0150	.0379	−.0822	.1405	−.1405	−.1167
2.18	.9854	.0146	.0371	−.0808	.1391	−.1416	−.1086
2.19	.9857	.0143	.0363	−.0794	.1377	−.1426	−.1006
2.20	.9861	.0139	.0355	−.0780	.1362	−.1436	−.0927
2.21	.9864	.0136	.0347	−.0767	.1348	−.1445	−.0850
2.22	.9868	.0132	.0339	−.0754	.1333	−.1453	−.0774
2.23	.9871	.0129	.0332	−.0740	.1319	−.1460	−.0700
2.24	.9875	.0125	.0325	−.0727	.1304	−.1467	−.0626
2.25	.9878	.0122	.0317	−.0714	.1289	−.1473	−.0554
2.26	.9881	.0119	.0310	−.0701	.1275	−.1478	−.0484
2.27	.9884	.0116	.0303	−.0689	.1260	−.1483	−.0414
2.28	.9887	.0113	.0297	−.0676	.1245	−.1486	−.0346
2.29	.9890	.0110	.0290	−.0664	.1230	−.1490	−.0279
2.30	.9893	.0107	.0283	−.0652	.1215	−.1492	−.0214
2.31	.9896	.0104	.0277	−.0639	.1200	−.1494	−.0150
2.32	.9898	.0102	.0270	−.0628	.1185	−.1495	−.0088
2.33	.9901	.0099	.0264	−.0616	.1170	−.1496	−.0027
2.34	.9904	.0096	.0258	−.0604	.1155	−.1496	.0033
2.35	.9906	.0094	.0252	−.0593	.1141	−.1495	.0092
2.36	.9909	.0091	.0246	−.0581	.1126	−.1494	.0149
2.37	.9911	.0089	.0241	−.0570	.1111	−.1492	.0204
2.38	.9913	.0087	.0235	−.0559	.1096	−.1490	.0258
2.39	.9916	.0084	.0229	−.0548	.1081	−.1487	.0311
2.40	.9918	.0082	.0224	−.0538	.1066	−.1483	.0362
2.41	.9920	.0080	.0219	−.0527	.1051	−.1480	.0412
2.42	.9922	.0078	.0213	−.0516	.1036	−.1475	.0461
2.43	.9925	.0075	.0208	−.0506	.1022	−.1470	.0508
2.44	.9927	.0073	.0203	−.0496	.1007	−.1465	.0554
2.45	.9929	.0071	.0198	−.0486	.0992	−.1459	.0598
2.46	.9931	.0069	.0194	−.0476	.0978	−.1453	.0641
2.47	.9932	.0068	.0189	−.0467	.0963	−.1446	.0683
2.48	.9934	.0066	.0184	−.0457	.0949	−.1439	.0723
2.49	.9936	.0064	.0180	−.0448	.0935	−.1432	.0762
2.50	.9938	.0062	.0175	−.0438	.0920	−.1424	.0800

Normal Distribution

NORMAL DISTRIBUTION AND RELATED FUNCTIONS

x	$F(x)$	$1 - F(x)$	$f(x)$	$f'(x)$	$f''(x)$	$f'''(x)$	$f^{IV}(x)$
2.50	.9938	.0062	.0175	−.0438	.0920	−.1424	.0800
2.51	.9940	.0060	.0171	−.0429	.0906	−.1416	.0836
2.52	.9941	.0059	.0167	−.0420	.0892	−.1408	.0871
2.53	.9943	.0057	.0163	−.0411	.0878	−.1399	.0905
2.54	.9945	.0055	.0158	−.0403	.0864	−.1389	.0937
2.55	.9946	.0054	.0155	−.0394	.0850	−.1380	.0968
2.56	.9948	.0052	.0151	−.0386	.0836	−.1370	.0998
2.57	.9949	.0051	.0147	−.0377	.0823	−.1360	.1027
2.58	.9951	.0049	.0143	−.0369	.0809	−.1350	.1054
2.59	.9952	.0048	.0139	−.0361	.0796	−.1339	.1080
2.60	.9953	.0047	.0136	−.0353	.0782	−.1328	.1105
2.61	.9955	.0045	.0132	−.0345	.0769	−.1317	.1129
2.62	.9956	.0044	.0129	−.0338	.0756	−.1305	.1152
2.63	.9957	.0043	.0126	−.0330	.0743	−.1294	.1173
2.64	.9959	.0041	.0122	−.0323	.0730	−.1282	.1194
2.65	.9960	.0040	.0119	−.0316	.0717	−.1270	.1213
2.66	.9961	.0039	.0116	−.0309	.0705	−.1258	.1231
2.67	.9962	.0038	.0113	−.0302	.0692	−.1245	.1248
2.68	.9963	.0037	.0110	−.0295	.0680	−.1233	.1264
2.69	.9964	.0036	.0107	−.0288	.0668	−.1220	.1279
2.70	.9965	.0035	.0104	−.0281	.0656	−.1207	.1293
2.71	.9966	.0034	.0101	−.0275	.0644	−.1194	.1306
2.72	.9967	.0033	.0099	−.0269	.0632	−.1181	.1317
2.73	.9968	.0032	.0096	−.0262	.0620	−.1168	.1328
2.74	.9969	.0031	.0093	−.0256	.0608	−.1154	.1338
2.75	.9970	.0030	.0091	−.0250	.0597	−.1141	.1347
2.76	.9971	.0029	.0088	−.0244	.0585	−.1127	.1356
2.77	.9972	.0028	.0086	−.0238	.0574	−.1114	.1363
2.78	.9973	.0027	.0084	−.0233	.0563	−.1100	.1369
2.79	.9974	.0026	.0081	−.0227	.0552	−.1087	.1375
2.80	.9974	.0026	.0079	−.0222	.0541	−.1073	.1379
2.81	.9975	.0025	.0077	−.0216	.0531	−.1059	.1383
2.82	.9976	.0024	.0075	−.0211	.0520	−.1045	.1386
2.83	.9977	.0023	.0073	−.0206	.0510	−.1031	.1389
2.84	.9977	.0023	.0071	−.0201	.0500	−.1017	.1390
2.85	.9978	.0022	.0069	−.0196	.0490	−.1003	.1391
2.86	.9979	.0021	.0067	−.0191	.0480	−.0990	.1391
2.87	.9979	.0021	.0065	−.0186	.0470	−.0976	.1391
2.88	.9980	.0020	.0063	−.0182	.0460	−.0962	.1389
2.89	.9981	.0019	.0061	−.0177	.0451	−.0948	.1388
2.90	.9981	.0019	.0060	−.0173	.0441	−.0934	.1385
2.91	.9982	.0018	.0058	−.0168	.0432	−.0920	.1382
2.92	.9982	.0018	.0056	−.0164	.0423	−.0906	.1378
2.93	.9983	.0017	.0055	−.0160	.0414	−.0893	.1374
2.94	.9984	.0016	.0053	−.0156	.0405	−.0879	.1369
2.95	.9984	.0016	.0051	−.0152	.0396	−.0865	.1364
2.96	.9985	.0015	.0050	−.0148	.0388	−.0852	.1358
2.97	.9985	.0015	.0048	−.0144	.0379	−.0838	.1352
2.98	.9986	.0014	.0047	−.0140	.0371	−.0825	.1345
2.99	.9986	.0014	.0046	−.0137	.0363	−.0811	.1337
3.00	.9987	.0013	.0044	−.0133	.0355	−.0798	.1330

NORMAL DISTRIBUTION AND RELATED FUNCTIONS

x	$F(x)$	$1 - F(x)$	$f(x)$	$f'(x)$	$f''(x)$	$f'''(x)$	$f^{\mathrm{IV}}(x)$
3.00	.9987	.0013	.0044	−.0133	.0355	−.0798	.1330
3.01	.9987	.0013	.0043	−.0130	.0347	−.0785	.1321
3.02	.9987	.0013	.0042	−.0126	.0339	−.0771	.1313
3.03	.9988	.0012	.0040	−.0123	.0331	−.0758	.1304
3.04	.9988	.0012	.0039	−.0119	.0324	−.0745	.1294
3.05	.9989	.0011	.0038	−.0116	.0316	−.0732	.1285
3.06	.9989	.0011	.0037	−.0113	.0309	−.0720	.1275
3.07	.9989	.0011	.0036	−.0110	.0302	−.0707	.1264
3.08	.9990	.0010	.0035	−.0107	.0295	−.0694	.1254
3.09	.9990	.0010	.0034	−.0104	.0288	−.0682	.1243
3.10	.9990	.0010	.0033	−.0101	.0281	−.0669	.1231
3.11	.9991	.0009	.0032	−.0099	.0275	−.0657	.1220
3.12	.9991	.0009	.0031	−.0096	.0268	−.0645	.1208
3.13	.9991	.0009	.0030	−.0093	.0262	−.0633	.1196
3.14	.9992	.0008	.0029	−.0091	.0256	−.0621	.1184
3.15	.9992	.0008	.0028	−.0088	.0249	−.0609	.1171
3.16	.9992	.0008	.0027	−.0086	.0243	−.0598	.1159
3.17	.9992	.0008	.0026	−.0083	.0237	−.0586	.1146
3.18	.9993	.0007	.0025	−.0081	.0232	−.0575	.1133
3.19	.9993	.0007	.0025	−.0079	.0226	−.0564	.1120
3.20	.9993	.0007	.0024	−.0076	.0220	−.0552	.1107
3.21	.9993	.0007	.0023	−.0074	.0215	−.0541	.1093
3.22	.9994	.0006	.0022	−.0072	.0210	−.0531	.1080
3.23	.9994	.0006	.0022	−.0070	.0204	−.0520	.1066
3.24	.9994	.0006	.0021	−.0068	.0199	−.0509	.1053
3.25	.9994	.0006	.0020	−.0066	.0194	−.0499	.1039
3.26	.9994	.0006	.0020	−.0064	.0189	−.0488	.1025
3.27	.9995	.0005	.0019	−.0062	.0184	−.0478	.1011
3.28	.9995	.0005	.0018	−.0060	.0180	−.0468	.0997
3.29	.9995	.0005	.0018	−.0059	.0175	−.0458	.0983
3.30	.9995	.0005	.0017	−.0057	.0170	−.0449	.0969
3.31	.9995	.0005	.0017	−.0055	.0166	−.0439	.0955
3.32	.9995	.0005	.0016	−.0054	.0162	−.0429	.0941
3.33	.9996	.0004	.0016	−.0052	.0157	−.0420	.0927
3.34	.9996	.0004	.0015	−.0050	.0153	−.0411	.0913
3.35	.9996	.0004	.0015	−.0049	.0149	−.0402	.0899
3.36	.9996	.0004	.0014	−.0047	.0145	−.0393	.0885
3.37	.9996	.0004	.0014	−.0046	.0141	−.0384	.0871
3.38	.9996	.0004	.0013	−.0045	.0138	−.0376	.0857
3.39	.9997	.0003	.0013	−.0043	.0134	−.0367	.0843
3.40	.9997	.0003	.0012	−.0042	.0130	−.0359	.0829
3.41	.9997	.0003	.0012	−.0041	.0127	−.0350	.0815
3.42	.9997	.0003	.0012	−.0039	.0123	−.0342	.0801
3.43	.9997	.0003	.0011	−.0038	.0120	−.0334	.0788
3.44	.9997	.0003	.0011	−.0037	.0116	−.0327	.0774
3.45	.9997	.0003	.0010	−.0036	.0113	−.0319	.0761
3.46	.9997	.0003	.0010	−.0035	.0110	−.0311	.0747
3.47	.9997	.0003	.0010	−.0034	.0107	−.0304	.0734
3.48	.9997	.0003	.0009	−.0033	.0104	−.0297	.0721
3.49	.9998	.0002	.0009	−.0032	.0101	−.0290	.0707
3.50	.9998	.0002	.0009	−.0031	.0098	−.0283	.0694

Normal Distribution

NORMAL DISTRIBUTION AND RELATED FUNCTIONS

x	$F(x)$	$1 - F(x)$	$f(x)$	$f'(x)$	$f''(x)$	$f'''(x)$	$f^{iv}(x)$
3.50	.9998	.0002	.0009	−.0031	.0098	−.0283	.0694
3.51	.9998	.0002	.0008	−.0030	.0095	−.0276	.0681
3.52	.9998	.0002	.0008	−.0029	.0093	−.0269	.0669
3.53	.9998	.0002	.0008	−.0028	.0090	−.0262	.0656
3.54	.9998	.0002	.0008	−.0027	.0087	−.0256	.0643
3.55	.9998	.0002	.0007	−.0026	.0085	−.0249	.0631
3.56	.9998	.0002	.0007	−.0025	.0082	−.0243	.0618
3.57	.9998	.0002	.0007	−.0024	.0080	−.0237	.0606
3.58	.9998	.0002	.0007	−.0024	.0078	−.0231	.0594
3.59	.9998	.0002	.0006	−.0023	.0075	−.0225	.0582
3.60	.9998	.0002	.0006	−.0022	.0073	−.0219	.0570
3.61	.9998	.0002	.0006	−.0021	.0071	−.0214	.0559
3.62	.9999	.0001	.0006	−.0021	.0069	−.0208	.0547
3.63	.9999	.0001	.0005	−.0020	.0067	−.0203	.0536
3.64	.9999	.0001	.0005	−.0019	.0065	−.0198	.0524
3.65	.9999	.0001	.0005	−.0019	.0063	−.0192	.0513
3.66	.9999	.0001	.0005	−.0018	.0061	−.0187	.0502
3.67	.9999	.0001	.0005	−.0017	.0059	−.0182	.0492
3.68	.9999	.0001	.0005	−.0017	.0057	−.0177	.0481
3.69	.9999	.0001	.0004	−.0016	.0056	−.0173	.0470
3.70	.9999	.0001	.0004	−.0016	.0054	−.0168	.0460
3.71	.9999	.0001	.0004	−.0015	.0052	−.0164	.0450
3.72	.9999	.0001	.0004	−.0015	.0051	−.0159	.0440
3.73	.9999	.0001	.0004	−.0014	.0049	−.0155	.0430
3.74	.9999	.0001	.0004	−.0014	.0048	−.0150	.0420
3.75	.9999	.0001	.0004	−.0013	.0046	−.0146	.0410
3.76	.9999	.0001	.0003	−.0013	.0045	−.0142	.0401
3.77	.9999	.0001	.0003	−.0012	.0043	−.0138	.0392
3.78	.9999	.0001	.0003	−.0012	.0042	−.0134	.0382
3.79	.9999	.0001	.0003	−.0012	.0041	−.0131	.0373
3.80	.9999	.0001	.0003	−.0011	.0039	−.0127	.0365
3.81	.9999	.0001	.0003	−.0011	.0038	−.0123	.0356
3.82	.9999	.0001	.0003	−.0010	.0037	−.0120	.0347
3.83	.9999	.0001	.0003	−.0010	.0036	−.0116	.0339
3.84	.9999	.0001	.0003	−.0010	.0034	−.0113	.0331
3.85	.9999	.0001	.0002	−.0009	.0033	−.0110	.0323
3.86	.9999	.0001	.0002	−.0009	.0032	−.0107	.0315
3.87	.9999	.0001	.0002	−.0009	.0031	−.0104	.0307
3.88	.9999	.0001	.0002	−.0008	.0030	−.0100	.0299
3.89	1.0000	.0000	.0002	−.0008	.0029	−.0098	.0292
3.90	1.0000	.0000	.0002	−.0008	.0028	−.0095	.0284
3.91	1.0000	.0000	.0002	−.0008	.0027	−.0092	.0277
3.92	1.0000	.0000	.0002	−.0007	.0026	−.0089	.0270
3.93	1.0000	.0000	.0002	−.0007	.0026	−.0086	.0263
3.94	1.0000	.0000	.0002	−.0007	.0025	−.0084	.0256
3.95	1.0000	.0000	.0002	−.0006	.0024	−.0081	.0250
3.96	1.0000	.0000	.0002	−.0006	.0023	−.0079	.0243
3.97	1.0000	.0000	.0002	−.0006	.0022	−.0076	.0237
3.98	1.0000	.0000	.0001	−.0006	.0022	−.0074	.0230
3.99	1.0000	.0000	.0001	−.0006	.0021	−.0072	.0224
4.00	1.0000	.0000	.0001	−.0005	.0020	−.0070	.0218

x	1.282	1.645	1.960	2.326	2.576	3.090
$F(x)$	.90	.95	.975	.99	.995	.999
$2[1 - F(x)]$	.20	.10	.05	.02	.01	.002

II.2 TOLERANCE FACTORS FOR NORMAL DISTRIBUTIONS

This table gives factors K such that the probability is γ that at least a proportion P of the distribution will be included between $\bar{x} - Ks$ and $\bar{x} + Ks$, where $\bar{x}$ and s are estimates of the mean and standard deviation computed from a sample of size N. Values of K are given for $P = 0.75, 0.90, 0.95, 0.99, 0.999$ and $\gamma = 0.75, 0.90, 0.95, 0.99$ and for various values of N. For example, if $\bar{x} = 10.0$ and $s = 1.0$, $N = 16$, the interval $\bar{x} \pm Ks = 10.0 \pm 3.812(1.0) = 10.0 \pm 3.812$, or the interval 6.188 to 13.812 will contain 99% of the population with confidence coefficient 0.95. The values of K are computed assuming that the observations are from normal populations.

Normal Distribution

TOLERANCE FACTORS FOR NORMAL DISTRIBUTIONS

$$\lambda = 0.75$$

P / N	0.75	0.90	0.95	0.99	0.999	P / N	0.75	0.90	0.95	0.99	0.999
2	4.498	6.301	7.414	9.531	11.920	55	1.249	1.785	2.127	2.795	3.571
3	2.501	3.538	4.187	5.431	6.844	60	1.243	1.778	2.118	2.784	3.556
4	2.035	2.892	3.431	4.471	5.657	65	1.239	1.771	2.110	2.773	3.543
5	1.825	2.599	3.088	4.033	5.117	70	1.235	1.765	2.104	2.764	3.531
6	1.704	2.429	2.889	3.779	4.802	75	1.231	1.760	2.098	2.757	3.521
7	1.624	2.318	2.757	3.611	4.593	80	1.228	1.756	2.092	2.749	3.512
8	1.568	2.238	2.663	3.491	4.444	85	1.225	1.752	2.087	2.743	3.504
9	1.525	2.178	2.593	3.400	4.330	90	1.223	1.748	2.083	2.737	3.497
10	1.492	2.131	2.537	3.328	4.241	95	1.220	1.745	2.079	2.732	3.490
11	1.465	2.093	2.493	3.271	4.169	100	1.218	1.742	2.075	2.727	3.484
12	1.443	2.062	2.456	3.223	4.110	110	1.214	1.736	2.069	2.719	3.473
13	1.425	2.036	2.424	3.183	4.059	120	1.211	1.732	2.063	2.712	3.464
14	1.409	2.013	2.398	3.148	4.016	130	1.208	1.728	2.059	2.705	3.456
15	1.395	1.994	2.375	3.118	3.979	140	1.206	1.724	2.054	2.700	3.449
16	1.383	1.977	2.355	3.092	3.946	150	1.204	1.721	2.051	2.695	3.443
17	1.372	1.962	2.337	3.069	3.917	160	1.202	1.718	2.047	2.691	3.437
18	1.363	1.948	2.321	3.048	3.891	170	1.200	1.716	2.044	2.687	3.432
19	1.355	1.936	2.307	3.030	3.867	180	1.198	1.713	2.042	2.683	3.427
20	1.347	1.925	2.294	3.013	3.846	190	1.197	1.711	2.039	2.680	3.423
21	1.340	1.915	2.282	2.998	3.827	200	1.195	1.709	2.037	2.677	3.419
22	1.334	1.906	2.271	2.984	3.809	250	1.190	1.702	2.028	2.665	3.404
23	1.328	1.898	2.261	2.971	3.793	300	1.186	1.696	2.021	2.656	3.393
24	1.322	1.891	2.252	2.959	3.778	400	1.181	1.688	2.012	2.644	3.378
25	1.317	1.883	2.244	2.948	3.764	500	1.177	1.683	2.006	2.636	3.368
26	1.313	1.877	2.236	2.938	3.751	600	1.175	1.680	2.002	2.631	3.360
27	1.309	1.871	2.229	2.929	3.740	700	1.173	1.677	1.998	2.626	3.355
30	1.297	1.855	2.210	2.904	3.708	800	1.171	1.675	1.996	2.623	3.350
35	1.283	1.834	2.185	2.871	3.667	900	1.170	1.673	1.993	2.620	3.347
40	1.271	1.818	2.166	2.846	3.635	1000	1.169	1.671	1.992	2.617	3.344
45	1.262	1.805	2.150	2.826	3.609	∞	1.150	1.645	1.960	2.576	3.291
50	1.255	1.794	2.138	2.809	3.588						

TOLERANCE FACTORS FOR NORMAL DISTRIBUTIONS

$\lambda = 0.90$

P / N	0.75	0.90	0.95	0.99	0.999	P / N	0.75	0.90	0.95	0.99	0.999
2	11.407	15.978	18.800	24.167	30.227	55	1.329	1.901	2.265	2.976	3.801
3	4.132	5.847	6.919	8.974	11.309	60	1.320	1.887	2.248	2.955	3.774
4	2.932	4.166	4.943	6.440	8.149	65	1.312	1.875	2.235	2.937	3.751
5	2.454	3.494	4.152	5.423	6.879	70	1.304	1.865	2.222	2.920	3.730
6	2.196	3.131	3.723	4.870	6.188	75	1.298	1.856	2.211	2.906	3.712
7	2.034	2.902	3.452	4.521	5.750	80	1.292	1.848	2.202	2.894	3.696
8	1.921	2.743	3.264	4.278	5.446	85	1.287	1.841	2.193	2.882	3.682
9	1.839	2.626	3.125	4.098	5.220	90	1.283	1.834	2.185	2.872	3.669
10	1.775	2.535	3.018	3.959	5.046	95	1.278	1.828	2.178	2.863	3.657
11	1.724	2.463	2.933	3.849	4.906	100	1.275	1.822	2.172	2.854	3.646
12	1.683	2.404	2.863	3.758	4.792	110	1.268	1.813	2.160	2.839	3.626
13	1.648	2.355	2.805	3.682	4.697	120	1.262	1.804	2.150	2.826	3.610
14	1.619	2.314	2.756	3.618	4.615	130	1.257	1.797	2.141	2.814	3.595
15	1.594	2.278	2.713	3.562	4.545	140	1.252	1.791	2.134	2.804	3.582
16	1.572	2.246	2.676	3.514	4.484	150	1.248	1.785	2.127	2.795	3.571
17	1.552	2.219	2.643	3.471	4.430	160	1.245	1.780	2.121	2.787	3.561
18	1.535	2.194	2.614	3.433	4.382	170	1.242	1.775	2.116	2.780	3.552
19	1.520	2.172	2.588	3.399	4.339	180	1.239	1.771	2.111	2.774	3.543
20	1.506	2.152	2.564	3.368	4.300	190	1.236	1.767	2.106	2.768	3.536
21	1.493	2.135	2.543	3.340	4.264	200	1.234	1.764	2.102	2.762	3.429
22	1.482	2.118	2.524	3.315	4.232	250	1.224	1.750	2.085	2.740	3.501
23	1.471	2.103	2.506	3.292	4.203	300	1.217	1.740	2.073	2.725	3.481
24	1.462	2.089	2.489	3.270	4.176	400	1.207	1.726	2.057	2.703	3.453
25	1.453	2.077	2.474	3.251	4.151	500	1.201	1.717	2.046	2.689	3.434
26	1.444	2.065	2.460	3.232	4.127	600	1.196	1.710	2.038	2.678	3.421
27	1.437	2.054	2.447	3.215	4.106	700	1.192	1.705	2.032	2.670	3.411
30	1.417	2.025	2.413	3.170	4.049	800	1.189	1.701	2.027	2.663	3.402
35	1.390	1.988	2.368	3.112	3.974	900	1.187	1.697	2.023	2.658	3.396
40	1.370	1.959	2.334	3.066	3.917	1000	1.185	1.695	2.019	2.654	3.390
45	1.354	1.935	2.306	3.030	3.871	∞	1.150	1.645	1.960	2.576	3.291
50	1.340	1.916	2.284	3.001	3.833						

Normal Distribution

TOLERANCE FACTORS FOR NORMAL DISTRIBUTIONS

$\lambda = 0.95$

P / N	0.75	0.90	0.95	0.99	0.999	P / N	0.75	0.90	0.95	0.99	0.999
2	22.858	32.019	37.674	48.430	60.573	55	1.382	1.976	2.354	3.094	3.951
3	5.922	8.380	9.916	12.861	16.208	60	1.369	1.958	2.333	3.066	3.916
4	3.779	5.369	6.370	8.299	10.502	65	1.359	1.943	2.315	3.042	3.886
5	3.002	4.275	5.079	6.634	8.415	70	1.349	1.929	2.299	3.021	3.859
6	2.604	3.712	4.414	5.775	7.337	75	1.341	1.917	2.285	3.002	3.835
7	2.361	3.369	4.007	5.248	6.676	80	1.334	1.907	2.272	2.986	3.814
8	2.197	3.136	3.732	4.891	6.226	85	1.327	1.897	2.261	2.971	3.795
9	2.078	2.967	3.532	4.631	5.899	90	1.321	1.889	2.251	2.958	3.778
10	1.987	2.839	3.379	4.433	5.649	95	1.315	1.881	2.241	2.945	3.763
11	1.916	2.737	3.259	4.277	5.452	100	1.311	1.874	2.233	2.934	3.748
12	1.858	2.655	3.162	4.150	5.291	110	1.302	1.861	2.218	2.915	3.723
13	1.810	2.587	3.081	4.044	5.158	120	1.294	1.850	2.205	2.898	3.702
14	1.770	2.529	3.012	3.955	5.045	130	1.288	1.841	2.194	2.883	3.683
15	1.735	2.480	2.954	3.878	4.949	140	1.282	1.833	2.184	2.870	3.666
16	1.705	2.437	2.903	3.812	4.865	150	1.277	1.825	2.175	2.859	3.652
17	1.679	2.400	2.858	3.754	4.791	160	1.272	1.819	2.167	2.848	3.638
18	1.655	2.366	2.819	3.702	4.725	170	1.268	1.813	2.160	2.839	3.627
19	1.635	2.337	2.784	3.656	4.667	180	1.264	1.808	2.154	2.831	3.616
20	1.616	2.310	2.752	3.615	4.614	190	1.261	1.803	2.148	2.823	3.606
21	1.599	2.286	2.723	3.577	4.567	200	1.258	1.798	2.143	2.816	3.597
22	1.584	2.264	2.697	3.543	4.523	250	1.245	1.780	2.121	2.788	3.561
23	1.570	2.244	2.673	3.512	4.484	300	1.236	1.767	2.106	2.767	3.535
24	1.557	2.225	2.651	3.483	4.447	400	1.223	1.749	2.084	2.739	3.499
25	1.545	2.208	2.631	3.457	4.413	500	1.215	1.737	2.070	2.721	3.475
26	1.534	2.193	2.612	3.432	4.382	600	1.209	1.729	2.060	2.707	3.458
27	1.523	2.178	2.595	3.409	4.353	700	1.204	1.722	2.052	2.697	3.445
30	1.497	2.140	2.549	3.350	4.278	800	1.201	1.717	2.046	2.688	3.434
35	1.462	2.090	2.490	3.272	4.179	900	1.198	1.712	2.040	2.682	3.426
40	1.435	2.052	2.445	3.213	4.104	1000	1.195	1.709	2.036	2.676	3.418
45	1.414	2.021	2.408	3.165	4.042	∞	1.150	1.645	1.960	2.576	3.291
50	1.396	1.996	2.379	3.126	3.993						

TOLERANCE FACTORS FOR NORMAL DISTRIBUTIONS

λ = 0.99

P / N	0.75	0.90	0.95	0.99	0.999	P / N	0.75	0.90	0.95	0.99	0.999
2	114.363	160.193	188.491	242.300	303.054	55	1.490	2.130	2.538	3.335	4.260
3	13.378	18.930	22.401	29.055	36.616	60	1.471	2.103	2.506	3.293	4.206
4	6.614	9.398	11.150	14.527	18.383	65	1.455	2.080	2.478	3.257	4.160
5	4.643	6.612	7.855	10.260	13.015	70	1.440	2.060	2.454	3.225	4.120
6	3.743	5.337	6.345	8.301	10.548	75	1.428	2.042	2.433	3.197	4.084
7	3.233	4.613	5.488	7.187	9.142	80	1.417	2.026	2.414	3.173	4.053
8	2.905	4.147	4.936	6.468	8.234	85	1.407	2.012	2.397	3.150	4.024
9	2.677	3.822	4.550	5.966	7.600	90	1.398	1.999	2.382	3.130	3.999
10	2.508	3.582	4.265	5.594	7.129	95	1.390	1.987	2.368	3.112	3.976
11	2.378	3.397	4.045	5.308	6.766	100	1.383	1.977	2.355	3.096	3.954
12	2.274	3.250	3.870	5.079	6.477	110	1.369	1.958	2.333	3.066	3.917
13	2.190	3.130	3.727	4.893	6.240	120	1.358	1.942	2.314	3.041	3.885
14	2.120	3.029	3.608	4.737	6.043	130	1.349	1.928	2.298	3.019	3.857
15	2.060	2.945	3.507	4.605	5.876	140	1.340	1.916	2.283	3.000	3.833
16	2.009	2.872	3.421	4.492	5.732	150	1.332	1.905	2.270	2.983	3.811
17	1.965	2.808	3.345	4.393	5.607	160	1.326	1.896	2.259	2.968	3.792
18	1.926	2.753	3.279	4.307	5.497	170	1.320	1.887	2.248	2.955	3.774
19	1.891	2.703	3.221	4.230	5.399	180	1.314	1.879	2.239	2.942	3.759
20	1.860	2.659	3.168	4.161	5.312	190	1.309	1.872	2.230	2.931	3.744
21	1.833	2.620	3.121	4.100	5.234	200	1.304	1.865	2.222	2.921	3.731
22	1.808	2.584	3.078	4.044	5.163	250	1.286	1.839	2.191	2.880	3.678
23	1.785	2.551	3.040	3.993	5.098	300	1.273	1.820	2.169	2.850	3.641
24	1.764	2.522	3.004	3.947	5.039	400	1.255	1.794	2.138	2.809	3.589
25	1.745	2.494	2.972	3.904	4.985	500	1.243	1.777	2.117	2.783	3.555
26	1.727	2.469	2.941	3.865	4.935	600	1.234	1.764	2.102	2.763	3.530
27	1.711	2.446	2.914	3.828	4.888	700	1.227	1.755	2.091	2.748	3.511
30	1.668	2.385	2.841	3.733	4.768	800	1.222	1.747	2.082	2.736	3.495
35	1.613	2.306	2.748	3.611	4.611	900	1.218	1.741	2.075	2.726	3.483
40	1.571	2.247	2.677	3.518	4.493	1000	1.214	1.736	2.068	2.718	3.472
45	1.539	2.200	2.621	3.444	4.399	∞	1.150	1.645	1.960	2.576	3.291
50	1.512	2.162	2.576	3.385	4.323						

II.3   FACTORS FOR COMPUTING PROBABLE ERRORS

The probable error of a series of n measures $a_1, a_2, a_3 \cdots a_n$, the mean of which is m is given by the expression,

$$e = \frac{0.6745}{\sqrt{n-1}} \sqrt{(m - a_1)^2 + (m - a_2)^2 + \cdots + (m - a_n)^2} \ .$$

The probable error of the mean is,

$$E = \frac{0.6745}{\sqrt{n(n-1)}} \sqrt{(m - a_1)^2 + (m - a_2)^2 + \cdots + (m - a_n)^2} \ .$$

The following approximate equations are convenient forms for computation,

$$e = 0.8453 \frac{\Sigma d}{\sqrt{n(n-1)}}$$

$$E = 0.8453 \frac{\Sigma d}{n \sqrt{n-1}} \ .$$

The symbol Σd represents the arithmetical sum of the deviations.

For convenience in computing the probable error the value of several of the factors involved is given for values of n from 2 to 100.

FACTORS FOR COMPUTING PROBABLE ERRORS

n	$\dfrac{1}{\sqrt{n}}$	$\dfrac{1}{\sqrt{n\,(n-1)}}$	$\dfrac{.6745}{\sqrt{n-1}}$	$\dfrac{.6745}{\sqrt{n\,(n-1)}}$	$\dfrac{.8453}{n\sqrt{n-1}}$	$\dfrac{.8453}{\sqrt{n\,(n-1)}}$
2	.707107	.707107	.6745	.4769	.4227	.5978
3	.577350	.408248	.4769	.2754	.1993	.3451
4	.500000	.288675	.3894	.1947	.1220	.2440
5	.447214	.223607	.3372	.1508	.0845	.1890
6	.408248	.182574	.3016	.1231	.0630	.1543
7	.377964	.154303	.2754	.1041	.0493	.1304
8	.353553	.133631	.2549	.0901	.0399	.1130
9	.333333	.117851	.2385	.0795	.0332	.0996
10	.316228	.105409	.2248	.0711	.0282	.0891
11	.301511	.095346	.2133	.0643	.0243	.0806
12	.288675	.087039	.2034	.0587	.0212	.0736
13	.277350	.080064	.1947	.0540	.0188	.0677
14	.267261	.074125	.1871	.0500	.0167	.0627
15	.258199	.069007	.1803	.0465	.0151	.0583
16	.250000	.064550	.1742	.0435	.0136	.0546
17	.242536	.060634	.1686	.0409	.0124	.0513
18	.235702	.057166	.1636	.0386	.0114	.0483
19	.229416	.054074	.1590	.0365	.0105	.0457
20	.223607	.051299	.1547	.0346	.0097	.0434
21	.218218	.048795	.1508	.0329	.0090	.0412
22	.213201	.046524	.1472	.0314	.0084	.0393
23	.208514	.044455	.1438	.0300	.0078	.0376
24	.204124	.042563	.1406	.0287	.0073	.0360
25	.200000	.040825	.1377	.0275	.0069	.0345
26	.196116	.039223	.1349	.0265	.0065	.0332
27	.192450	.037743	.1323	.0255	.0061	.0319
28	.188982	.036370	.1298	.0245	.0058	.0307
29	.185695	.035093	.1275	.0237	.0055	.0297
30	.182574	.033903	.1252	.0229	.0052	.0287
31	.179605	.032791	.1231	.0221	.0050	.0277
32	.176777	.031750	.1211	.0214	.0047	.0268
33	.174078	.030773	.1192	.0208	.0045	.0260
34	.171499	.029854	.1174	.0201	.0043	.0252
35	.169031	.028989	.1157	.0196	.0041	.0245
36	.166667	.028172	.1140	.0190	.0040	.0238
37	.164399	.027400	.1124	.0185	.0038	.0232
38	.162221	.026669	.1109	.0180	.0037	.0225
39	.160128	.025976	.1094	.0175	.0035	.0220
40	.158114	.025318	.1080	.0171	.0034	.0214
41	.156174	.024693	.1066	.0167	.0033	.0209
42	.154303	.024098	.1053	.0163	.0031	.0204
43	.152499	.023531	.1041	.0159	.0030	.0199
44	.150756	.022990	.1029	.0155	.0029	.0194
45	.149071	.022473	.1017	.0152	.0028	.0190
46	.147442	.021979	.1005	.0148	.0027	.0186
47	.145865	.021507	.0994	.0145	.0027	.0182
48	.144338	.021054	.0984	.0142	.0026	.0178
49	.142857	.020620	.0974	.0139	.0025	.0174
50	.141421	.020203	.0964	.0136	.0024	.0171

Normal Distribution

FACTORS FOR COMPUTING PROBABLE ERRORS

n	$\dfrac{1}{\sqrt{n}}$	$\dfrac{1}{\sqrt{n\,(n-1)}}$	$\dfrac{.6745}{\sqrt{n-1}}$	$\dfrac{.6745}{\sqrt{n(n-1)}}$	$\dfrac{.8453}{n\sqrt{n-1}}$	$\dfrac{.8453}{\sqrt{n(n-1)}}$
50	.141421	.020203	.0964	.0136	.0024	.0171
51	.140028	.019803	.0954	.0134	.0023	.0167
52	.138675	.019418	.0945	.0131	.0023	.0164
53	.137361	.019048	.0935	.0129	.0022	.0161
54	.136083	.018692	.0927	.0126	.0022	.0158
55	.134840	.018349	.0918	.0124	.0021	.0155
56	.133631	.018019	.0910	.0122	.0020	.0152
57	.132453	.017700	.0901	.0119	.0020	.0150
58	.131306	.017392	.0893	.0117	.0019	.0147
59	.130189	.017095	.0886	.0115	.0019	.0145
60	.129099	.016807	.0878	.0113	.0018	.0142
61	.128037	.016529	.0871	.0112	.0018	.0140
62	.127000	.016261	.0864	.0110	.0018	.0138
63	.125988	.016001	.0857	.0108	.0017	.0135
64	.125000	.015749	.0850	.0106	.0017	.0133
65	.124035	.015504	.0843	.0105	.0016	.0131
66	.123091	.015268	.0837	.0103	.0016	.0129
67	.122169	.015038	.0830	.0101	.0016	.0127
68	.121268	.014815	.0824	.0100	.0015	.0125
69	.120386	.014599	.0818	.0099	.0015	.0123
70	.119523	.014389	.0812	.0097	.0015	.0122
71	.118678	.014185	.0806	.0096	.0014	.0120
72	.117851	.013986	.0801	.0094	.0014	.0118
73	.117041	.013793	.0795	.0093	.0014	.0117
74	.116248	.013606	.0789	.0092	.0013	.0115
75	.115470	.013423	.0784	.0091	.0013	.0113
76	.114708	.013245	.0779	.0089	.0013	.0112
77	.113961	.013072	.0773	.0088	.0013	.0111
78	.113228	.012904	.0769	.0087	.0012	.0109
79	.112509	.012739	.0764	.0086	.0012	.0108
80	.111803	.012579	.0759	.0085	.0012	.0106
81	.111111	.012423	.0754	.0084	.0012	.0105
82	.110432	.012270	.0749	.0083	.0012	.0104
83	.109764	.012121	.0745	.0082	.0011	.0103
84	.109109	.011976	.0740	.0081	.0011	.0101
85	.108465	.011835	.0736	.0080	.0011	.0100
86	.107833	.011696	.0732	.0079	.0011	.0099
87	.107211	.011561	.0727	.0078	.0011	.0098
88	.106600	.011429	.0723	.0077	.0010	.0097
89	.106000	.011300	.0719	.0076	.0010	.0096
90	.105409	.011173	.0715	.0075	.0010	.0094
91	.104828	.011050	.0711	.0075	.0010	.0093
92	.104257	.010929	.0707	.0074	.0010	.0092
93	.103695	.010811	.0703	.0073	.0010	.0091
94	.103142	.010695	.0699	.0072	.0009	.0090
95	.102598	.010582	.0696	.0071	.0009	.0089
96	.102062	.010471	.0692	.0071	.0009	.0089
97	.101535	.010363	.0688	.0070	.0009	.0088
98	.101015	.010257	.0685	.0069	.0009	.0087
99	.100504	.010152	.0681	.0069	.0009	.0086
100	.100000	.010050	.0678	.0068	.0008	.0085

II.4 PROBABILITY OF OCCURRENCE OF DEVIATIONS

The significance of deviations is indicated by this table. The probability of occurrence of deviations as great as or greater than any specific value is given for various ratios of deviation to probable error and also with respect to the standard deviation. The probability of occurrence is stated in per cent or chances in 100. The odds against occurrence are also stated. The probable error is 0.6745 × the standard deviation.

Ratio, dev. to P.E.	Probable occurrence %	Odds against, to 1	Ratio dev. to std. dev.	Probable occurrence %	Odds against, to 1
1.0	50.00	1.00	0.67449	50.00	1.00
1.1	45.81	1.18	0.7	48.39	1.07
1.2	41.83	1.39	0.8	42.37	1.36
1.3	38.06	1.63	0.9	36.81	1.72
1.4	34.50	1.90	1.0	31.73	2.15
1.5	31.17	2.21	1.1	27.13	2.69
1.6	28.05	2.57	1.2	23.01	3.35
1.7	25.15	2.98	1.3	19.36	4.17
1.8	22.47	3.45	1.4	16.15	5.19
1.9	20.00	4.00	1.5	13.36	6.48
2.0	17.73	4.64	1.6	10.96	8.12
2.1	15.67	5.38	1.7	8.91	10.22
2.2	13.78	6.25	1.8	7.19	12.92
2.3	12.08	7.28	1.9	5.74	16.41
2.4	10.55	8.48	2.0	4.55	20.98
2.5	9.18	9.90	2.1	3.57	26.99
2.6	7.95	11.58	2.2	2.78	34.96
2.7	6.86	13.58	2.3	2.14	45.62
2.8	5.89	15.96	2.4	1.64	59.99
2.9	5.05	18.82	2.5	1.24	79.52
3.0	4.30	22.24	2.6	.932	106.3
3.1	3.65	26.37	2.7	.693	143.2
3.2	3.09	31.36	2.8	.511	194.7
3.3	2.60	37.42	2.9	.373	267.0
3.4	2.18	44.80	3.0	.270	369.4
3.5	1.82	53.82	3.1	.194	515.7
3.6	1.52	64.89	3.2	.137	726.7
3.7	1.26	78.53	3.3	.0967	1033.
3.8	1.04	95.38	3.4	.0674	1483.
3.9	.853	116.3	3.5	.0465	2149.
4.0	.698	142.3	3.6	.0318	3142.
4.1	.569	174.9	3.7	.0216	4637.
4.2	.461	215.8	3.8	.0145	6915.
4.3	.373	267.2	3.9	.00962	10394.
4.4	.300	332.4	4.0	.00634	15772.
4.5	.240	415.0	5.0	5.73×10^{-5}	1.744×10^6
4.6	.192	520.4	6.0	2.0×10^{-7}	5.0×10^8
4.7	.152	655.3	7.0	2.6×10^{-10}	3.9×10^{11}
4.8	.121	828.3			
4.9	.0950	1052.			
5.0	.0745	1341.			
6.0	.0052	19300.			
7.0	.00023	4.27×10^5			
8.0	6.8×10^{-6}	1.47×10^7			
9.0	1.3×10^{-7}	7.30×10^8			
10.0	1.5×10^{-9}	6.5×10^{10}			

Valid for samples of size 30 or greater.

III. Binomial, Poisson, Hypergeometric, and Negative Binomial Distributions

III.1 INDIVIDUAL TERMS, BINOMIAL DISTRIBUTION

The $(x + 1)^{st}$ term in the expansion of the binomial $(q + p)^n$ is given by

$$f(x) = \binom{n}{x} p^x q^{n-x}, \qquad x = 0, 1, 2, \ldots, n; q + p = 1 .$$

This is the probability of exactly x successes in n independent binomial trials with probability of success on a single trial equal to p. This table contains the individual terms of $f(x)$ for specified choices of x, n, and p.

For $p > 0.5$, the value of $\binom{n}{x} p^x q^{n-x}$ is found by using the table entry for $\binom{n}{n-x} q^{n-x} p^x$.

INDIVIDUAL TERMS, BINOMIAL DISTRIBUTION

n	x	.05	.10	.15	.20	.25	.30	.35	.40	.45	.50
1	0	.9500	.9000	.8500	.8000	.7500	.7000	.6500	.6000	.5500	.5000
	1	.0500	.1000	.1500	.2000	.2500	.3000	.3500	.4000	.4500	.5000
2	0	.9025	.8100	.7225	.6400	.5625	.4900	.4225	.3600	.3025	.2500
	1	.0950	.1800	.2550	.3200	.3750	.4200	.4550	.4800	.4950	.5000
	2	.0025	.0100	.0225	.0400	.0625	.0900	.1225	.1600	.2025	.2500
3	0	.8574	.7290	.6141	.5120	.4219	.3430	.2746	.2160	.1664	.1250
	1	.1354	.2430	.3251	.3840	.4219	.4410	.4436	.4320	.4084	.3750
	2	.0071	.0270	.0574	.0960	.1406	.1890	.2389	.2880	.3341	.3750
	3	.0001	.0010	.0034	.0080	.0156	.0270	.0429	.0640	.0911	.1250
4	0	.8145	.6561	.5220	.4096	.3164	.2401	.1785	.1296	.0915	.0625
	1	.1715	.2916	.3685	.4096	.4219	.4116	.3845	.3456	.2995	.2500
	2	.0135	.0486	.0975	.1536	.2109	.2646	.3105	.3456	.3675	.3750
	3	.0005	.0036	.0115	.0256	.0469	.0756	.1115	.1536	.2005	.2500
	4	.0000	.0001	.0005	.0016	.0039	.0081	.0150	.0256	.0410	.0625
5	0	.7738	.5905	.4437	.3277	.2373	.1681	.1160	.0778	.0503	.0312
	1	.2036	.3280	.3915	.4096	.3955	.3602	.3124	.2592	.2059	.1562
	2	.0214	.0729	.1382	.2048	.2637	.3087	.3364	.3456	.3369	.3125
	3	.0011	.0081	.0244	.0512	.0879	.1323	.1811	.2304	.2757	.3125
	4	.0000	.0004	.0022	.0064	.0146	.0284	.0488	.0768	.1128	.1562
	5	.0000	.0000	.0001	.0003	.0010	.0024	.0053	.0102	.0185	.0312
6	0	.7351	.5314	.3771	.2621	.1780	.1176	.0754	.0467	.0277	.0156
	1	.2321	.3543	.3993	.3932	.3560	.3025	.2437	.1866	.1359	.0938
	2	.0305	.0984	.1762	.2458	.2966	.3241	.3280	.3110	.2780	.2344
	3	.0021	.0146	.0415	.0819	.1318	.1852	.2355	.2765	.3032	.3125
	4	.0001	.0012	.0055	.0154	.0330	.0595	.0951	.1382	.1861	.2344
	5	.0000	.0001	.0004	.0015	.0044	.0102	.0205	.0369	.0609	.0938
	6	.0000	.0000	.0000	.0001	.0002	.0007	.0018	.0041	.0083	.0156
7	0	.6983	.4783	.3206	.2097	.1335	.0824	.0490	.0280	.0152	.0078
	1	.2573	.3720	.3960	.3670	.3115	.2471	.1848	.1306	.0872	.0547
	2	.0406	.1240	.2097	.2753	.3115	.3177	.2985	.2613	.2140	.1641
	3	.0036	.0230	.0617	.1147	.1730	.2269	.2679	.2903	.2918	.2734
	4	.0002	.0026	.0109	.0287	.0577	.0972	.1442	.1935	.2388	.2734
	5	.0000	.0002	.0012	.0043	.0115	.0250	.0466	.0774	.1172	.1641
	6	.0000	.0000	.0001	.0004	.0013	.0036	.0084	.0172	.0320	.0547
	7	.0000	.0000	.0000	.0000	.0001	.0002	.0006	.0016	.0037	.0078
8	0	.6634	.4305	.2725	.1678	.1001	.0576	.0319	.0168	.0084	.0039
	1	.2793	.3826	.3847	.3355	.2670	.1977	.1373	.0896	.0548	.0312
	2	.0515	.1488	.2376	.2936	.3115	.2965	.2587	.2090	.1569	.1094
	3	.0054	.0331	.0839	.1468	.2076	.2541	.2786	.2787	.2568	.2188
	4	.0004	.0046	.0185	.0459	.0865	.1361	.1875	.2322	.2627	.2734
	5	.0000	.0004	.0026	.0092	.0231	.0467	.0808	.1239	.1719	.2188
	6	.0000	.0000	.0002	.0011	.0038	.0100	.0217	.0413	.0703	.1094
	7	.0000	.0000	.0000	.0001	.0004	.0012	.0033	.0079	.0164	.0312
	8	.0000	.0000	.0000	.0000	.0000	.0001	.0002	.0007	.0017	.0039

Linear interpolations with respect to p will in general be accurate at most to two decimal places.

Binomial, Poisson, and Hypergeometric Distributions

INDIVIDUAL TERMS, BINOMIAL DISTRIBUTION

n	x	.05	.10	.15	.20	.25	.30	.35	.40	.45	.50
9	0	.6302	.3874	.2316	.1342	.0751	.0404	.0207	.0101	.0046	.0020
	1	.2985	.3874	.3679	.3020	.2253	.1556	.1004	.0605	.0339	.0176
	2	.0629	.1722	.2597	.3020	.3003	.2668	.2162	.1612	.1110	.0703
	3	.0077	.0446	.1069	.1762	.2336	.2668	.2716	.2508	.2119	.1641
	4	.0006	.0074	.0283	.0661	.1168	.1715	.2194	.2508	.2600	.2461
	5	.0000	.0008	.0050	.0165	.0389	.0735	.1181	.1672	.2128	.2461
	6	.0000	.0001	.0006	.0028	.0087	.0210	.0424	.0743	.1160	.1641
	7	.0000	.0000	.0000	.0003	.0012	.0039	.0098	.0212	.0407	.0703
	8	.0000	.0000	.0000	.0000	.0001	.0004	.0013	.0035	.0083	.0176
	9	.0000	.0000	.0000	.0000	.0000	.0000	.0001	.0003	.0008	.0020
10	0	.5987	.3487	.1969	.1074	.0563	.0282	.0135	.0060	.0025	.0010
	1	.3151	.3874	.3474	.2684	.1877	.1211	.0725	.0403	.0207	.0098
	2	.0746	.1937	.2759	.3020	.2816	.2335	.1757	.1209	.0763	.0439
	3	.0105	.0574	.1298	.2013	.2503	.2668	.2522	.2150	.1665	.1172
	4	.0010	.0112	.0401	.0881	.1460	.2001	.2377	.2508	.2384	.2051
	5	.0001	.0015	.0085	.0264	.0584	.1029	.1536	.2007	.2340	.2461
	6	.0000	.0001	.0012	.0055	.0162	.0368	.0689	.1115	.1596	.2051
	7	.0000	.0000	.0001	.0008	.0031	.0090	.0212	.0425	.0746	.1172
	8	.0000	.0000	.0000	.0001	.0004	.0014	.0043	.0106	.0229	.0439
	9	.0000	.0000	.0000	.0000	.0000	.0001	.0005	.0016	.0042	.0098
	10	.0000	.0000	.0000	.0000	.0000	.0000	.0000	.0001	.0003	.0010
11	0	.5688	.3138	.1673	.0859	.0422	.0198	.0088	.0036	.0014	.0004
	1	.3293	.3835	.3248	.2362	.1549	.0932	.0518	.0266	.0125	.0055
	2	.0867	.2131	.2866	.2953	.2581	.1998	.1395	.0887	.0513	.0269
	3	.0137	.0710	.1517	.2215	.2581	.2568	.2254	.1774	.1259	.0806
	4	.0014	.0158	.0536	.1107	.1721	.2201	.2428	.2365	.2060	.1611
	5	.0001	.0025	.0132	.0388	.0803	.1321	.1830	.2207	.2360	.2256
	6	.0000	.0003	.0023	.0097	.0268	.0566	.0985	.1471	.1931	.2256
	7	.0000	.0000	.0003	.0017	.0064	.0173	.0379	.0701	.1128	.1611
	8	.0000	.0000	.0000	.0002	.0011	.0037	.0102	.0234	.0462	.0806
	9	.0000	.0000	.0000	.0000	.0001	.0005	.0018	.0052	.0126	.0269
	10	.0000	.0000	.0000	.0000	.0000	.0000	.0002	.0007	.0021	.0054
	11	.0000	.0000	.0000	.0000	.0000	.0000	.0000	.0000	.0002	.0005
12	0	.5404	.2824	.1422	.0687	.0317	.0138	.0057	.0022	.0008	.0002
	1	.3413	.3766	.3012	.2062	.1267	.0712	.0368	.0174	.0075	.0029
	2	.0988	.2301	.2924	.2835	.2323	.1678	.1088	.0639	.0339	.0161
	3	.0173	.0852	.1720	.2362	.2581	.2397	.1954	.1419	.0923	.0537
	4	.0021	.0213	.0683	.1329	.1936	.2311	.2367	.2128	.1700	.1208
	5	.0002	.0038	.0193	.0532	.1032	.1585	.2039	.2270	.2225	.1934
	6	.0000	.0005	.0040	.0155	.0401	.0792	.1281	.1766	.2124	.2256
	7	.0000	.0000	.0006	.0033	.0115	.0291	.0591	.1009	.1489	.1934
	8	.0000	.0000	.0001	.0005	.0024	.0078	.0199	.0420	.0762	.1208
	9	.0000	.0000	.0000	.0001	.0004	.0015	.0048	.0125	.0277	.0537
	10	.0000	.0000	.0000	.0000	.0000	.0002	.0008	.0025	.0068	.0161
	11	.0000	.0000	.0000	.0000	.0000	.0000	.0001	.0003	.0010	.0029
	12	.0000	.0000	.0000	.0000	.0000	.0000	.0000	.0000	.0001	.0002

INDIVIDUAL TERMS, BINOMIAL DISTRIBUTION

x	.05	.10	.15	.20	.25	.30	.35	.40	.45	.50
3 0	.5133	.2542	.1209	.0550	.0238	.0097	.0037	.0013	.0004	.0001
1	.3512	.3672	.2774	.1787	.1029	.0540	.0259	.0113	.0045	.0016
2	.1109	.2448	.2937	.2680	.2059	.1388	.0836	.0453	.0220	.0095
3	.0214	.0997	.1900	.2457	.2517	.2181	.1651	.1107	.0660	.0349
4	.0028	.0277	.0838	.1535	.2097	.2337	.2222	.1845	.1350	.0873
5	.0003	.0055	.0266	.0691	.1258	.1803	.2154	.2214	.1989	.1571
6	.0000	.0008	.0063	.0230	.0559	.1030	.1546	.1968	.2169	.2095
7	.0000	.0001	.0011	.0058	.0186	.0442	.0833	.1312	.1775	.2095
8	.0000	.0000	.0001	.0011	.0047	.0142	.0336	.0656	.1089	.1571
9	.0000	.0000	.0000	.0001	.0009	.0034	.0101	.0243	.0495	.0873
10	.0000	.0000	.0000	.0000	.0001	.0006	.0022	.0065	.0162	.0349
11	.0000	.0000	.0000	.0000	.0000	.0001	.0003	.0012	.0036	.0095
12	.0000	.0000	.0000	.0000	.0000	.0000	.0000	.0001	.0005	.0016
13	.0000	.0000	.0000	.0000	.0000	.0000	.0000	.0000	.0000	.0001
4 0	.4877	.2288	.1028	.0440	.0178	.0068	.0024	.0008	.0002	.0001
1	.3593	.3559	.2539	.1539	.0832	.0407	.0181	.0073	.0027	.0009
2	.1229	.2570	.2912	.2501	.1802	.1134	.0634	.0317	.0141	.0056
3	.0259	.1142	.2056	.2501	.2402	.1943	.1366	.0845	.0462	.0222
4	.0037	.0349	.0998	.1720	.2202	.2290	.2022	.1549	.1040	.0611
5	.0004	.0078	.0352	.0860	.1468	.1963	.2178	.2066	.1701	.1222
6	.0000	.0013	.0093	.0322	.0734	.1262	.1759	.2066	.2088	.1833
7	.0000	.0002	.0019	.0092	.0280	.0618	.1082	.1574	.1952	.2095
8	.0000	.0000	.0003	.0020	.0082	.0232	.0510	.0918	.1398	.1833
9	.0000	.0000	.0000	.0003	.0018	.0066	.0183	.0408	.0762	.1222
10	.0000	.0000	.0000	.0000	.0003	.0014	.0049	.0136	.0312	.0611
11	.0000	.0000	.0000	.0000	.0000	.0002	.0010	.0033	.0093	.0222
12	.0000	.0000	.0000	.0000	.0000	.0000	.0001	.0005	.0019	.0056
13	.0000	.0000	.0000	.0000	.0000	.0000	.0000	.0001	.0002	.0009
14	.0000	.0000	.0000	.0000	.0000	.0000	.0000	.0000	.0000	.0001
5 0	.4633	.2059	.0874	.0352	.0134	.0047	.0016	.0005	.0001	.0000
1	.3658	.3432	.2312	.1319	.0668	.0305	.0126	.0047	.0016	.0005
2	.1348	.2669	.2856	.2309	.1559	.0916	.0476	.0219	.0090	.0032
3	.0307	.1285	.2184	.2501	.2252	.1700	.1110	.0634	.0318	.0139
4	.0049	.0428	.1156	.1876	.2252	.2186	.1792	.1268	.0780	.0417
5	.0006	.0105	.0449	.1032	.1651	.2061	.2123	.1859	.1404	.0916
6	.0000	.0019	.0132	.0430	.0917	.1472	.1906	.2066	.1914	.1527
7	.0000	.0003	.0030	.0138	.0393	.0811	.1319	.1771	.2013	.1964
8	.0000	.0000	.0005	.0035	.0131	.0348	.0710	.1181	.1647	.1964
9	.0000	.0000	.0001	.0007	.0034	.0116	.0298	.0612	.1048	.1527
10	.0000	.0000	.0000	.0001	.0007	.0030	.0096	.0245	.0515	.0916
11	.0000	.0000	.0000	.0000	.0001	.0006	.0024	.0074	.0191	.0417
12	.0000	.0000	.0000	.0000	.0000	.0001	.0004	.0016	.0052	.0139
13	.0000	.0000	.0000	.0000	.0000	.0000	.0001	.0003	.0010	.0032
14	.0000	.0000	.0000	.0000	.0000	.0000	.0000	.0000	.0001	.0005
15	.0000	.0000	.0000	.0000	.0000	.0000	.0000	.0000	.0000	.0000

INDIVIDUAL TERMS, BINOMIAL DISTRIBUTION

n	x	.05	.10	.15	.20	p .25	.30	.35	.40	.45	.50
16	0	.4401	.1853	.0743	.0281	.0100	.0033	.0010	.0003	.0001	.0000
	1	.3706	.3294	.2097	.1126	.0535	.0228	.0087	.0030	.0009	.0002
	2	.1463	.2745	.2775	.2111	.1336	.0732	.0353	.0150	.0056	.0018
	3	.0359	.1423	.2285	.2463	.2079	.1465	.0888	.0468	.0215	.0085
	4	.0061	.0514	.1311	.2001	.2252	.2040	.1553	.1014	.0572	.0278
	5	.0008	.0137	.0555	.1201	.1802	.2099	.2008	.1623	.1123	.0667
	6	.0001	.0028	.0180	.0550	.1101	.1649	.1982	.1983	.1684	.1222
	7	.0000	.0004	.0045	.0197	.0524	.1010	.1524	.1889	.1969	.1746
	8	.0000	.0001	.0009	.0055	.0197	.0487	.0923	.1417	.1812	.1964
	9	.0000	.0000	.0001	.0012	.0058	.0185	.0442	.0840	.1318	.1746
	10	.0000	.0000	.0000	.0002	.0014	.0056	.0167	.0392	.0755	.1222
	11	.0000	.0000	.0000	.0000	.0002	.0013	.0049	.0142	.0337	.0667
	12	.0000	.0000	.0000	.0000	.0000	.0002	.0011	.0040	.0115	.0278
	13	.0000	.0000	.0000	.0000	.0000	.0000	.0002	.0008	.0029	.0085
	14	.0000	.0000	.0000	.0000	.0000	.0000	.0000	.0001	.0005	.0018
	15	.0000	.0000	.0000	.0000	.0000	.0000	.0000	.0000	.0001	.0002
	16	.0000	.0000	.0000	.0000	.0000	.0000	.0000	.0000	.0000	.0000
17	0	.4181	.1668	.0631	.0225	.0075	.0023	.0007	.0002	.0000	.0000
	1	.3741	.3150	.1893	.0957	.0426	.0169	.0060	.0019	.0005	.0001
	2	.1575	.2800	.2673	.1914	.1136	.0581	.0260	.0102	.0035	.0010
	3	.0415	.1556	.2359	.2393	.1893	.1245	.0701	.0341	.0144	.0052
	4	.9076	.0605	.1457	.2093	.2209	.1868	.1320	.0796	.0411	.0182
	5	.0010	.0175	.0668	.1361	.1914	.2081	.1849	.1379	.0875	.0472
	6	.0001	.0039	.0236	.0680	.1276	.1784	.1991	.1839	.1432	.0944
	7	.0000	.0007	.0065	.0267	.0668	.1201	.1685	.1927	.1841	.1484
	8	.0000	.0001	.0014	.0084	.0279	.0644	.1134	.1606	.1883	.1855
	9	.0000	.0000	.0003	.0021	.0093	.0276	.0611	.1070	.1540	.1855
	10	.0000	.0000	.0000	.0004	.0025	.0095	.0263	.0571	.1008	.1484
	11	.0000	.0000	.0000	.0001	.0005	.0026	.0090	.0242	.0525	.0944
	12	.0000	.0000	.0000	.0000	.0001	.0006	.0024	.0081	.0215	.0472
	13	.0000	.0000	.0000	.0000	.0000	.0001	.0005	.0021	.0068	.0182
	14	.0000	.0000	.0000	.0000	.0000	.0000	.0001	.0004	.0016	.0052
	15	.0000	.0000	.0000	.0000	.0000	.0000	.0000	.0001	.0003	.0010
	16	.0000	.0000	.0000	.0000	.0000	.0000	.0000	.0000	.0000	.0001
	17	.0000	.0000	.0000	.0000	.0000	.0000	.0000	.0000	.0000	.0000
18	0	.3972	.1501	.0536	.0180	.0056	.0016	.0004	.0001	.0000	.0000
	1	.3763	.3002	.1704	.0811	.0338	.0126	.0042	.0012	.0003	.0001
	2	.1683	.2835	.2556	.1723	.0958	.0458	.0190	.0069	.0022	.0006
	3	.0473	.1680	.2406	.2297	.1704	.1046	.0547	.0246	.0095	.0031
	4	.0093	.0700	.1592	.2153	.2130	.1681	.1104	.0614	.0291	.0117
	5	.0014	.0218	.0787	.1507	.1988	.2017	.1664	.1146	.0666	.0327
	6	.0002	.0052	.0301	.0816	.1436	.1873	.1941	.1655	.1181	.0708
	7	.0000	.0010	.0091	.0350	.0820	.1376	.1792	.1892	.1657	.1214
	8	.0000	.0002	.0022	.0120	.0376	.0811	.1327	.1734	.1864	.1669
	9	.0000	.0000	.0004	.0033	.0139	.0386	.0794	.1284	.1694	.1855
	10	.0000	.0000	.0001	.0008	.0042	.0149	.0385	.0771	.1248	.1669
	11	.0000	.0000	.0000	.0001	.0010	.0046	.0151	.0374	.0742	.1214

INDIVIDUAL TERMS, BINOMIAL DISTRIBUTION

n	x	.05	.10	.15	.20	.25	.30	.35	.40	.45	.50
8	12	.0000	.0000	.0000	.0000	.0002	.0012	.0047	.0145	.0354	.0708
	13	.0000	.0000	.0000	.0000	.0000	.0002	.0012	.0045	.0134	.0327
	14	.0000	.0000	.0000	.0000	.0000	.0000	.0002	.0011	.0039	.0117
	15	.0000	.0000	.0000	.0000	.0000	.0000	.0000	.0002	.0009	.0031
	16	.0000	.0000	.0000	.0000	.0000	.0000	.0000	.0000	.0001	.0006
	17	.0000	.0000	.0000	.0000	.0000	.0000	.0000	.0000	.0000	.0001
	18	.0000	.0000	.0000	.0000	.0000	.0000	.0000	.0000	.0000	.0000
9	0	.3774	.1351	.0456	.0144	.0042	.0011	.0003	.0001	.0000	.0000
	1	.3774	.2852	.1529	.0685	.0268	.0093	.0029	.0008	.0002	.0000
	2	.1787	.2852	.2428	.1540	.0803	.0358	.0138	.0046	.0013	.0003
	3	.0533	.1796	.2428	.2182	.1517	.0869	.0422	.0175	.0062	.0018
	4	.0112	.0798	.1714	.2182	.2023	.1491	.0909	.0467	.0203	.0074
	5	.0018	.0266	.0907	.1636	.2023	.1916	.1468	.0933	.0497	.0222
	6	.0002	.0069	.0374	.0955	.1574	.1916	.1844	.1451	.0949	.0518
	7	.0000	.0014	.0122	.0443	.0974	.1525	.1844	.1797	.1443	.0961
	8	.0000	.0002	.0032	.0166	.0487	.0981	.1489	.1797	.1771	.1442
	9	.0000	.0000	.0007	.0051	.0198	.0514	.0980	.1464	.1771	.1762
	10	.0000	.0000	.0001	.0013	.0066	.0220	.0528	.0976	.1449	.1762
	11	.0000	.0000	.0000	.0003	.0018	.0077	.0233	.0532	.0970	.1442
	12	.0000	.0000	.0000	.0000	.0004	.0022	.0083	.0237	.0529	.0961
	13	.0000	.0000	.0000	.0000	.0001	.0005	.0024	.0085	.0233	.0518
	14	.0000	.0000	.0000	.0000	.0000	.0001	.0006	.0024	.0082	.0222
	15	.0000	.0000	.0000	.0000	.0000	.0000	.0001	.0005	.0022	.0074
	16	.0000	.0000	.0000	.0000	.0000	.0000	.0000	.0001	.0005	.0018
	17	.0000	.0000	.0000	.0000	.0000	.0000	.0000	.0000	.0001	.0003
	18	.0000	.0000	.0000	.0000	.0000	.0000	.0000	.0000	.0000	.0000
	19	.0000	.0000	.0000	.0000	.0000	.0000	.0000	.0000	.0000	.0000
20	0	.3585	.1216	.0388	.0115	.0032	.0008	.0002	.0000	.0000	.0000
	1	.3774	.2702	.1368	.0576	.0211	.0068	.0020	.0005	.0001	.0000
	2	.1887	.2852	.2293	.1369	.0669	.0278	.0100	.0031	.0008	.0002
	3	.0596	.1901	.2428	.2054	.1339	.0716	.0323	.0123	.0040	.0011
	4	.0133	.0898	.1821	.2182	.1897	.1304	.0738	.0350	.0139	.0046
	5	.0022	.0319	.1028	.1746	.2023	.1789	.1272	.0746	.0365	.0148
	6	.0003	.0089	.0454	.1091	.1686	.1916	.1712	.1244	.0746	.0370
	7	.0000	.0020	.0160	.0545	.1124	.1643	.1844	.1659	.1221	.0739
	8	.0000	.0004	.0046	.0222	.0609	.1144	.1614	.1797	.1623	.1201
	9	.0000	.0001	.0011	.0074	.0271	.0654	.1158	.1597	.1771	.1602
	10	.0000	.0000	.0002	.0020	.0099	.0308	.0686	.1171	.1593	.1762
	11	.0000	.0000	.0000	.0005	.0030	.0120	.0336	.0710	.1185	.1602
	12	.0000	.0000	.0000	.0001	.0008	.0039	.0136	.0355	.0727	.1201
	13	.0000	.0000	.0000	.0000	.0002	.0010	.0045	.0146	.0366	.0739
	14	.0000	.0000	.0000	.0000	.0000	.0002	.0012	.0049	.0150	.0370
	15	.0000	.0000	.0000	.0000	.0000	.0000	.0003	.0013	.0049	.0148
	16	.0000	.0000	.0000	.0000	.0000	.0000	.0000	.0003	.0013	.0046
	17	.0000	.0000	.0000	.0000	.0000	.0000	.0000	.0000	.0002	.0011
	18	.0000	.0000	.0000	.0000	.0000	.0000	.0000	.0000	.0000	.0002
	19	.0000	.0000	.0000	.0000	.0000	.0000	.0000	.0000	.0000	.0000
	20	.0000	.0000	.0000	.0000	.0000	.0000	.0000	.0000	.0000	.0000

III.2　CUMULATIVE TERMS, BINOMIAL DISTRIBUTION

For the binomial probability function $f(x)$, the probability of observing x' or more successes is given by

$$\sum_{x=x'}^{n} \binom{n}{x} p^x q^{n-x}, \quad \text{where } p + q = 1.$$

This table contains the values of $\sum_{x=x'}^{n} \binom{n}{x} p^x q^{n-x}$ for specified values of n, x', and p. I

$p > 0.5$, the values for $\sum_{x=x'}^{n} \binom{n}{x} p^x q^{n-x}$ are obtained using the corresponding results ob

tained from

$$1 - \sum_{x=n-x'+1}^{n} \binom{n}{x} q^x p^{n-x}.$$

The cumulative binomial distribution is related to the incomplete beta function as follows:

$$\sum_{x=x'}^{n} \binom{n}{x} p^x q^{n-x} = I_p(x', n - x' + 1) ,$$

where

$$\begin{aligned}
I_x(a, b) &= \frac{B_x(a, b)}{B(a, b)} \\
&= \frac{\Gamma(a + b)}{\Gamma(a)\Gamma(b)} \int_0^x u^{a-1}(1 - u)^{b-1}\, du .
\end{aligned}$$

Thus

$$\begin{aligned}
\sum_{x=0}^{x'-1} \binom{n}{x} p^x q^{n-x} &= 1 - I_p(x', n - x' + 1) \\
&= 1 - \int_0^p u^{x'-1}(1 - u)^{n-x'}\, du \Big/ \int_0^1 u^{x'-1}(1 - u)^{n-x'}\, du .
\end{aligned}$$

The cumulative binomial distribution is related to the cumulative negative binomial distribution as follows:

$$1 - \sum_{x'=0}^{r-1} \binom{x + r}{x'} p^{x'} q^{x+r-x'} = \sum_{x'=0}^{x} \binom{x' + r - 1}{r - 1} p^r q^{x'}$$

or

$$\sum_{x'=r}^{x+r} \binom{x + r}{x'} p^{x'} q^{x+r-x'} = \sum_{x'=0}^{x} \binom{x' + r - 1}{r - 1} p^r q^{x'} .$$

CUMULATIVE TERMS, BINOMIAL DISTRIBUTION

n	x'	.05	.10	.15	.20	.25	.30	.35	.40	.45	.50
						p					
2	1	.0975	.1900	.2775	.3600	.4375	.5100	.5775	.6400	.6975	.7500
	2	.0025	.0100	.0225	.0400	.0625	.0900	.1225	.1600	.2025	.2500
3	1	.1426	.2710	.3859	.4880	.5781	.6570	.7254	.7840	.8336	.8750
	2	.0072	.0280	.0608	.1040	.1562	.2160	.2818	.3520	.4252	.5000
	3	.0001	.0010	.0034	.0080	.0156	.0270	.0429	.0640	.0911	.1250
4	1	.1855	.3439	.4780	.5904	.6836	.7599	.8215	.8704	.9085	.9375
	2	.0140	.0523	.1095	.1808	.2617	.3483	.4370	.5248	.6090	.6875
	3	.0005	.0037	.0120	.0272	.0508	.0837	.1265	.1792	.2415	.3125
	4	.0000	.0001	.0005	.0016	.0039	.0081	.0150	.0256	.0410	.0625
5	1	.2262	.4095	.5563	.6723	.7627	.8319	.8840	.9222	.9497	.9688
	2	.0226	.0815	.1648	.2627	.3672	.4718	.5716	.6630	.7438	.8125
	3	.0012	.0086	.0266	.0579	.1035	.1631	.2352	.3174	.4069	.5000
	4	.0000	.0005	.0022	.0067	.0156	.0308	.0540	.0870	.1312	.1875
	5	.0000	.0000	.0001	.0003	.0010	.0024	.0053	.0102	.0185	.0312
6	1	.2649	.4686	.6229	.7379	.8220	.8824	.9246	.9533	.9723	.9844
	2	.0328	.1143	.2235	.3447	.4661	.5798	.6809	.7667	.8364	.8906
	3	.0022	.0158	.0473	.0989	.1694	.2557	.3529	.4557	.5585	.6562
	4	.0001	.0013	.0059	.0170	.0376	.0705	.1174	.1792	.2553	.3438
	5	.0000	.0001	.0004	.0016	.0046	.0109	.0223	.0410	.0692	.1094
	6	.0000	.0000	.0000	.0001	.0002	.0007	.0018	.0041	.0083	.0156
7	1	.3017	.5217	.6794	.7903	.8665	.9176	.9510	.9720	.9848	.9922
	2	.0444	.1497	.2834	.4233	.5551	.6706	.7662	.8414	.8976	.9375
	3	.0038	.0257	.0738	.1480	.2436	.3529	.4677	.5801	.6836	.7734
	4	.0002	.0027	.0121	.0333	.0706	.1260	.1998	.2898	.3917	.5000
	5	.0000	.0002	.0012	.0047	.0129	.0288	.0556	.0963	.1529	.2266
	6	.0000	.0000	.0001	.0004	.0013	.0038	.0090	.0188	.0357	.0625
	7	.0000	.0000	.0000	.0000	.0001	.0002	.0006	.0016	.0037	.0078
8	1	.3366	.5695	.7275	.8322	.8999	.9424	.9681	.9832	.9916	.9961
	2	.0572	.1869	.3428	.4967	.6329	.7447	.8309	.8936	.9368	.9648
	3	.0058	.0381	.1052	.2031	.3215	.4482	.5722	.6846	.7799	.8555
	4	.0004	.0050	.0214	.0563	.1138	.1941	.2936	.4059	.5230	.6367
	5	.0000	.0004	.0029	.0104	.0273	.0580	.1061	.1737	.2604	.3633
	6	.0000	.0000	.0002	.0012	.0042	.0113	.0253	.0498	.0885	.1445
	7	.0000	.0000	.0000	.0001	.0004	.0013	.0036	.0085	.0181	.0352
	8	.0000	.0000	.0000	.0000	.0000	.0001	.0002	.0007	.0017	.0039
9	1	.3698	.6126	.7684	.8658	.9249	.9596	.9793	.9899	.9954	.9980
	2	.0712	.2252	.4005	.5638	.6997	.8040	.8789	.9295	.9615	.9805
	3	.0084	.0530	.1409	.2618	.3993	.5372	.6627	.7682	.8505	.9102
	4	.0006	.0083	.0339	.0856	.1657	.2703	.3911	.5174	.6386	.7461
	5	.0000	.0009	.0056	.0196	.0489	.0988	.1717	.2666	.3786	.5000
	6	.0000	.0001	.0006	.0031	.0100	.0253	.0536	.0994	.1658	.2539
	7	.0000	.0000	.0000	.0003	.0013	.0043	.0112	.0250	.0498	.0898
	8	.0000	.0000	.0000	.0000	.0001	.0004	.0014	.0038	.0091	.0195
	9	.0000	.0000	.0000	.0000	.0000	.0000	.0001	.0003	.0008	.0020

Linear interpolation will be accurate at most to two decimal places.

Binomial, Poisson, and Hypergeometric Distributions

CUMULATIVE TERMS, BINOMIAL DISTRIBUTION

n	x'	.05	.10	.15	.20	.25	.30	.35	.40	.45	.50
10	1	.4013	.6513	.8031	.8926	.9437	.9718	.9865	.9940	.9975	.9990
	2	.0861	.2639	.4557	.6242	.7560	.8507	.9140	.9536	.9767	.9893
	3	.0115	.0702	.1798	.3222	.4744	.6172	.7384	.8327	.9004	.9453
	4	.0010	.0128	.0500	.1209	.2241	.3504	.4862	.6177	.7340	.8281
	5	.0001	.0016	.0099	.0328	.0781	.1503	.2485	.3669	.4956	.6230
	6	.0000	.0001	.0014	.0064	.0197	.0473	.0949	.1662	.2616	.3770
	7	.0000	.0000	.0001	.0009	.0035	.0106	.0260	.0548	.1020	.1719
	8	.0000	.0000	.0000	.0001	.0004	.0016	.0048	.0123	.0274	.0547
	9	.0000	.0000	.0000	.0000	.0000	.0001	.0005	.0017	.0045	.0107
	10	.0000	.0000	.0000	.0000	.0000	.0000	.0000	.0001	.0003	.0010
11	1	.4312	.6862	.8327	.9141	.9578	.9802	.9912	.9964	.9986	.9995
	2	.1019	.3026	.5078	.6779	.8029	.8870	.9394	.9698	.9861	.9941
	3	.0152	.0896	.2212	.3826	.5448	.6873	.7999	.8811	.9348	.9673
	4	.0016	.0185	.0694	.1611	.2867	.4304	.5744	.7037	.8089	.8867
	5	.0001	.0028	.0159	.0504	.1146	.2103	.3317	.4672	.6029	.7256
	6	.0000	.0003	.0027	.0117	.0343	.0782	.1487	.2465	.3669	.5000
	7	.0000	.0000	.0003	.0020	.0076	.0216	.0501	.0994	.1738	.2744
	8	.0000	.0000	.0000	.0002	.0012	.0043	.0122	.0293	.0610	.1133
	9	.0000	.0000	.0000	.0000	.0001	.0006	.0020	.0059	.0148	.0327
	10	.0000	.0000	.0000	.0000	.0000	.0000	.0002	.0007	.0022	.0059
	11	.0000	.0000	.0000	.0000	.0000	.0000	.0000	.0000	.0002	.0005
12	1	.4596	.7176	.8578	.9313	.9683	.9862	.9943	.9978	.9992	.9998
	2	.1184	.3410	.5565	.7251	.8416	.9150	.9576	.9804	.9917	.9968
	3	.0196	.1109	.2642	.4417	.6093	.7472	.8487	.9166	.9579	.9807
	4	.0022	.0256	.0922	.2054	.3512	.5075	.6533	.7747	.8655	.9270
	5	.0002	.0043	.0239	.0726	.1576	.2763	.4167	.5618	.6956	.8062
	6	.0000	.0005	.0046	.0194	.0544	.1178	.2127	.3348	.4731	.6128
	7	.0000	.0001	.0007	.0039	.0143	.0386	.0846	.1582	.2607	.3872
	8	.0000	.0000	.0001	.0006	.0028	.0095	.0255	.0573	.1117	.1938
	9	.0000	.0000	.0000	.0001	.0004	.0017	.0056	.0153	.0356	.0730
	10	.0000	.0000	.0000	.0000	.0000	.0002	.0008	.0028	.0079	.0193
	11	.0000	.0000	.0000	.0000	.0000	.0000	.0001	.0003	.0011	.0032
	12	.0000	.0000	.0000	.0000	.0000	.0000	.0000	.0000	.0001	.0002
13	1	.4867	.7458	.8791	.9450	.9762	.9903	.9963	.9987	.9996	.9999
	2	.1354	.3787	.6017	.7664	.8733	.9363	.9704	.9874	.9951	.9983
	3	.0245	.1339	.2704	.4983	.6674	.7975	.8868	.9421	.9731	.9888
	4	.0031	.0342	.0967	.2527	.4157	.5794	.7217	.8314	.9071	.9539
	5	.0003	.0065	.0260	.0991	.2060	.3457	.4995	.6470	.7721	.8666
	6	.0000	.0009	.0053	.0300	.0802	.1654	.2841	.4256	.5732	.7095
	7	.0000	.0001	.0013	.0070	.0243	.0624	.1295	.2288	.3563	.5000
	8	.0000	.0000	.0002	.0012	.0056	.0182	.0462	.0977	.1788	.2905
	9	.0000	.0000	.0000	.0002	.0010	.0040	.0126	.0321	.0698	.1334
	10	.0000	.0000	.0000	.0000	.0001	.0007	.0025	.0078	.0203	.0461
	11	.0000	.0000	.0000	.0000	.0000	.0001	.0003	.0013	.0041	.0112
	12	.0000	.0000	.0000	.0000	.0000	.0000	.0000	.0001	.0005	.0017
	13	.0000	.0000	.0000	.0000	.0000	.0000	.0000	.0000	.0000	.0001

CUMULATIVE TERMS, BINOMIAL DISTRIBUTION

n	x′	.05	.10	.15	.20	.25	.30	.35	.40	.45	.50
4	1	.5123	.7712	.8972	.9560	.9822	.9932	.9976	.9992	.9998	.9999
	2	.1530	.4154	.6433	.8021	.8990	.9525	.9795	.9919	.9971	.9991
	3	.0301	.1584	.3521	.5519	.7189	.8392	.9161	.9602	.9830	.9935
	4	.0042	.0441	.1465	.3018	.4787	.6448	.7795	.8757	.9368	.9713
	5	.0004	.0092	.0467	.1298	.2585	.4158	.5773	.7207	.8328	.9102
	6	.0000	.0015	.0115	.0439	.1117	.2195	.3595	.5141	.6627	.7880
	7	.0000	.0002	.0022	.0116	.0383	.0933	.1836	.3075	.4539	.6047
	8	.0000	.0000	.0003	.0024	.0103	.0315	.0753	.1501	.2586	.3953
	9	.0000	.0000	.0000	.0004	.0022	.0083	.0243	.0583	.1189	.2120
	10	.0000	.0000	.0000	.0000	.0003	.0017	.0060	.0175	.0426	.0898
	11	.0000	.0000	.0000	.0000	.0000	.0002	.0011	.0039	.0114	.0287
	12	.0000	.0000	.0000	.0000	.0000	.0000	.0001	.0006	.0022	.0065
	13	.0000	.0000	.0000	.0000	.0000	.0000	.0000	.0001	.0003	.0009
	14	.0000	.0000	.0000	.0000	.0000	.0000	.0000	.0000	.0000	.0001
15	1	.5367	.7941	.9126	.9648	.9866	.9953	.9984	.9995	.9999	1.0000
	2	.1710	.4510	.6814	.8329	.9198	.9647	.9858	.9948	.9983	.9995
	3	.0362	.1841	.3958	.6020	.7639	.8732	.9383	.9729	.9893	.9963
	4	.0055	.0556	.1773	.3518	.5387	.7031	.8273	.9095	.9576	.9824
	5	.0006	.0127	.0617	.1642	.3135	.4845	.6481	.7827	.8796	.9408
	6	.0001	.0022	.0168	.0611	.1484	.2784	.4357	.5968	.7392	.8491
	7	.0000	.0003	.0036	.0181	.0566	.1311	.2452	.3902	.5478	.6964
	8	.0000	.0000	.0006	.0042	.0173	.0500	.1132	.2131	.3465	.5000
	9	.0000	.0000	.0001	.0008	.0042	.0152	.0422	.0950	.1818	.3036
	10	.0000	.0000	.0000	.0001	.0008	.0037	.0124	.0338	.0769	.1509
	11	.0000	.0000	.0000	.0000	.0001	.0007	.0028	.0093	.0255	.0592
	12	.0000	.0000	.0000	.0000	.0000	.0001	.0005	.0019	.0063	.0176
	13	.0000	.0000	.0000	.0000	.0000	.0000	.0001	.0003	.0011	.0037
	14	.0000	.0000	.0000	.0000	.0000	.0000	.0000	.0000	.0001	.0005
	15	.0000	.0000	.0000	.0000	.0000	.0000	.0000	.0000	.0000	.0000
16	1	.5599	.8147	.9257	.9719	.9900	.9967	.9990	.9997	.9999	1.0000
	2	.1892	.4853	.7161	.8593	.9365	.9739	.9902	.9967	.9990	.9997
	3	.0429	.2108	.4386	.6482	.8029	.9006	.9549	.9817	.9934	.9979
	4	.0070	.0684	.2101	.4019	.5950	.7541	.8661	.9349	.9719	.9894
	5	.0009	.0170	.0791	.2018	.3698	.5501	.7108	.8334	.9147	.9616
	6	.0001	.0033	.0235	.0817	.1897	.3402	.5100	.6712	.8024	.8949
	7	.0000	.0005	.0056	.0267	.0796	.1753	.3119	.4728	.6340	.7228
	8	.0000	.0001	.0011	.0070	.0271	.0744	.1594	.2839	.4371	.5982
	9	.0000	.0000	.0002	.0015	.0075	.0257	.0671	.1423	.2559	.4018
	10	.0000	.0000	.0000	.0002	.0016	.0071	.0229	.0583	.1241	.2272
	11	.0000	.0000	.0000	.0000	.0003	.0016	.0062	.0191	.0486	.1051
	12	.0000	.0000	.0000	.0000	.0000	.0003	.0013	.0049	.0149	.0384
	13	.0000	.0000	.0000	.0000	.0000	.0000	.0002	.0009	.0035	.0106
	14	.0000	.0000	.0000	.0000	.0000	.0000	.0000	.0001	.0006	.0021
	15	.0000	.0000	.0000	.0000	.0000	.0000	.0000	.0000	.0001	.0003
	16	.0000	.0000	.0000	.0000	.0000	.0000	.0000	.0000	.0000	.0000

CUMULATIVE TERMS, BINOMIAL DISTRIBUTION

n	x'	.05	.10	.15	.20	p .25	.30	.35	.40	.45	.50
17	1	.5819	.8332	.9369	.9775	.9925	.9977	.9993	.9998	1.0000	1.0000
	2	.2078	.5182	.7475	.8818	.9499	.9807	.9933	.9979	.9994	.9999
	3	.0503	.2382	.4802	.6904	.8363	.9226	.9673	.9877	.9959	.9988
	4	.0088	.0826	.2444	.4511	.6470	.7981	.8972	.9536	.9816	.9936
	5	.0012	.0221	.0987	.2418	.4261	.6113	.7652	.8740	.9404	.9755
	6	.0001	.0047	.0319	.1057	.2347	.4032	.5803	.7361	.8529	.9283
	7	.0000	.0008	.0083	.0377	.1071	.2248	.3812	.5522	.7098	.8338
	8	.0000	.0001	.0017	.0109	.0402	.1046	.2128	.3595	.5257	.6855
	9	.0000	.0000	.0003	.0026	.0124	.0403	.0994	.1989	.3374	.5000
	10	.0000	.0000	.0000	.0005	.0031	.0127	.0383	.0919	.1834	.3145
	11	.0000	.0000	.0000	.0001	.0006	.0032	.0120	.0348	.0826	.1662
	12	.0000	.0000	.0000	.0000	.0001	.0007	.0030	.0106	.0301	.0717
	13	.0000	.0000	.0000	.0000	.0000	.0001	.0006	.0025	.0086	.0245
	14	.0000	.0000	.0000	.0000	.0000	.0000	.0000	.0005	.0019	.0064
	15	.0000	.0000	.0000	.0000	.0000	.0000	.0000	.0001	.0003	.0012
	16	.0000	.0000	.0000	.0000	.0000	.0000	.0000	.0000	.0000	.0001
	17	.0000	.0000	.0000	.0000	.0000	.0000	.0000	.0000	.0000	.0000
18	1	.6028	.8499	.9464	.9820	.9944	.9984	.9996	.9999	1.0000	1.0000
	2	.2265	.5497	.7759	.9009	.9605	.9858	.9954	.9987	.9997	.9999
	3	.0581	.2662	.5203	.7287	.8647	.9400	.9764	.9918	.9975	.9993
	4	.0109	.0982	.2798	.4990	.6943	.8354	.9217	.9672	.9880	.9962
	5	.0015	.0282	.1206	.2836	.4813	.6673	.8114	.9058	.9589	.9846
	6	.0002	.0064	.0419	.1329	.2825	.4656	.6450	.7912	.8923	.9519
	7	.0000	.0012	.0118	.0513	.1390	.2783	.4509	.6257	.7742	.8811
	8	.0000	.0002	.0027	.0163	.0569	.1407	.2717	.4366	.6085	.7597
	9	.0000	.0000	.0005	.0043	.0193	.0596	.1391	.2632	.4222	.5927
	10	.0000	.0000	.0001	.0009	.0054	.0210	.0597	.1347	.2527	.4073
	11	.0000	.0000	.0000	.0002	.0012	.0061	.0212	.0576	.1280	.2403
	12	.0000	.0000	.0000	.0000	.0002	.0014	.0062	.0203	.0537	.1189
	13	.0000	.0000	.0000	.0000	.0000	.0003	.0014	.0058	.0183	.0481
	14	.0000	.0000	.0000	.0000	.0000	.0000	.0003	.0013	.0049	.0154
	15	.0000	.0000	.0000	.0000	.0000	.0000	.0000	.0002	.0010	.0038
	16	.0000	.0000	.0000	.0000	.0000	.0000	.0000	.0000	.0001	.0007
	17	.0000	.0000	.0000	.0000	.0000	.0000	.0000	.0000	.0000	.0001
	18	.0000	.0000	.0000	.0000	.0000	.0000	.0000	.0000	.0000	.0000
19	1	.6226	.8649	.9544	.9856	.9958	.9989	.9997	.9999	1.0000	1.0000
	2	.2453	.5797	.8015	.9171	.9690	.9896	.9969	.9992	.9998	1.0000
	3	.0665	.2946	.5587	.7631	.8887	.9538	.9830	.9945	.9985	.9996
	4	.0132	.1150	.3159	.5449	.7369	.8668	.9409	.9770	.9923	.9978
	5	.0020	.0352	.1444	.3267	.5346	.7178	.8500	.9304	.9720	.9904
	6	.0002	.0086	.0537	.1631	.3322	.5261	.7032	.8371	.9223	.9682
	7	.0000	.0017	.0163	.0676	.1749	.3345	.5188	.6919	.8273	.9165
	8	.0000	.0003	.0041	.0233	.0775	.1820	.3344	.5122	.6831	.8204
	9	.0000	.0000	.0008	.0067	.0287	.0839	.1855	.3325	.5060	.6762
	10	.0000	.0000	.0001	.0016	.0089	.0326	.0875	.1861	.3290	.5000

CUMULATIVE TERMS, BINOMIAL DISTRIBUTION

x'	.05	.10	.15	.20	.25	.30	.35	.40	.45	.50
11	.0000	.0000	.0000	.0003	.0023	.0105	.0347	.0885	.1841	.3238
12	.0000	.0000	.0000	.0000	.0005	.0028	.0114	.0352	.0871	.1796
13	.0000	.0000	.0000	.0000	.0001	.0006	.0031	.0116	.0342	.0835
14	.0000	.0000	.0000	.0000	.0000	.0001	.0007	.0031	.0109	.0318
15	.0000	.0000	.0000	.0000	.0000	.0000	.0001	.0006	.0028	.0096
16	.0000	.0000	.0000	.0000	.0000	.0000	.0000	.0001	.0005	.0022
17	.0000	.0000	.0000	.0000	.0000	.0000	.0000	.0000	.0001	.0004
18	.0000	.0000	.0000	.0000	.0000	.0000	.0000	.0000	.0000	.0000
19	.0000	.0000	.0000	.0000	.0000	.0000	.0000	.0000	.0000	.0000
1	.6415	.8784	.9612	.9885	.9968	.9992	.9998	1.0000	1.0000	1.0000
2	.2642	.6083	.8244	.9308	.9757	.9924	.9979	.9995	.9999	1.0000
3	.0755	.3231	.5951	.7939	.9087	.9645	.9879	.9964	.9991	.9998
4	.0159	.1330	.3523	.5886	.7748	.8929	.9556	.9840	.9951	.9987
5	.0026	.0432	.1702	.3704	.5852	.7625	.8818	.9490	.9811	.9941
6	.0003	.0113	.0673	.1958	.3828	.5836	.7546	.8744	.9447	.9793
7	.0000	.0024	.0219	.0867	.2142	.3920	.5834	.7500	.8701	.9423
8	.0000	.0004	.0059	.0321	.1018	.2277	.3990	.5841	.7480	.8684
9	.0000	.0001	.0013	.0100	.0409	.1133	.2376	.4044	.5857	.7483
10	.0000	.0000	.0002	.0026	.0139	.0480	.1218	.2447	.4086	.5881
11	.0000	.0000	.0000	.0006	.0039	.0171	.0532	.1275	.2493	.4119
12	.0000	.0000	.0000	.0001	.0009	.0051	.0196	.0565	.1308	.2517
13	.0000	.0000	.0000	.0000	.0002	.0013	.0060	.0210	.0580	.1316
14	.0000	.0000	.0000	.0000	.0000	.0003	.0015	.0065	.0214	.0577
15	.0000	.0000	.0000	.0000	.0000	.0000	.0003	.0016	.0064	.0207
16	.0000	.0000	.0000	.0000	.0000	.0000	.0000	.0003	.0015	.0059
17	.0000	.0000	.0000	.0000	.0000	.0000	.0000	.0000	.0003	.0013
18	.0000	.0000	.0000	.0000	.0000	.0000	.0000	.0000	.0000	.0002
19	.0000	.0000	.0000	.0000	.0000	.0000	.0000	.0000	.0000	.0000
20	.0000	.0000	.0000	.0000	.0000	.0000	.0000	.0000	.0000	.0000

III.3   INDIVIDUAL TERMS, POISSON DISTRIBUTION

The Poisson probability function is given by

$$f(x) = \frac{m^x e^{-m}}{x!}, \qquad m > 0, \, x = 0, 1, 2, \ldots \ .$$

This table contains the individual terms of $f(x)$ for specified values of x and m.

INDIVIDUAL TERMS, POISSON DISTRIBUTION

m

	0.1	0.2	0.3	0.4	0.5	0.6	0.7	0.8	0.9	1.0
0	.9048	.8187	.7408	.6703	.6065	.5488	.4966	.4493	.4066	.3679
1	.0905	.1637	.2222	.2681	.3033	.3293	.3476	.3595	.3659	.3679
2	.0045	.0164	.0333	.0536	.0758	.0988	.1217	.1438	.1647	.1839
3	.0002	.0011	.0033	.0072	.0126	.0198	.0284	.0383	.0494	.0613
4	.0000	.0001	.0003	.0007	.0016	.0030	.0050	.0077	.0111	.0153
5	.0000	.0000	.0000	.0001	.0002	.0004	.0007	.0012	.0020	.0031
6	.0000	.0000	.0000	.0000	.0000	.0000	.0001	.0002	.0003	.0005
7	.0000	.0000	.0000	.0000	.0000	.0000	.0000	.0000	.0000	.0001

m

	1.1	1.2	1.3	1.4	1.5	1.6	1.7	1.8	1.9	2.0
0	.3329	.3012	.2725	.2466	.2231	.2019	.1827	.1653	.1496	.1353
1	.3662	.3614	.3543	.3452	.3347	.3230	.3106	.2975	.2842	.2707
2	.2014	.2169	.2303	.2417	.2510	.2584	.2640	.2678	.2700	.2707
3	.0738	.0867	.0998	.1128	.1255	.1378	.1496	.1607	.1710	.1804
4	.0203	.0260	.0324	.0395	.0471	.0551	.0636	.0723	.0812	.0902
5	.0045	.0062	.0084	.0111	.0141	.0176	.0216	.0260	.0309	.0361
6	.0008	.0012	.0018	.0026	.0035	.0047	.0061	.0078	.0098	.0120
7	.0001	.0002	.0003	.0005	.0008	.0011	.0015	.0020	.0027	.0034
8	.0000	.0000	.0001	.0001	.0001	.0002	.0003	.0005	.0006	.0009
9	.0000	.0000	.0000	.0000	.0000	.0000	.0001	.0001	.0001	.0002

m

	2.1	2.2	2.3	2.4	2.5	2.6	2.7	2.8	2.9	3.0
0	.1225	.1108	.1003	.0907	.0821	.0743	.0672	.0608	.0550	.0498
1	.2572	.2438	.2306	.2177	.2052	.1931	.1815	.1703	.1596	.1494
2	.2700	.2681	.2652	.2613	.2565	.2510	.2450	.2384	.2314	.2240
3	.1890	.1966	.2033	.2090	.2138	.2176	.2205	.2225	.2237	.2240
4	.0992	.1082	.1169	.1254	.1336	.1414	.1488	.1557	.1622	.1680
5	.0417	.0476	.0538	.0602	.0668	.0735	.0804	.0872	.0940	.1008
6	.0146	.0174	.0206	.0241	.0278	.0319	.0362	.0407	.0455	.0504
7	.0044	.0055	.0068	.0083	.0099	.0118	.0139	.0163	.0188	.0216
8	.0011	.0015	.0019	.0025	.0031	.0038	.0047	.0057	.0068	.0081
9	.0003	.0004	.0005	.0007	.0009	.0011	.0014	.0018	.0022	.0027
10	.0001	.0001	.0001	.0002	.0002	.0003	.0004	.0005	.0006	.0008
11	.0000	.0000	.0000	.0000	.0000	.0001	.0001	.0001	.0002	.0002
12	.0000	.0000	.0000	.0000	.0000	.0000	.0000	.0000	.0000	.0001

m

	3.1	3.2	3.3	3.4	3.5	3.6	3.7	3.8	3.9	4.0
0	.0450	.0408	.0369	.0334	.0302	.0273	.0247	.0224	.0202	.0183
1	.1397	.1304	.1217	.1135	.1057	.0984	.0915	.0850	.0789	.0733
2	.2165	.2087	.2008	.1929	.1850	.1771	.1692	.1615	.1539	.1465
3	.2237	.2226	.2209	.2186	.2158	.2125	.2087	.2046	.2001	.1954
4	.1734	.1781	.1823	.1858	.1888	.1912	.1931	.1944	.1951	.1954
5	.1075	.1140	.1203	.1264	.1322	.1377	.1429	.1477	.1522	.1563
6	.0555	.0608	.0662	.0716	.0771	.0826	.0881	.0936	.0989	.1042
7	.0246	.0278	.0312	.0348	.0385	.0425	.0466	.0508	.0551	.0595
8	.0095	.0111	.0129	.0148	.0169	.0191	.0215	.0241	.0269	.0298
9	.0033	.0040	.0047	.0056	.0066	.0076	.0089	.0102	.0116	.0132

INDIVIDUAL TERMS, POISSON DISTRIBUTION

x	3.1	3.2	3.3	3.4	3.5	3.6	3.7	3.8	3.9	4.0
10	.0010	.0013	.0016	.0019	.0023	.0028	.0033	.0039	.0045	.005
11	.0003	.0004	.0005	.0006	.0007	.0009	.0011	.0013	.0016	.001
12	.0001	.0001	.0001	.0002	.0002	.0003	.0003	.0004	.0005	.000
13	.0000	.0000	.0000	.0000	.0001	.0001	.0001	.0001	.0002	.000
14	.0000	.0000	.0000	.0000	.0000	.0000	.0000	.0000	.0000	.000

x	4.1	4.2	4.3	4.4	4.5	4.6	4.7	4.8	4.9	5.0
0	.0166	.0150	.0136	.0123	.0111	.0101	.0091	.0082	.0074	.006
1	.0679	.0630	.0583	.0540	.0500	.0462	.0427	.0395	.0365	.033
2	.1393	.1323	.1254	.1188	.1125	.1063	.1005	.0948	.0894	.084
3	.1904	.1852	.1798	.1743	.1687	.1631	.1574	.1517	.1460	.140
4	.1951	.1944	.1933	.1917	.1898	.1875	.1849	.1820	.1789	.175
5	.1600	.1633	.1662	.1687	.1708	.1725	.1738	.1747	.1753	.175
6	.1093	.1143	.1191	.1237	.1281	.1323	.1362	.1398	.1432	.146
7	.0640	.0686	.0732	.0778	.0824	.0869	.0914	.0959	.1002	.104
8	.0328	.0360	.0393	.0428	.0463	.0500	.0537	.0575	.0614	.065
9	.0150	.0168	.0188	.0209	.0232	.0255	.0280	.0307	.0334	.036
10	.0061	.0071	.0081	.0092	.0104	.0118	.0132	.0147	.0164	.018
11	.0023	.0027	.0032	.0037	.0043	.0049	.0056	.0064	.0073	.008
12	.0008	.0009	.0011	.0014	.0016	.0019	.0022	.0026	.0030	.003
13	.0002	.0003	.0004	.0005	.0006	.0007	.0008	.0009	.0011	.001
14	.0001	.0001	.0001	.0001	.0002	.0002	.0003	.0003	.0004	.000
15	.0000	.0000	.0000	.0000	.0001	.0001	.0001	.0001	.0001	.000

x	5.1	5.2	5.3	5.4	5.5	5.6	5.7	5.8	5.9	6.0
0	.0061	.0055	.0050	.0045	.0041	.0037	.0033	.0030	.0027	.002
1	.0311	.0287	.0265	.0244	.0225	.0207	.0191	.0176	.0162	.014
2	.0793	.0746	.0701	.0659	.0618	.0580	.0544	.0509	.0477	.044
3	.1348	.1293	.1239	.1185	.1133	.1082	.1033	.0985	.0938	.089
4	.1719	.1681	.1641	.1600	.1558	.1515	.1472	.1428	.1383	.133
5	.1753	.1748	.1740	.1728	.1714	.1697	.1678	.1656	.1632	.160
6	.1490	.1515	.1537	.1555	.1571	.1584	.1594	.1601	.1605	.160
7	.1086	.1125	.1163	.1200	.1234	.1267	.1298	.1326	.1353	.137
8	.0692	.0731	.0771	.0810	.0849	.0887	.0925	.0962	.0998	.103
9	.0392	.0423	.0454	.0486	.0519	.0552	.0586	.0620	.0654	.068
10	.0200	.0220	.0241	.0262	.0285	.0309	.0334	.0359	.0386	.041
11	.0093	.0104	.0116	.0129	.0143	.0157	.0173	.0190	.0207	.022
12	.0039	.0045	.0051	.0058	.0065	.0073	.0082	.0092	.0102	.011
13	.0015	.0018	.0021	.0024	.0028	.0032	.0036	.0041	.0046	.005
14	.0006	.0007	.0008	.0009	.0011	.0013	.0015	.0017	.0019	.002
15	.0002	.0002	.0003	.0003	.0004	.0005	.0006	.0007	.0008	.000
16	.0001	.0001	.0001	.0001	.0001	.0002	.0002	.0002	.0003	.000
17	.0000	.0000	.0000	.0000	.0000	.0000	.0001	.0001	.0001	.000

INDIVIDUAL TERMS, POISSON DISTRIBUTION

x	6.1	6.2	6.3	6.4	6.5	m 6.6	6.7	6.8	6.9	7.0
0	.0022	.0020	.0018	.0017	.0015	.0014	.0012	.0011	.0010	.0009
1	.0137	.0126	.0116	.0106	.0098	.0090	.0082	.0076	.0070	.0064
2	.0417	.0390	.0364	.0340	.0318	.0296	.0276	.0258	.0240	.0223
3	.0848	.0806	.0765	.0726	.0688	.0652	.0617	.0584	.0552	.0521
4	.1294	.1249	.1205	.1162	.1118	.1076	.1034	.0992	.0952	.0912
5	.1579	.1549	.1519	.1487	.1454	.1420	.1385	.1349	.1314	.1277
6	.1605	.1601	.1595	.1586	.1575	.1562	.1546	.1529	.1511	.1490
7	.1399	.1418	.1435	.1450	.1462	.1472	.1480	.1486	.1489	.1490
8	.1066	.1099	.1130	.1160	.1188	.1215	.1240	.1263	.1284	.1304
9	.0723	.0757	.0791	.0825	.0858	.0891	.0923	.0954	.0985	.1014
10	.0441	.0469	.0498	.0528	.0558	.0588	.0618	.0649	.0679	.0710
11	.0245	.0265	.0285	.0307	.0330	.0353	.0377	.0401	.0426	.0452
12	.0124	.0137	.0150	.0164	.0179	.0194	.0210	.0227	.0245	.0264
13	.0058	.0065	.0073	.0081	.0089	.0098	.0108	.0119	.0130	.0142
14	.0025	.0029	.0033	.0037	.0041	.0046	.0052	.0058	.0064	.0071
15	.0010	.0012	.0014	.0016	.0018	.0020	.0023	.0026	.0029	.0033
16	.0004	.0005	.0005	.0006	.0007	.0008	.0010	.0011	.0013	.0014
17	.0001	.0002	.0002	.0002	.0003	.0003	.0004	.0004	.0005	.0006
18	.0000	.0001	.0001	.0001	.0001	.0001	.0001	.0002	.0002	.0002
19	.0000	.0000	.0000	.0000	.0000	.0000	.0000	.0001	.0001	.0001

x	7.1	7.2	7.3	7.4	7.5	m 7.6	7.7	7.8	7.9	8.0
0	.0008	.0007	.0007	.0006	.0006	.0005	.0005	.0004	.0004	.0003
1	.0059	.0054	.0049	.0045	.0041	.0038	.0035	.0032	.0029	.0027
2	.0208	.0194	.0180	.0167	.0156	.0145	.0134	.0125	.0116	.0107
3	.0492	.0464	.0438	.0413	.0389	.0366	.0345	.0324	.0305	.0286
4	.0874	.0836	.0799	.0764	.0729	.0696	.0663	.0632	.0602	.0573
5	.1241	.1204	.1167	.1130	.1094	.1057	.1021	.0986	.0951	.0916
6	.1468	.1445	.1420	.1394	.1367	.1339	.1311	.1282	.1252	.1221
7	.1489	.1486	.1481	.1474	.1465	.1454	.1442	.1428	.1413	.1396
8	.1321	.1337	.1351	.1363	.1373	.1382	.1388	.1392	.1395	.1396
9	.1042	.1070	.1096	.1121	.1144	.1167	.1187	.1207	.1224	.1241
10	.0740	.0770	.0800	.0829	.0858	.0887	.0914	.0941	.0967	.0993
11	.0478	.0504	.0531	.0558	.0585	.0613	.0640	.0667	.0695	.0722
12	.0283	.0303	.0323	.0344	.0366	.0388	.0411	.0434	.0457	.0481
13	.0154	.0168	.0181	.0196	.0211	.0227	.0243	.0260	.0278	.0296
14	.0078	.0086	.0095	.0104	.0113	.0123	.0134	.0145	.0157	.0169
15	.0037	.0041	.0046	.0051	.0057	.0062	.0069	.0075	.0083	.0090
16	.0016	.0019	.0021	.0024	.0026	.0030	.0033	.0037	.0041	.0045
17	.0007	.0008	.0009	.0010	.0012	.0013	.0015	.0017	.0019	.0021
18	.0003	.0003	.0004	.0004	.0005	.0006	.0006	.0007	.0008	.0009
19	.0001	.0001	.0001	.0002	.0002	.0002	.0003	.0003	.0003	.0004
20	.0000	.0000	.0001	.0001	.0001	.0001	.0001	.0001	.0001	.0002
21	.0000	.0000	.0000	.0000	.0000	.0000	.0000	.0000	.0001	.0001

INDIVIDUAL TERMS, POISSON DISTRIBUTION

x	8.1	8.2	8.3	8.4	8.5	8.6	8.7	8.8	8.9	9.0
0	.0003	.0003	.0002	.0002	.0002	.0002	.0002	.0002	.0001	.0001
1	.0025	.0023	.0021	.0019	.0017	.0016	.0014	.0013	.0012	.0011
2	.0100	.0092	.0086	.0079	.0074	.0068	.0063	.0058	.0054	.0050
3	.0269	.0252	.0237	.0222	.0208	.0195	.0183	.0171	.0160	.0150
4	.0544	.0517	.0491	.0466	.0443	.0420	.0398	.0377	.0357	.0337
5	.0882	.0849	.0816	.0784	.0752	.0722	.0692	.0663	.0635	.0607
6	.1191	.1160	.1128	.1097	.1066	.1034	.1003	.0972	.0941	.0911
7	.1378	.1358	.1338	.1317	.1294	.1271	.1247	.1222	.1197	.1171
8	.1395	.1392	.1388	.1382	.1375	.1366	.1356	.1344	.1332	.1318
9	.1256	.1269	.1280	.1290	.1299	.1306	.1311	.1315	.1317	.1318
10	.1017	.1040	.1063	.1084	.1104	.1123	.1140	.1157	.1172	.1186
11	.0749	.0776	.0802	.0828	.0853	.0878	.0902	.0925	.0948	.0970
12	.0505	.0530	.0555	.0579	.0604	.0629	.0654	.0679	.0703	.0728
13	.0315	.0334	.0354	.0374	.0395	.0416	.0438	.0459	.0481	.0504
14	.0182	.0196	.0210	.0225	.0240	.0256	.0272	.0289	.0306	.0324
15	.0098	.0107	.0116	.0126	.0136	.0147	.0158	.0169	.0182	.0194
16	.0050	.0055	.0060	.0066	.0072	.0079	.0086	.0093	.0101	.0109
17	.0024	.0026	.0029	.0033	.0036	.0040	.0044	.0048	.0053	.0058
18	.0011	.0012	.0014	.0015	.0017	.0019	.0021	.0024	.0026	.0029
19	.0005	.0005	.0006	.0007	.0008	.0009	.0010	.0011	.0012	.0014
20	.0002	.0002	.0002	.0003	.0003	.0004	.0004	.0005	.0005	.0006
21	.0001	.0001	.0001	.0001	.0001	.0002	.0002	.0002	.0002	.0003
22	.0000	.0000	.0000	.0000	.0001	.0001	.0001	.0001	.0001	.0001

x	9.1	9.2	9.3	9.4	9.5	9.6	9.7	9.8	9.9	10
0	.0001	.0001	.0001	.0001	.0001	.0001	.0001	.0001	.0001	.0000
1	.0010	.0009	.0009	.0008	.0007	.0007	.0006	.0005	.0005	.0005
2	.0046	.0043	.0040	.0037	.0034	.0031	.0029	.0027	.0025	.0023
3	.0140	.0131	.0123	.0115	.0107	.0100	.0093	.0087	.0081	.0076
4	.0319	.0302	.0285	.0269	.0254	.0240	.0226	.0213	.0201	.0189
5	.0581	.0555	.0530	.0506	.0483	.0460	.0439	.0418	.0398	.0378
6	.0881	.0851	.0822	.0793	.0764	.0736	.0709	.0682	.0656	.0631
7	.1145	.1118	.1091	.1064	.1037	.1010	.0982	.0955	.0928	.0901
8	.1302	.1286	.1269	.1251	.1232	.1212	.1191	.1170	.1148	.1126
9	.1317	.1315	.1311	.1306	.1300	.1293	.1284	.1274	.1263	.1251
10	.1198	.1210	.1219	.1228	.1235	.1241	.1245	.1249	.1250	.1251
11	.0991	.1012	.1031	.1049	.1067	.1083	.1098	.1112	.1125	.1137
12	.0752	.0776	.0799	.0822	.0844	.0866	.0888	.0908	.0928	.0948
13	.0526	.0549	.0572	.0594	.0617	.0640	.0662	.0685	.0707	.0729
14	.0342	.0361	.0380	.0399	.0419	.0439	.0459	.0479	.0500	.0521
15	.0208	.0221	.0235	.0250	.0265	.0281	.0297	.0313	.0330	.0347
16	.0118	.0127	.0137	.0147	.0157	.0168	.0180	.0192	.0204	.0217
17	.0063	.0069	.0075	.0081	.0088	.0095	.0103	.0111	.0119	.0128
18	.0032	.0035	.0039	.0042	.0046	.0051	.0055	.0060	.0065	.0071
19	.0015	.0017	.0019	.0021	.0023	.0026	.0028	.0031	.0034	.0037

INDIVIDUAL TERMS, POISSON DISTRIBUTION

x	m 9.1	9.2	9.3	9.4	9.5	9.6	9.7	9.8	9.9	10
20	.0007	.0008	.0009	.0010	.0011	.0012	.0014	.0015	.0017	.0019
21	.0003	.0003	.0004	.0004	.0005	.0006	.0006	.0007	.0008	.0009
22	.0001	.0001	.0002	.0002	.0002	.0002	.0003	.0003	.0004	.0004
23	.0000	.0001	.0001	.0001	.0001	.0001	.0001	.0001	.0002	.0002
24	.0000	.0000	.0000	.0000	.0000	.0000	.0000	.0001	.0001	.0001

x	m 11	12	13	14	15	16	17	18	19	20
0	.0000	.0000	.0000	.0000	.0000	.0000	.0000	.0000	.0000	.0000
1	.0002	.0001	.0000	.0000	.0000	.0000	.0000	.0000	.0000	.0000
2	.0010	.0004	.0002	.0001	.0000	.0000	.0000	.0000	.0000	.0000
3	.0037	.0018	.0008	.0004	.0002	.0001	.0000	.0000	.0000	.0000
4	.0102	.0053	.0027	.0013	.0006	.0003	.0001	.0001	.0000	.0000
5	.0224	.0127	.0070	.0037	.0019	.0010	.0005	.0002	.0001	.0001
6	.0411	.0255	.0152	.0087	.0048	.0026	.0014	.0007	.0004	.0002
7	.0646	.0437	.0281	.0174	.0104	.0060	.0034	.0018	.0010	.0005
8	.0888	.0655	.0457	.0304	.0194	.0120	.0072	.0042	.0024	.0013
9	.1085	.0874	.0661	.0473	.0324	.0213	.0135	.0083	.0050	.0029
10	.1194	.1048	.0859	.0663	.0486	.0341	.0230	.0150	.0095	.0058
11	.1194	.1144	.1015	.0844	.0663	.0496	.0355	.0245	.0164	.0106
12	.1094	.1144	.1099	.0984	.0829	.0661	.0504	.0368	.0259	.0176
13	.0926	.1056	.1099	.1060	.0956	.0814	.0658	.0509	.0378	.0271
14	.0728	.0905	.1021	.1060	.1024	.0930	.0800	.0655	.0514	.0387
15	.0534	.0724	.0885	.0989	.1024	.0992	.0906	.0786	.0650	.0516
16	.0367	.0543	.0719	.0866	.0960	.0992	.0963	.0884	.0772	.0646
17	.0237	.0383	.0550	.0713	.0847	.0934	.0963	.0936	.0863	.0760
18	.0145	.0256	.0397	.0554	.0706	.0830	.0909	.0936	.0911	.0844
19	.0084	.0161	.0272	.0409	.0557	.0699	.0814	.0887	.0911	.0888
20	.0046	.0097	.0177	.0286	.0418	.0559	.0692	.0798	.0866	.0888
21	.0024	.0055	.0109	.0191	.0299	.0426	.0560	.0684	.0783	.0846
22	.0012	.0030	.0065	.0121	.0204	.0310	.0433	.0560	.0676	.0769
23	.0006	.0016	.0037	.0074	.0133	.0216	.0320	.0438	.0559	.0669
24	.0003	.0008	.0020	.0043	.0083	.0144	.0226	.0328	.0442	.0557
25	.0001	.0004	.0010	.0024	.0050	.0092	.0154	.0237	.0336	.0446
26	.0000	.0002	.0005	.0013	.0029	.0057	.0101	.0164	.0246	.0343
27	.0000	.0001	.0002	.0007	.0016	.0034	.0063	.0109	.0173	.0254
28	.0000	.0000	.0001	.0003	.0009	.0019	.0038	.0070	.0117	.0181
29	.0000	.0000	.0001	.0002	.0004	.0011	.0023	.0044	.0077	.0125
30	.0000	.0000	.0000	.0001	.0002	.0006	.0013	.0026	.0049	.0083
31	.0000	.0000	.0000	.0000	.0001	.0003	.0007	.0015	.0030	.0054
32	.0000	.0000	.0000	.0000	.0001	.0001	.0004	.0009	.0018	.0034
33	.0000	.0000	.0000	.0000	.0000	.0001	.0002	.0005	.0010	.0020
34	.0000	.0000	.0000	.0000	.0000	.0000	.0001	.0002	.0006	.0012
35	.0000	.0000	.0000	.0000	.0000	.0000	.0000	.0001	.0003	.0007
36	.0000	.0000	.0000	.0000	.0000	.0000	.0000	.0001	.0002	.0004
37	.0000	.0000	.0000	.0000	.0000	.0000	.0000	.0000	.0001	.0002
38	.0000	.0000	.0000	.0000	.0000	.0000	.0000	.0000	.0001	.0001
39	.0000	.0000	.0000	.0000	.0000	.0000	.0000	.0000	.0000	.0001

III.4 CUMULATIVE TERMS, POISSON DISTRIBUTION

This table contains the values of

$$\sum_{x=x'}^{\infty} \frac{e^{-m}m^x}{x!}$$

for specified values of x' and m. The cumulative Poisson distribution and the cumulative chi-square (χ^2) distribution are related as follows:

$$\sum_{x=0}^{x'-1} \frac{e^{-m}m^x}{x!} = 1 - F(\chi^2)$$

$$= \frac{1}{2^{\frac{n}{2}}\Gamma\left(\frac{n}{2}\right)} \int_{\chi^2}^{\infty} x^{\frac{n}{2}-1} e^{-\frac{x}{2}} \, dx$$

where $m = \frac{1}{2}\chi^2$ and $x' = \frac{1}{2}n$.

CUMULATIVE TERMS, POISSON DISTRIBUTION

					m					
x'	0.1	0.2	0.3	0.4	0.5	0.6	0.7	0.8	0.9	1.0
0	1.0000	1.0000	1.0000	1.0000	1.0000	1.0000	1.0000	1.0000	1.0000	1.0000
1	.0952	.1813	.2592	.3297	.3935	.4512	.5034	.5507	.5934	.6321
2	.0047	.0175	.0369	.0616	.0902	.1219	.1558	.1912	.2275	.2642
3	.0002	.0011	.0036	.0079	.0144	.0231	.0341	.0474	.0629	.0803
4	.0000	.0001	.0003	.0008	.0018	.0034	.0058	.0091	.0135	.0190
5	.0000	.0000	.0000	.0001	.0002	.0004	.0008	.0014	.0023	.0037
6	.0000	.0000	.0000	.0000	.0000	.0000	.0001	.0002	.0003	.0006
7	.0000	.0000	.0000	.0000	.0000	.0000	.0000	.0000	.0000	.0001

					m					
x'	1.1	1.2	1.3	1.4	1.5	1.6	1.7	1.8	1.9	2.0
0	1.0000	1.0000	1.0000	1.0000	1.0000	1.0000	1.0000	1.0000	1.0000	1.0000
1	.6671	.6988	.7275	.7534	.7769	.7981	.8173	.8347	.8504	.8647
2	.3010	.3374	.3732	.4082	.4422	.4751	.5068	.5372	.5663	.5940
3	.0996	.1205	.1429	.1665	.1912	.2166	.2428	.2694	.2963	.3233
4	.0257	.0338	.0431	.0537	.0656	.0788	.0932	.1087	.1253	.1429
5	.0054	.0077	.0107	.0143	.0186	.0237	.0296	.0364	.0441	.0527
6	.0010	.0015	.0022	.0032	.0045	.0060	.0080	.0104	.0132	.0166
7	.0001	.0003	.0004	.0006	.0009	.0013	.0019	.0026	.0034	.0045
8	.0000	.0000	.0001	.0001	.0002	.0003	.0004	.0006	.0008	.0011
9	.0000	.0000	.0000	.0000	.0000	.0000	.0001	.0001	.0002	.0002

					m					
x'	2.1	2.2	2.3	2.4	2.5	2.6	2.7	2.8	2.9	3.0
0	1.0000	1.0000	1.0000	1.0000	1.0000	1.0000	1.0000	1.0000	1.0000	1.0000
1	.8775	.8892	.8997	.9093	.9179	.9257	.9328	.9392	.9450	.9502
2	.6204	.6454	.6691	.6916	.7127	.7326	.7513	.7689	.7854	.8009
3	.3504	.3773	.4040	.4303	.4562	.4816	.5064	.5305	.5540	.5768
4	.1614	.1806	.2007	.2213	.2424	.2640	.2859	.3081	.3304	.3528
5	.0621	.0725	.0838	.0959	.1088	.1226	.1371	.1523	.1682	.1847
6	.0204	.0249	.0300	.0357	.0420	.0490	.0567	.0651	.0742	.0839
7	.0059	.0075	.0094	.0116	.0142	.0172	.0206	.0244	.0287	.0335
8	.0015	.0020	.0026	.0033	.0042	.0053	.0066	.0081	.0099	.0119
9	.0003	.0005	.0006	.0009	.0011	.0015	.0019	.0024	.0031	.0038
10	.0001	.0001	.0001	.0002	.0003	.0004	.0005	.0007	.0009	.0011
11	.0000	.0000	.0000	.0000	.0001	.0001	.0001	.0002	.0002	.0003
12	.0000	.0000	.0000	.0000	.0000	.0000	.0000	.0000	.0001	.0001

					m					
x'	3.1	3.2	3.3	3.4	3.5	3.6	3.7	3.8	3.9	4.0
0	1.0000	1.0000	1.0000	1.0000	1.0000	1.0000	1.0000	1.0000	1.0000	1.0000
1	.9550	.9592	.9631	.9666	.9698	.9727	.9753	.9776	.9798	.9817
2	.8153	.8288	.8414	.8532	.8641	.8743	.8838	.8926	.9008	.9084
3	.5988	.6201	.6406	.6603	.6792	.6973	.7146	.7311	.7469	.7619
4	.3752	.3975	.4197	.4416	.4634	.4848	.5058	.5265	.5468	.5665

Binomial, Poisson, and Hypergeometric Distributions

CUMULATIVE TERMS, POISSON DISTRIBUTION

x'	3.1	3.2	3.3	3.4	m 3.5	3.6	3.7	3.8	3.9	4.0
5	.2018	.2194	.2374	.2558	.2746	.2936	.3128	.3322	.3516	.3712
6	.0943	.1054	.1171	.1295	.1424	.1559	.1699	.1844	.1994	.2149
7	.0388	.0446	.0510	.0579	.0653	.0733	.0818	.0909	.1005	.1107
8	.0142	.0168	.0198	.0231	.0267	.0308	.0352	.0401	.0454	.0511
9	.0047	.0057	.0069	.0083	.0099	.0117	.0137	.0160	.0185	.0214
10	.0014	.0018	.0022	.0027	.0033	.0040	.0048	.0058	.0069	.0081
11	.0004	.0005	.0006	.0008	.0010	.0013	.0016	.0019	.0023	.0028
12	.0001	.0001	.0002	.0002	.0003	.0004	.0005	.0006	.0007	.0009
13	.0000	.0000	.0000	.0001	.0001	.0001	.0001	.0002	.0002	.0003
14	.0000	.0000	.0000	.0000	.0000	.0000	.0000	.0000	.0001	.0001

x'	4.1	4.2	4.3	4.4	m 4.5	4.6	4.7	4.8	4.9	5.0
0	1.0000	1.0000	1.0000	1.0000	1.0000	1.0000	1.0000	1.0000	1.0000	1.0000
1	.9834	.9850	.9864	.9877	.9889	.9899	.9909	.9918	.9926	.9933
2	.9155	.9220	.9281	.9337	.9389	.9437	.9482	.9523	.9561	.9596
3	.7762	.7898	.8026	.8149	.8264	.8374	.8477	.8575	.8667	.8753
4	.5858	.6046	.6228	.6406	.6577	.6743	.6903	.7058	.7207	.7350
5	.3907	.4102	.4296	.4488	.4679	.4868	.5054	.5237	.5418	.5595
6	.2307	.2469	.2633	.2801	.2971	.3142	.3316	.3490	.3665	.3840
7	.1214	.1325	.1442	.1564	.1689	.1820	.1954	.2092	.2233	.2378
8	.0573	.0639	.0710	.0786	.0866	.0951	.1040	.1133	.1231	.1334
9	.0245	.0279	.0317	.0358	.0403	.0451	.0503	.0558	.0618	.0681
10	.0095	.0111	.0129	.0149	.0171	.0195	.0222	.0251	.0283	.0318
11	.0034	.0041	.0048	.0057	.0067	.0078	.0090	.0104	.0120	.0137
12	.0011	.0014	.0017	.0020	.0024	.0029	.0034	.0040	.0047	.0055
13	.0003	.0004	.0005	.0007	.0008	.0010	.0012	.0014	.0017	.0020
14	.0001	.0001	.0002	.0002	.0003	.0003	.0004	.0005	.0006	.0007
15	.0000	.0000	.0000	.0001	.0001	.0001	.0001	.0001	.0002	.0002
16	.0000	.0000	.0000	.0000	.0000	.0000	.0000	.0000	.0001	.0001

x'	5.1	5.2	5.3	5.4	m 5.5	5.6	5.7	5.8	5.9	6.0
0	1.0000	1.0000	1.0000	1.0000	1.0000	1.0000	1.0000	1.0000	1.0000	1.0000
1	.9939	.9945	.9950	.9955	.9959	.9963	.9967	.9970	.9973	.9975
2	.9628	.9658	.9686	.9711	.9734	.9756	.9776	.9794	.9811	.9826
3	.8835	.8912	.8984	.9052	.9116	.9176	.9232	.9285	.9334	.9380
4	.7487	.7619	.7746	.7867	.7983	.8094	.8200	.8300	.8396	.8488
5	.5769	.5939	.6105	.6267	.6425	.6579	.6728	.6873	.7013	.7149
6	.4016	.4191	.4365	.4539	.4711	.4881	.5050	.5217	.5381	.5543
7	.2526	.2676	.2829	.2983	.3140	.3297	.3456	.3616	.3776	.3937
8	.1440	.1551	.1665	.1783	.1905	.2030	.2159	.2290	.2424	.2560
9	.0748	.0819	.0894	.0974	.1056	.1143	.1234	.1328	.1426	.1528

CUMULATIVE TERMS, POISSON DISTRIBUTION

x	5.1	5.2	5.3	5.4	5.5	5.6	5.7	5.8	5.9	6.0
					m					
10	.0356	.0397	.0441	.0488	.0538	.0591	.0648	.0708	.0772	.0839
11	.0156	.0177	.0200	.0225	.0253	.0282	.0314	.0349	.0386	.0426
12	.0063	.0073	.0084	.0096	.0110	.0125	.0141	.0160	.0179	.0201
13	.0024	.0028	.0033	.0038	.0045	.0051	.0059	.0068	.0078	.0088
14	.0008	.0010	.0012	.0014	.0017	.0020	.0023	.0027	.0031	.0036
15	.0003	.0003	.0004	.0005	.0006	.0007	.0009	.0010	.0012	.0014
16	.0001	.0001	.0001	.0002	.0002	.0002	.0003	.0004	.0004	.0005
17	.0000	.0000	.0000	.0001	.0001	.0001	.0001	.0001	.0001	.0002
18	.0000	.0000	.0000	.0000	.0000	.0000	.0000	.0000	.0000	.0001

x'	6.1	6.2	6.3	6.4	6.5	6.6	6.7	6.8	6.9	7.0
					m					
0	1.0000	1.0000	1.0000	1.0000	1.0000	1.0000	1.0000	1.0000	1.0000	1.0000
1	.9978	.9980	.9982	.9983	.9985	.9986	.9988	.9989	.9990	.9991
2	.9841	.9854	.9866	.9877	.9887	.9897	.9905	.9913	.9920	.9927
3	.9423	.9464	.9502	.9537	.9570	.9600	.9629	.9656	.9680	.9704
4	.8575	.8658	.8736	.8811	.8882	.8948	.9012	.9072	.9129	.9182
5	.7281	.7408	.7531	.7649	.7763	.7873	.7978	.8080	.8177	.8270
6	.5702	.5859	.6012	.6163	.6310	.6453	.6594	.6730	.6863	.6993
7	.4098	.4258	.4418	.4577	.4735	.4892	.5047	.5201	.5353	.5503
8	.2699	.2840	.2983	.3127	.3272	.3419	.3567	.3715	.3864	.4013
9	.1633	.1741	.1852	.1967	.2084	.2204	.2327	.2452	.2580	.2709
10	.0910	.0984	.1061	.1142	.1226	.1314	.1404	.1498	.1505	.1695
11	.0469	.0514	.0563	.0614	.0668	.0726	.0786	.0849	.0916	.0985
12	.0224	.0250	.0277	.0307	.0339	.0373	.0409	.0448	.0490	.0534
13	.0100	.0113	.0127	.0143	.0160	.0179	.0199	.0221	.0245	.0270
14	.0042	.0048	.0055	.0063	.0071	.0080	.0091	.0102	.0115	.0128
15	.0016	.0019	.0022	.0026	.0030	.0034	.0039	.0044	.0050	.0057
16	.0006	.0007	.0008	.0010	.0012	.0014	.0016	.0018	.0021	.0024
17	.0002	.0003	.0003	.0004	.0004	.0005	.0006	.0007	.0008	.0010
18	.0001	.0001	.0001	.0001	.0002	.0002	.0002	.0003	.0003	.0004
19	.0000	.0000	.0000	.0000	.0001	.0001	.0001	.0001	.0001	.0001

x'	7.1	7.2	7.3	7.4	7.5	7.6	7.7	7.8	7.9	8.0
					m					
0	1.0000	1.0000	1.0000	1.0000	1.0000	1.0000	1.0000	1.0000	1.0000	1.0000
1	.9992	.9993	.9993	.9994	.9994	.9995	.9995	.9996	.9996	.9997
2	.9933	.9939	.9944	.9949	.9953	.9957	.9961	.9964	.9967	.9970
3	.9725	.9745	.9764	.0781	.9797	.9812	.9826	.9839	.9851	.9862
4	.9233	.9281	.9326	.9368	.9409	.9446	.9482	.9515	.9547	.9576
5	.8359	.8445	.8527	.8605	.8679	.8751	.8819	.8883	.8945	.9004
6	.7119	.7241	.7360	.7474	.7586	.7693	.7797.	.7897	.7994	.8088
7	.5651	.5796	.5940	.6080	.6218	.6354	.6486	.6616	.6743	.6866
8	.4162	.4311	.4459	.4607	.4754	.4900	.5044	.5188	.5330	.5470
9	.2840	.2973	.3108	.3243	.3380	.3518	.3657	.3796	.3935	.4075
10	.1798	.1904	.2012	.2123	.2236	.2351	.2469	.2589	.2710	.2834
11	.1058	.1133	.1212	.1293	.1378	.1465	.1555	.1648	.1743	.1841
12	.0580	.0629	.0681	.0735	.0792	.0852	.0915	.0980	.1048	.1119
13	.0297	.0327	.0358	.0391	.0427	.0464	.0504	.0546	.0591	.0638
14	.0143	.0159	.0176	.0195	.0216	.0238	.0261	.0286	.0313	.0342

CUMULATIVE TERMS, POISSON DISTRIBUTION

					m					
x'	7.1	7.2	7.3	7.4	7.5	7.6	7.7	7.8	7.9	8.0
15	.0065	.0073	.0082	.0092	.0103	.0114	.0127	.0141	.0156	.017
16	.0028	.0031	.0036	.0041	.0046	.0052	.0059	.0066	.0074	.008
17	.0011	.0013	.0015	.0017	.0020	.0022	.0026	.0029	.0033	.003
18	.0004	.0005	.0006	.0007	.0008	.0009	.0011	.0012	.0014	.001
19	.0002	.0002	.0002	.0003	.0003	.0004	.0004	.0005	.0006	.000
20	.0001	.0001	.0001	.0001	.0001	.0001	.0002	.0002	.0002	.000
21	.0000	.0000	.0000	.0000	.0000	.0000	.0001	.0001	.0001	.000

					m					
x'	8.1	8.2	8.3	8.4	8.5	8.6	8.7	8.8	8.9	9.0
0	1.0000	1.0000	1.0000	1.0000	1.0000	1.0000	1.0000	1.0000	1.0000	1.000
1	.9997	.9997	.9998	.9998	.9998	.9998	.9998	.9998	.9999	.999
2	.9972	.9975	.9977	.9979	.9981	.9982	.9984	.9985	.9987	.998
3	.9873	.9882	.9891	.9900	.9907	.9914	.9921	.9927	.9932	.993
4	.9604	.9630	.9654	.9677	.9699	.9719	.9738	.9756	.9772	.978
5	.9060	.9113	.9163	.9211	.9256	.9299	.9340	.9379	.9416	.945
6	.8178	.8264	.8347	.8427	.8504	.8578	.8648	.8716	.8781	.884
7	.6987	.7104	.7219	.7330	.7438	.7543	.7645	.7744	.7840	.793
8	.5609	.5746	.5881	.6013	.6144	.6272	.6398	.6522	.6643	.676
9	.4214	.4353	.4493	.4631	.4769	.4906	.5042	.5177	.5311	.544
10	.2959	.3085	.3212	.3341	.3470	.3600	.3731	.3863	.3994	.412
11	.1942	.2045	.2150	.2257	.2366	.2478	.2591	.2706	.2822	.294
12	.1193	.1269	.1348	.1429	.1513	.1600	.1689	.1780	.1874	.197
13	.0687	.0739	.0793	.0850	.0909	.0971	.1035	.1102	.1171	.124
14	.0372	.0405	.0439	.0476	.0514	.0555	.0597	.0642	.0689	.073
15	.0190	.0209	.0229	.0251	.0274	.0299	.0325	.0353	.0383	.041
16	.0092	.0102	.0113	.0125	.0138	.0152	.0168	.0184	.0202	.022
17	.0042	.0047	.0053	.0059	.0066	.0074	.0082	.0091	.0101	.011
18	.0018	.0021	.0023	.0027	.0030	.0034	.0038	.0043	.0048	.005
19	.0008	.0009	.0010	.0011	.0013	.0015	.0017	.0019	.0022	.002
20	.0003	.0003	.0004	.0005	.0005	.0006	.0007	.0008	.0009	.001
21	.0001	.0001	.0002	.0002	.0002	.0002	.0003	.0003	.0004	.000
22	.0000	.0000	.0001	.0001	.0001	.0001	.0001	.0001	.0002	.000
23	.0000	.0000	.0000	.0000	.0000	.0000	.0000	.0000	.0001	.000

					m					
x'	9.1	9.2	9.3	9.4	9.5	9.6	9.7	9.8	9.9	10
0	1.0000	1.0000	1.0000	1.0000	1.0000	1.0000	1.0000	1.0000	1.0000	1.000
1	.9999	.9999	.9999	.9999	.9999	.9999	.9999	.9999	1.0000	1.0000
2	.9989	.9990	.9991	.9991	.9992	.9993	.9993	.9994	.9995	.999
3	.9942	.9947	.9951	.9955	.9958	.9962	.9965	.9967	.9970	.997
4	.9802	.9816	.9828	.9840	.9851	.9862	.9871	.9880	.9889	.989
5	.9483	.9514	.9544	.9571	.9597	.9622	.9645	.9667	.9688	.9707
6	.8902	.8959	.9014	.9065	.9115	.9162	.9207	.9250	.9290	.9329
7	.8022	.8108	.8192	.8273	.8351	.8426	.8498	.8567	.8634	.8699
8	.6877	.6990	.7101	.7208	.7313	.7416	.7515	.7612	.7706	.7798
9	.5574	.5704	.5832	.5958	.6082	.6204	.6324	.6442	.6558	.6672

CUMULATIVE TERMS, POISSON DISTRIBUTION

c'	m 9.1	9.2	9.3	9.4	9.5	9.6	9.7	9.8	9.9	10
10	.4258	.4389	.4521	.4651	.4782	.4911	.5040	.5168	.5295	.5421
11	.3059	.3180	.3301	.3424	.3547	.3671	.3795	.3920	.4045	.4170
12	.2068	.2168	.2270	.2374	.2480	.2588	.2697	.2807	.2919	.3032
13	.1316	.1393	.1471	.1552	.1636	.1721	.1809	.1899	.1991	.2084
14	.0790	.0844	.0900	.0958	.1019	.1081	.1147	.1214	.1284	.1355
15	.0448	.0483	.0520	.0559	.0600	.0643	.0688	.0735	.0784	.0835
16	.0240	.0262	.0285	.0309	.0335	.0362	.0391	.0421	.0454	.0487
17	.0122	.0135	.0148	.0162	.0177	.0194	.0211	.0230	.0249	.0270
18	.0059	.0066	.0073	.0081	.0089	.0098	.0108	.0119	.0130	.0143
19	.0027	.0031	.0034	.0038	.0043	.0048	.0053	.0059	.0065	.0072
20	.0012	.0014	.0015	.0017	.0020	.0022	.0025	.0028	.0031	.0035
21	.0005	.0006	.0007	.0008	.0009	.0010	.0011	.0013	.0014	.0016
22	.0002	.0002	.0003	.0003	.0004	.0004	.0005	.0005	.0006	.0007
23	.0001	.0001	.0001	.0001	.0001	.0002	.0002	.0002	.0003	.0003
24	.0000	.0000	.0000	.0000	.0001	.0001	.0001	.0001	.0001	.0001

c'	m 11	12	13	14	15	16	17	18	19	20
0	1.0000	1.0000	1.0000	1.0000	1.0000	1.0000	1.0000	1.0000	1.0000	1.0000
1	1.0000	1.0000	1.0000	1.0000	1.0000	1.0000	1.0000	1.0000	1.0000	1.0000
2	.9998	.9999	1.0000	1.0000	1.0000	1.0000	1.0000	1.0000	1.0000	1.0000
3	.9988	.9995	.9998	.9999	1.0000	1.0000	1.0000	1.0000	1.0000	1.0000
4	.9951	.9977	.9990	.9995	.9998	.9999	1.0000	1.0000	1.0000	1.0000
5	.9849	.9924	.9963	.9982	.9991	.9996	.9998	.9999	1.0000	1.0000
6	.9625	.9797	.9893	.9945	.9972	.9986	.9993	.9997	.9998	.9999
7	.9214	.9542	.9741	.9858	.9924	.9960	.9979	.9990	.9995	.9997
8	.8568	.9105	.9460	.9684	.9820	.9900	.9946	.9971	.9985	.9992
9	.7680	.8450	.9002	.9379	.9626	.9780	.9874	.9929	.9961	.9979
10	.6595	.7576	.8342	.8906	.9301	.9567	.9739	.9846	.9911	.9950
11	.5401	.6528	.7483	.8243	.8815	.9226	.9509	.9696	.9817	.9892
12	.4207	.5384	.6468	.7400	.8152	.8730	.9153	.9451	.9653	.9786
13	.3113	.4240	.5369	.6415	.7324	.8069	.8650	.9083	.9394	.9610
14	.2187	.3185	.4270	.5356	.6368	.7255	.7991	.8574	.9016	.9339
15	.1460	.2280	.3249	.4296	.5343	.6325	.7192	.7919	.8503	.8951
16	.0926	.1556	.2364	.3306	.4319	.5333	.6285	.7133	.7852	.8435
17	.0559	.1013	.1645	.2441	.3359	.4340	.5323	.6250	.7080	.7789
18	.0322	.0630	.1095	.1728	.2511	.3407	.4360	.5314	.6216	.7030
19	.0177	.0374	.0698	.1174	.1805	.2577	.3450	.4378	.5305	.6186
20	.0093	.0213	.0427	.0765	.1248	.1878	.2637	.3491	.4394	.5297
21	.0047	.0116	.0250	.0479	.0830	.1318	.1945	.2693	.3528	.4409
22	.0023	.0061	.0141	.0288	.0531	.0892	.1385	.2009	.2745	.3563
23	.0010	.0030	.0076	.0167	.0327	.0582	.0953	.1449	.2069	.2794
24	.0005	.0015	.0040	.0093	.0195	.0367	.0633	.1011	.1510	.2125
25	.0002	.0007	.0020	.0050	.0112	.0223	.0406	.0683	.1067	.1568
26	.0001	.0003	.0010	.0026	.0062	.0131	.0252	.0446	.0731	.1122
27	.0000	.0001	.0005	.0013	.0033	.0075	.0152	.0282	.0486	.0779
28	.0000	.0001	.0002	.0006	.0017	.0041	.0088	.0173	.0313	.0525
29	.0000	.0000	.0001	.0003	.0009	.0022	.0050	.0103	.0195	.0343

CUMULATIVE TERMS, POISSON DISTRIBUTION

x'	11	12	13	14	m 15	16	17	18	19	20
30	.0000	.0000	.0000	.0001	.0004	.0011	.0027	.0059	.0118	.0218
31	.0000	.0000	.0000	.0001	.0002	.0006	.0014	.0033	.0070	.0135
32	.0000	.0000	.0000	.0000	.0001	.0003	.0007	.0018	.0040	.0081
33	.0000	.0000	.0000	.0000	.0000	.0001	.0004	.0010	.0022	.0047
34	.0000	.0000	.0000	.0000	.0000	.0001	.0002	.0005	.0012	.0027
35	.0000	.0000	.0000	.0000	.0000	.0000	.0001	.0002	.0006	.0015
36	.0000	.0000	.0000	.0000	.0000	.0000	.0000	.0001	.0003	.0008
37	.0000	.0000	.0000	.0000	.0000	.0000	.0000	.0001	.0002	.0004
38	.0000	.0000	.0000	.0000	.0000	.0000	.0000	.0000	.0001	.0002
39	.0000	.0000	.0000	.0000	.0000	.0000	.0000	.0000	.0000	.0001
40	.0000	.0000	.0000	.0000	.0000	.0000	.0000	.0000	.0000	.0001

III.5 CONFIDENCE LIMITS FOR PROPORTIONS

The general term of the binomial expansion is given by

$$f(x;n,p) = \binom{n}{x} p^x q^{n-x}, \qquad x = 0, 1, 2, \ldots, n; \ p + q = 1.$$

For known n, and for a given value of x', the values of the confidence limits p_a and p_b ($p_a < p_b$) are defined by

$$\sum_{x=x'}^{n} f(x;n,p_a) = \alpha \qquad \text{and} \qquad \sum_{x=0}^{x'} f(x;n,p_b) = \alpha,$$

where the cumulative sums may be evaluated conveniently from the cumulative binomial distribution tables.

The charts show confidence limits for p for

$$1\text{–}2\alpha = .95, \text{ and } .99$$

or

$$\alpha = .025, \text{ and } .005$$

and for $n = 5, 10, 15, 20, 30, 50, 100, 250, 1000$.

CONFIDENCE LIMITS FOR PROPORTIONS
(Confidence coefficient .95)

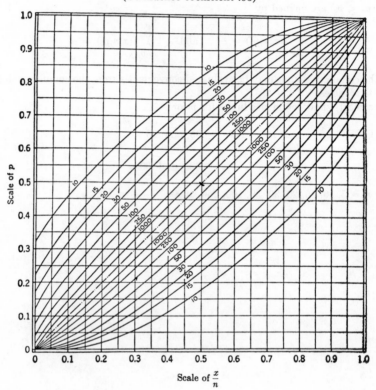

Scale of $\frac{x}{n}$

CONFIDENCE LIMITS FOR PROPORTIONS
(Confidence coefficient .99)

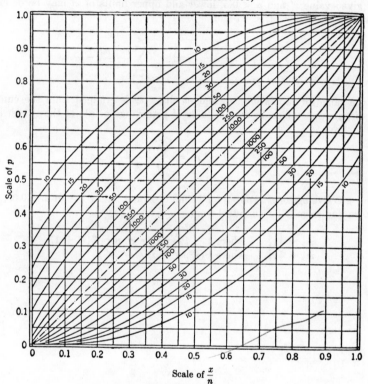

III.6 CONFIDENCE LIMITS FOR THE EXPECTED VALUE OF A POISSON DISTRIBUTION

The general term of a Poisson distributed variable is given by

$$f(x) = \frac{e^{-m}m^x}{x!}, \qquad x = 0, 1, 2, \ldots .$$

For any given value x' and $\alpha < 0.5$, lower and upper limits of m may be determined, say m_a and m_b, such that $m_a < m_b$ and

$$\sum_{x=x'}^{\infty} \frac{e^{-m_a}m_a^x}{x!} = \alpha \qquad \text{and} \qquad \sum_{x=0}^{x'} \frac{e^{-m_b}m_b^x}{x!} = \alpha.$$

Within the range of tabulation, m_a and m_b may be determined from the cumulative Poisson distribution tables, or from a table of percentage points of the χ^2 distribution, since

$$1 - P(\chi^2;n) = \sum_{x=0}^{x'-1} \frac{e^{-m}m^x}{x!}.$$

This table gives values of m_a and m_b for values of x' and for $2\alpha = 0.01$ and 0.05. Beyond $x' = 50$, m_a and m_b may be computed from

$$m_b = \tfrac{1}{2}\chi_1^2 \qquad \text{where } 1 - P(\chi^2;n) = \alpha, \ n = 2(x' + 1)$$
$$m_a = \tfrac{1}{2}\chi_2^2 \qquad \text{where } P(\chi^2;n) = \alpha, \ n = 2x' .$$

CONFIDENCE LIMITS FOR THE EXPECTED VALUE OF A POISSON DISTRIBUTION

Total observed count $x' = \Sigma x_i$	Significance level				Total observed count $x' = \Sigma x_i$	Significance level			
	$2\alpha = 0.01$		$2\alpha = 0.05$			$2\alpha = 0.01$		$2\alpha = 0.05$	
	Lower Limit	Upper Limit	Lower Limit	Upper Limit		Lower Limit	Upper Limit	Lower Limit	Upper Limit
0	0.0	5.3	0.0	3.7					
1	0.0	7.4	0.1	5.6	26	14.7	42.2	17.0	38.0
2	0.1	9.3	0.2	7.2	27	15.4	43.5	17.8	39.2
3	0.3	11.0	0.6	8.8	28	16.2	44.8	18.6	40.4
4	0.6	12.6	1.0	10.2	29	17.0	46.0	19.4	41.6
5	1.0	14.1	1.6	11.7	30	17.7	47.2	20.2	42.8
6	1.5	15.6	2.2	13.1	31	18.5	48.4	21.0	44.0
7	2.0	17.1	2.8	14.4	32	19.3	49.6	21.8	45.1
8	2.5	18.5	3.4	15.8	33	20.0	50.8	22.7	46.3
9	3.1	20.0	4.0	17.1	34	20.8	52.1	23.5	47.5
10	3.7	21.3	4.7	18.4	35	21.6	53.3	24.3	48.7
11	4.3	22.6	5.4	19.7	36	22.4	54.5	25.1	49.8
12	4.9	24.0	6.2	21.0	37	23.2	55.7	26.0	51.0
13	5.5	25.4	6.9	22.3	38	24.0	56.9	26.8	52.2
14	6.2	26.7	7.7	23.5	39	24.8	58.1	27.7	53.3
15	6.8	28.1	8.4	24.8	40	25.6	59.3	28.6	54.5
16	7.5	29.4	9.4	26.0	41	26.4	60.5	29.4	55.6
17	8.2	30.7	9.9	27.2	42	27.2	61.7	30.3	56.8
18	8.9	32.0	10.7	28.4	43	28.0	62.9	31.1	57.9
19	9.6	33.3	11.5	29.6	44	28.8	64.1	32.0	59.0
20	10.3	34.6	12.2	30.8	45	29.6	65.3	32.8	60.2
21	11.0	35.9	13.0	32.0	46	30.4	66.5	33.6	61.3
22	11.8	37.2	13.8	33.2	47	31.2	67.7	34.5	62.5
23	12.5	38.4	14.6	34.4	48	32.0	68.9	35.3	63.6
24	13.2	39.7	15.4	35.6	49	32.8	70.1	36.1	64.8
25	14.0	41.0	16.2	36.8	50	33.6	71.3	37.0	65.9

III.7   VARIOUS FUNCTIONS OF p AND $q = 1 - p$

The columns of this table give values of the various functions

$$pq, \ \sqrt{pq}, \ 1 - p^2, \ \sqrt{1 - p^2}, \ \frac{1}{\sqrt{1 - p^2}}, \ p^2 + q^2, \ 1 - q^2, \ \Gamma(1 + p), \ \log_{10}\Gamma(1 + p),$$

$$2 \text{ arc sin } \sqrt{p}, \ 2 \text{ arc sin } \sqrt{q}$$

The transformation, $y = 2 \text{ arc sin } \sqrt{p}$, is approximately normally distributed with mean $2 \text{ arc sin } \sqrt{\theta}$ and variance $\frac{1}{n}$ for $n\theta(1 - \theta) > 9$. Here p is the probability of success in a sample of n drawn from an infinite population where the probability of success is θ.

VARIOUS FUNCTIONS OF p AND $q = 1 - p$

p	$q = 1 - p$	pq	$\sqrt{(pq)}$	$1 - p^2$	$\sqrt{(1 - p^2)}$	$\dfrac{1}{\sqrt{(1 - p^2)}}$
0.00	1.00	0.0000	0.00000	1.0000	1.00000	1.00000
0.01	0.99	0.0099	0.09950	0.9999	0.99995	1.00005
0.02	0.98	0.0196	0.14000	0.9996	0.99980	1.00020
0.03	0.97	0.0291	0.17059	0.9991	0.99955	1.00045
0.04	0.96	0.0384	0.19596	0.9984	0.99920	1.00080
0.05	0.95	0.0475	0.21794	0.9975	0.99875	1.00125
0.06	0.94	0.0564	0.23749	0.9964	0.99820	1.00180
0.07	0.93	0.0651	0.25515	0.9951	0.99755	1.00246
0.08	0.92	0.0736	0.27129	0.9936	0.99679	1.00322
0.09	0.91	0.0819	0.28618	0.9919	0.99594	1.00407
0.10	0.90	0.0900	0.30000	0.9900	0.99499	1.00504
0.11	0.89	0.0979	0.31289	0.9879	0.99393	1.00611
0.12	0.88	0.1056	0.32496	0.9856	0.99277	1.00728
0.13	0.87	0.1131	0.33630	0.9831	0.99151	1.00856
0.14	0.86	0.1204	0.34699	0.9804	0.99015	1.00995
0.15	0.85	0.1275	0.35707	0.9775	0.98869	1.01144
0.16	0.84	0.1344	0.36661	0.9744	0.98712	1.01305
0.17	0.83	0.1411	0.37563	0.9711	0.98544	1.01477
0.18	0.82	0.1476	0.38419	0.9676	0.98367	1.01660
0.19	0.81	0.1539	0.39230	0.9639	0.98178	1.01855
0.20	0.80	0.1600	0.40000	0.9600	0.97980	1.02062
0.21	0.79	0.1659	0.40731	0.9559	0.97770	1.02281
0.22	0.78	0.1716	0.41425	0.9516	0.97550	1.02512
0.23	0.77	0.1771	0.42083	0.9471	0.97319	1.02755
0.24	0.76	0.1824	0.42708	0.9424	0.97077	1.03011
0.25	0.75	0.1875	0.43301	0.9375	0.96825	1.03280
0.26	0.74	0.1924	0.43863	0.9324	0.96561	1.03562
0.27	0.73	0.1971	0.44396	0.9271	0.96286	1.03857
0.28	0.72	0.2016	0.44900	0.9216	0.96000	1.04167
0.29	0.71	0.2059	0.45376	0.9159	0.95703	1.04490
0.30	0.70	0.2100	0.45826	0.9100	0.95394	1.04828
0.31	0.69	0.2139	0.46249	0.9039	0.95074	1.05182
0.32	0.68	0.2176	0.46648	0.8976	0.94742	1.05550
0.33	0.67	0.2211	0.47021	0.8911	0.94398	1.05934
0.34	0.66	0.2244	0.47371	0.8844	0.94043	1.06335
0.35	0.65	0.2275	0.47697	0.8775	0.93675	1.06752
0.36	0.64	0.2304	0.48000	0.8704	0.93295	1.07187
0.37	0.63	0.2331	0.48280	0.8631	0.92903	1.07639
0.38	0.62	0.2356	0.48539	0.8556	0.92499	1.08110
0.39	0.61	0.2379	0.48775	0.8479	0.92081	1.08599
0.40	0.60	0.2400	0.48990	0.8400	0.91652	1.09109
0.41	0.59	0.2419	0.49183	0.8319	0.91209	1.09639
0.42	0.58	0.2436	0.49356	0.8236	0.90752	1.10190
0.43	0.57	0.2451	0.49508	0.8151	0.90283	1.10763
0.44	0.56	0.2464	0.49639	0.8064	0.89800	1.11359
0.45	0.55	0.2475	0.49749	0.7975	0.89303	1.11979
0.46	0.54	0.2484	0.49840	0.7884	0.88792	1.12623
0.47	0.53	0.2491	0.49910	0.7791	0.88267	1.13293
0.48	0.52	0.2496	0.49960	0.7696	0.87727	1.13990
0.49	0.51	0.2499	0.49990	0.7599	0.87172	1.14715
0.50	0.50	0.2500	0.50000	0.7500	0.86603	1.15470

VARIOUS FUNCTIONS OF p AND $q = 1 - p$

p	$q = 1 - p$	$p^2 + q^2$	$1 - q^2$	$\Gamma(1 + p)$	$\log_{10} \Gamma(1 + p)$	$2 \text{ arc sin } \sqrt{p}$	$2 \text{ arc sin } \sqrt{}$
0.00	1.00	1.0000	0.0000	1.0000 000	0.0000 000	0.0000	3.1416
0.01	0.99	0.9802	0.0199	0.9943 259	$\bar{1}$.9975 287	0.2003	2.9413
0.02	0.98	0.9608	0.0396	0.9888 442	$\bar{1}$.9951 279	0.2838	2.8578
0.03	0.97	0.9418	0.0591	0.9835 500	$\bar{1}$.9927 964	0.3482	2.7934
0.04	0.96	0.9232	0.0784	0.9784 382	$\bar{1}$.9905 334	0.4027	2.7389
0.05	0.95	0.9050	0.0975	0.9735 043	$\bar{1}$.9883 379	0.4510	2.6906
0.06	0.94	0.8872	0.1164	0.9687 436	$\bar{1}$.9862 089	0.4949	2.6467
0.07	0.93	0.8698	0.1351	0.9641 520	$\bar{1}$.9841 455	0.5355	2.6061
0.08	0.92	0.8528	0.1536	0.9597 253	$\bar{1}$.9821 469	0.5735	2.5681
0.09	0.91	0.8362	0.1719	0.9554 595	$\bar{1}$.9802 123	0.6094	2.5322
0.10	0.90	0.8200	0.1900	0.9513 508	$\bar{1}$.9783 407	0.6435	2.4981
0.11	0.89	0.8042	0.2079	0.9473 955	$\bar{1}$.9765 313	0.6761	2.4655
0.12	0.88	0.7888	0.2256	0.9435 902	$\bar{1}$.9747 834	0.7075	2.4341
0.13	0.87	0.7738	0.2431	0.9399 314	$\bar{1}$.9730 962	0.7377	2.4039
0.14	0.86	0.7592	0.2604	0.9364 161	$\bar{1}$.9714 689	0.7670	2.3746
0.15	0.85	0.7450	0.2775	0.9330 409	$\bar{1}$.9699 007	0.7954	2.3462
0.16	0.84	0.7312	0.2944	0.9298 031	$\bar{1}$.9683 910	0.8230	2.3186
0.17	0.83	0.7178	0.3111	0.9266 996	$\bar{1}$.9669 390	0.8500	2.2916
0.18	0.82	0.7048	0.3276	0.9237 278	$\bar{1}$.9655 440	0.8763	2.2653
0.19	0.81	0.6922	0.3439	0.9208 850	$\bar{1}$.9642 054	0.9021	2.2395
0.20	0.80	0.6800	0.3600	0.9181 687	$\bar{1}$.9629 225	0.9273	2.2143
0.21	0.79	0.6682	0.3759	0.9155 765	$\bar{1}$.9616 946	0.9521	2.1895
0.22	0.78	0.6568	0.3916	0.9131 059	$\bar{1}$.9605 212	0.9764	2.1652
0.23	0.77	0.6458	0.4071	0.9107 549	$\bar{1}$.9594 015	1.0004	2.1412
0.24	0.76	0.6352	0.4224	0.9085 211	$\bar{1}$.9583 350	1.0239	2.1176
0.25	0.75	0.6250	0.4375	0.9064 025	$\bar{1}$.9573 211	1.0472	2.0944
0.26	0.74	0.6152	0.4524	0.9043 971	$\bar{1}$.9563 592	1.0701	2.0715
0.27	0.73	0.6058	0.4671	0.9025 031	$\bar{1}$.9554 487	1.0928	2.0488
0.28	0.72	0.5968	0.4816	0.9007 185	$\bar{1}$.9545 891	1.1152	2.0264
0.29	0.71	0.5882	0.4959	0.8990 416	$\bar{1}$.9537 798	1.1374	2.0042
0.30	0.70	0.5800	0.5100	0.8974 707	$\bar{1}$.9530 203	1.1593	1.9823
0.31	0.69	0.5722	0.5239	0.8960 042	$\bar{1}$.9523 100	1.1810	1.9606
0.32	0.68	0.5648	0.5376	0.8946 405	$\bar{1}$.9516 485	1.2025	1.9391
0.33	0.67	0.5578	0.5511	0.8933 781	$\bar{1}$.9510 353	1.2239	1.9177
0.34	0.66	0.5512	0.5644	0.8922 155	$\bar{1}$.9504 698	1.2451	1.8965
0.35	0.65	0.5450	0.5775	0.8911 514	$\bar{1}$.9499 515	1.2661	1.8755
0.36	0.64	0.5392	0.5904	0.8901 845	$\bar{1}$.9494 800	1.2870	1.8546
0.37	0.63	0.5338	0.6031	0.8893 135	$\bar{1}$.9490 549	1.3078	1.8338
0.38	0.62	0.5288	0.6156	0.8885 371	$\bar{1}$.9486 756	1.3284	1.8132
0.39	0.61	0.5242	0.6279	0.8878 543	$\bar{1}$.9483 417	1.3490	1.7926
0.40	0.60	0.5200	0.6400	0.8872 638	$\bar{1}$.9480 528	1.3694	1.7722
0.41	0.59	0.5162	0.6519	0.8867 647	$\bar{1}$.9478 084	1.3898	1.7518
0.42	0.58	0.5128	0.6636	0.8863 558	$\bar{1}$.9476 081	1.4101	1.7315
0.43	0.57	0.5098	0.6751	0.8860 362	$\bar{1}$.9474 515	1.4303	1.7113
0.44	0.56	0.5072	0.6864	0.8858 051	$\bar{1}$.9473 382	1.4505	1.6911
0.45	0.55	0.5050	0.6975	0.8856 614	$\bar{1}$.9472 677	1.4706	1.6710
0.46	0.54	0.5032	0.7084	0.8856 043	$\bar{1}$.9472 397	1.4907	1.6509
0.47	0.53	0.5018	0.7191	0.8856 331	$\bar{1}$.9472 539	1.5108	1.6308
0.48	0.52	0.5008	0.7296	0.8857 470	$\bar{1}$.9473 097	1.5308	1.6108
0.49	0.51	0.5002	0.7399	0.8859 451	$\bar{1}$.9474 068	1.5508	1.5908
0.50	0.50	0.5000	0.7500	0.8862 269	$\bar{1}$.9475 449	1.5708	1.5708

VARIOUS FUNCTIONS OF p AND $q = 1 - p$

p	$q = 1 - p$	pq	$\sqrt{(pq)}$	$1 - p^2$	$\sqrt{(1 - p^2)}$	$\dfrac{1}{\sqrt{(1 - p^2)}}$
.50	0.50	0.2500	0.50000	0.7500	0.86603	1.15470
.51	0.49	0.2499	0.49990	0.7399	0.86017	1.16255
.52	0.48	0.2496	0.49960	0.7296	0.85417	1.17073
.53	0.47	0.2491	0.49910	0.7191	0.84800	1.17925
.54	0.46	0.2484	0.49840	0.7084	0.84167	1.18812
.55	0.45	0.2475	0.49749	0.6975	0.83516	1.19737
.56	0.44	0.2464	0.49639	0.6864	0.82849	1.20701
.57	0.43	0.2451	0.49508	0.6751	0.82164	1.21707
.58	0.42	0.2436	0.49356	0.6636	0.81462	1.22757
.59	0.41	0.2419	0.49183	0.6519	0.80740	1.23854
.60	0.40	0.2400	0.48990	0.6400	0.80000	1.25000
.61	0.39	0.2379	0.48775	0.6279	0.79240	1.26199
.62	0.38	0.2356	0.48539	0.6156	0.78460	1.27453
.63	0.37	0.2331	0.48280	0.6031	0.77660	1.28767
.64	0.36	0.2304	0.48000	0.5904	0.76837	1.30145
.65	0.35	0.2275	0.47697	0.5775	0.75993	1.31590
.66	0.34	0.2244	0.47371	0.5644	0.75127	1.33109
.67	0.33	0.2211	0.47021	0.5511	0.74236	1.34705
.68	0.32	0.2176	0.46648	0.5376	0.73321	1.36386
.69	0.31	0.2139	0.46249	0.5239	0.72381	1.38158
.70	0.30	0.2100	0.45826	0.5100	0.71414	1.40028
.71	0.29	0.2059	0.45376	0.4959	0.70420	1.42005
.72	0.28	0.2016	0.44900	0.4816	0.69397	1.44098
.73	0.27	0.1971	0.44396	0.4671	0.68345	1.46317
.74	0.26	0.1924	0.43863	0.4524	0.67261	1.48675
.75	0.25	0.1875	0.43301	0.4375	0.66144	1.51186
.76	0.24	0.1824	0.42708	0.4224	0.64992	1.53864
.77	0.23	0.1771	0.42083	0.4071	0.63804	1.56729
.78	0.22	0.1716	0.41425	0.3916	0.62578	1.59801
.79	0.21	0.1659	0.40731	0.3759	0.61311	1.63104
.80	0.20	0.1600	0.40000	0.3600	0.60000	1.66667
.81	0.19	0.1539	0.39230	0.3439	0.58643	1.70523
.82	0.18	0.1476	0.38419	0.3276	0.57236	1.74714
.83	0.17	0.1411	0.37563	0.3111	0.55776	1.79287
.84	0.16	0.1344	0.36661	0.2944	0.54259	1.84302
.85	0.15	0.1275	0.35707	0.2775	0.52678	1.89832
.86	0.14	0.1204	0.34699	0.2604	0.51029	1.95965
.87	0.13	0.1131	0.33630	0.2431	0.49305	2.02818
.88	0.12	0.1056	0.32496	0.2256	0.47497	2.10538
.89	0.11	0.0979	0.31289	0.2079	0.45596	2.19317
.90	0.10	0.0900	0.30000	0.1900	0.43589	2.29416
.91	0.09	0.0819	0.28618	0.1719	0.41461	2.41192
.92	0.08	0.0736	0.27129	0.1536	0.39192	2.55155
.93	0.07	0.0651	0.25515	0.1351	0.36756	2.72065
.94	0.06	0.0564	0.23749	0.1164	0.34117	2.93105
.95	0.05	0.0475	0.21794	0.0975	0.31225	3.20256
.96	0.04	0.0384	0.19596	0.0784	0.28000	3.57143
.97	0.03	0.0291	0.17059	0.0591	0.24310	4.11345
.98	0.02	0.0196	0.14000	0.0396	0.19900	5.02519
.99	0.01	0.0099	0.09950	0.0199	0.14107	7.08881
.00	0.00	0.0000	0.00000	0.0000	0.00000	∞

Binomial, Poisson, and Hypergeometric Distributions

VARIOUS FUNCTIONS OF p AND $q = 1 - p$

p	$q = 1 - p$	$p^2 + q^2$	$1 - q^2$	$\Gamma(1 + p)$	$\log_{10} \Gamma(1 + p)$	$2 \text{ arc sin } \sqrt{p}$	$2 \text{ arc sin } \sqrt{q}$
0.50	0.50	0.5000	0.7500	0.8862 269	$\bar{1}$.9475 449	1.5708	1.5708
0.51	0.49	0.5002	0.7599	0.8865 917	$\bar{1}$.9477 237	1.5908	1.5508
0.52	0.48	0.5008	0.7696	0.8870 388	$\bar{1}$.9479 426	1.6108	1.5308
0.53	0.47	0.5018	0.7791	0.8875 676	$\bar{1}$.9482 015	1.6308	1.5108
0.54	0.46	0.5032	0.7884	0.8881 777	$\bar{1}$.9484 998	1.6509	1.4907
0.55	0.45	0.5050	0.7975	0.8888 683	$\bar{1}$.9488 374	1.6710	1.4706
0.56	0.44	0.5072	0.8064	0.8896 392	$\bar{1}$.9492 139	1.6911	1.4505
0.57	0.43	0.5098	0.8151	0.8904 897	$\bar{1}$.9496 289	1.7113	1.4303
0.58	0.42	0.5128	0.8236	0.8914 196	$\bar{1}$.9500 822	1.7315	1.4101
0.59	0.41	0.5162	0.8319	0.8924 282	$\bar{1}$.9505 733	7.7518	1.3898
0.60	0.40	0.5200	0.8400	0.8935 153	$\bar{1}$.9511 020	1.7722	1.3694
0.61	0.39	0.5242	0.8479	0.8946 806	$\bar{1}$.9516 680	1.7926	1.3490
0.62	0.38	0.5288	0.8556	0.8959 237	$\bar{1}$.9522 710	1.8132	1.3284
0.63	0.37	0.5338	0.8631	0.8972 442	$\bar{1}$.9529 107	1.8338	1.3078
0.64	0.36	0.5392	0.8704	0.8986 420	$\bar{1}$.9535 867	1.8546	1.2870
0.65	0.35	0.5450	0.8775	0.9001 168	$\bar{1}$.9542 989	1.8755	1.2661
0.66	0.34	0.5512	0.8844	0.9016 684	$\bar{1}$.9550 468	1.8965	1.2451
0.67	0.33	0.5578	0.8911	0.9032 965	$\bar{1}$.9558 303	1.9177	1.2239
0.68	0.32	0.5648	0.8976	0.9050 010	$\bar{1}$.9566 491	1.9391	1.2025
0.69	0.31	0.5722	0.9039	0.9067 818	$\bar{1}$.9575 028	1.9606	1.1810
0.70	0.30	0.5800	0.9100	0.9086 387	$\bar{1}$.9583 912	1.9823	1.1593
0.71	0.29	0.5882	0.9159	0.9105 717	$\bar{1}$.9593 141	2.0042	1.1374
0.72	0.28	0.5968	0.9216	0.9125 806	$\bar{1}$.9602 712	2.0264	1.1152
0.73	0.27	0.6058	0.9271	0.9146 654	$\bar{1}$.9612 622	2.0488	1.0928
0.74	0.26	0.6152	0.9324	0.9168 260	$\bar{1}$.9622 869	2.0715	1.0701
0.75	0.25	0.6250	0.9375	0.9190 625	$\bar{1}$.9633 451	2.0944	1.0472
0.76	0.24	0.6352	0.9424	0.9213 749	$\bar{1}$.9644 364	2.1176	1.0239
0.77	0.23	0.6458	0.9471	0.9237 631	$\bar{1}$.9655 606	2.1412	1.0004
0.78	0.22	0.6568	0.9516	0.9262 273	$\bar{1}$.9667 176	2.1652	0.9764
0.79	0.21	0.6682	0.9559	0.9287 675	$\bar{1}$.9679 070	2.1895	0.9521
0.80	0.20	0.6800	0.9600	0.9313 838	$\bar{1}$.9691 287	2.2143	0.9273
0.81	0.19	0.6922	0.9639	0.9340 763	$\bar{1}$.9703 823	2.2395	0.9021
0.82	0.18	0.7048	0.9676	0.9368 451	$\bar{1}$.9716 678	2.2653	0.8763
0.83	0.17	0.7178	0.9711	0.9396 904	$\bar{1}$.9729 848	2.2916	0.8500
0.84	0.16	0.7312	0.9744	0.9426 124	$\bar{1}$.9743 331	2.3186	0.8230
0.85	0.15	0.7450	0.9775	0.9456 112	$\bar{1}$.9757 126	2.3462	0.7954
0.86	0.14	0.7592	0.9804	0.9486 870	$\bar{1}$.9771 230	2.3746	0.7670
0.87	0.13	0.7738	0.9831	0.9518 402	$\bar{1}$.9785 640	2.4039	0.7377
0.88	0.12	0.7888	0.9856	0.9550 709	$\bar{1}$.9800 356	2.4341	0.7075
0.89	0.11	0.8042	0.9879	0.9583 793	$\bar{1}$.9815 374	2.4655	0.6761
0.90	0.10	0.8200	0.9900	0.9617 658	$\bar{1}$.9830 693	2.4981	0.6435
0.91	0.09	0.8362	0.9919	0.9652 307	$\bar{1}$.9846 311	2.5322	0.6094
0.92	0.08	0.8528	0.9936	0.9687 743	$\bar{1}$.9862 226	2.5681	0.5735
0.93	0.07	0.8698	0.9951	0.9723 969	$\bar{1}$.9878 436	2.6061	0.5355
0.94	0.06	0.8872	0.9964	0.9760 989	$\bar{1}$.9894 938	2.6467	0.4949
0.95	0.05	0.9050	0.9975	0.9798 807	$\bar{1}$.9911 732	2.6906	0.4510
0.96	0.04	0.9232	0.9984	0.9837 425	$\bar{1}$.9928 815	2.7389	0.4027
0.97	0.03	0.9418	0.9991	0.9876 850	$\bar{1}$.9946 185	2.7934	0.3482
0.98	0.02	0.9608	0.9996	0.9917 084	$\bar{1}$.9963 840	2.8578	0.2838
0.99	0.01	0.9802	0.9999	0.9958 133	$\bar{1}$.9981 779	2.9413	0.2003
1.00	0.00	1.0000	1.0000	1.0000 000	0.0000 000	3.1416	0.0000

III.8 HYPERGEOMETRIC DISTRIBUTION

The hypergeometric probability function is given by

$$f(x) = \frac{\dbinom{k}{x}\dbinom{N-k}{n-x}}{\dbinom{N}{n}} = \frac{\dfrac{k!}{x!(k-x)!}\dfrac{(N-k)!}{(n-x)!(N-k-n+x)!}}{\dfrac{N!}{n!(N-n)!}}$$

$$= \frac{k!n!}{x!(k-x)!(n-x)!}\frac{(N-k)!(N-n)!}{N!(N-k-n+x)!},$$

here N = number of items in a finite population consisting of A successes and B failures
$\quad\quad (A + B = N)$

$\quad n$ = number of items drawn in sample without replacement, from the N items

$\quad k$ = number of failures in finite population = B

$\quad x$ = number of failures in sample.

$f(x)$ gives the probability of exactly x failures and $n - x$ successes in the sample of
items.

$$F(x) = \sum_{r=0}^{x} \frac{\dbinom{k}{r}\dbinom{N-k}{n-r}}{\dbinom{N}{n}}.$$

$F(x)$ gives the probability of x or fewer failures in the sample of n items.

Binomial Poisson and Hypergeometric Distributions

HYPERGEOMETRIC PROBABILITY AND DISTRIBUTION FUNCTIONS

$$f(x) = \frac{\binom{k}{x}\binom{N-k}{n-x}}{\binom{N}{n}}, \quad F(x) = \sum_{r=0}^{x} \frac{\binom{k}{r}\binom{N-k}{n-r}}{\binom{N}{n}}$$

N	n	k	x	F(x)	f(x)	N	n	k	x	F(x)	f(x)
2	1	1	0	0.500000	0.500000	6	2	2	2	1.000000	0.066667
2	1	1	1	1.000000	0.500000	6	3	1	0	0.500000	0.500000
3	1	1	0	0.666667	0.666667	6	3	1	1	1.000000	0.500000
3	1	1	1	1.000000	0.333333	6	3	2	0	0.200000	0.200000
3	2	1	0	0.333333	0.333333	6	3	2	1	0.800000	0.600000
3	2	1	1	1.000000	0.666667	6	3	2	2	1.000000	0.200000
3	2	2	1	0.666667	0.666667	6	3	3	0	0.050000	0.050000
3	2	2	2	1.000000	0.333333	6	3	3	1	0.500000	0.450000
4	1	1	0	0.750000	0.750000	6	3	3	2	0.950000	0.450000
4	1	1	1	1.000000	0.250000	6	3	3	3	1.000000	0.050000
4	2	1	0	0.500000	0.500000	6	4	1	0	0.333333	0.333333
4	2	1	1	1.000000	0.500000	6	4	1	1	1.000000	0.666667
4	2	2	0	0.166667	0.166667	6	4	2	0	0.066667	0.066667
4	2	2	1	0.833333	0.666667	6	4	2	1	0.600000	0.533333
4	2	2	2	1.000000	0.166667	6	4	2	2	1.000000	0.400000
4	3	1	0	0.250000	0.250000	6	4	3	1	0.200000	0.200000
4	3	1	1	1.000000	0.750000	6	4	3	2	0.800000	0.600000
4	3	2	1	0.500000	0.500000	6	4	3	3	1.000000	0.200000
4	3	2	2	1.000000	0.500000	6	4	4	2	0.400000	0.400000
4	3	3	2	0.750000	0.750000	6	4	4	3	0.933333	0.533333
4	3	3	3	1.000000	0.250000	6	4	4	4	1.000000	0.066667
5	1	1	0	0.800000	0.800000	6	5	1	0	0.166667	0.166667
5	1	1	1	1.000000	0.200000	6	5	1	1	1.000000	0.833333
5	2	1	0	0.600000	0.600000	6	5	2	1	0.333333	0.333333
5	2	1	1	1.000000	0.400000	6	5	2	2	1.000000	0.666667
5	2	2	0	0.300000	0.300000	6	5	3	2	0.500000	0.500000
5	2	2	1	0.900000	0.600000	6	5	3	3	1.000000	0.500000
5	2	2	2	1.000000	0.100000	6	5	4	3	0.666667	0.666667
5	3	1	0	0.400000	0.400000	6	5	4	4	1.000000	0.333333
5	3	1	1	1.000000	0.600000	6	5	5	4	0.833333	0.833333
5	3	2	0	0.100000	0.100000	6	5	5	5	1.000000	0.166667
5	3	2	1	0.700000	0.600000	7	1	1	0	0.857143	0.857143
5	3	2	2	1.000000	0.300000	7	1	1	1	1.000000	0.142857
5	3	3	1	0.300000	0.300000	7	2	1	0	0.714286	0.714286
5	3	3	2	0.900000	0.600000	7	2	1	1	1.000000	0.285714
5	3	3	3	1.000000	0.100000	7	2	2	0	0.476190	0.476190
5	4	1	0	0.200000	0.200000	7	2	2	1	0.952381	0.476190
5	4	1	1	1.000000	0.800000	7	2	2	2	1.000000	0.047619
5	4	2	1	0.400000	0.400000	7	3	1	0	0.571429	0.571429
5	4	2	2	0.000000	0.600000	7	3	1	1	1.000000	0.428571
5	4	3	2	0.600000	0.600000	7	3	2	0	0.285714	0.285714
5	4	3	3	1.000000	0.400000	7	3	2	1	0.857143	0.571429
5	4	4	3	0.800000	0.800000	7	3	2	2	1.000000	0.142857
5	4	4	4	1.000000	0.200000	7	3	3	0	0.114286	0.114286
6	1	1	0	0.833333	0.833333	7	3	3	1	0.628571	0.514286
6	1	1	1	1.000000	0.166667	7	3	3	2	0.971428	0.342857
6	2	1	0	0.666667	0.666667	7	3	3	3	1.000000	0.028571
6	2	1	1	1.000000	0.333333	7	4	1	0	0.428571	0.428571
6	2	2	0	0.400000	0.400000	7	4	1	1	1.000000	0.571429
6	2	2	1	0.933333	0.533333	7	4	2	0	0.142857	0.142857

HYPERGEOMETRIC PROBABILITY AND DISTRIBUTION FUNCTIONS

N	n	k	x	F(x)	f(x)	N	n	k	x	F(x)	f(x)
7	4	2	1	0.714286	0.571429	8	3	3	2	0.982143	0.267857
7	4	2	2	1.000000	0.285714	8	3	3	3	1.000000	0.017857
7	4	3	0	0.028571	0.028571	8	4	1	0	0.500000	0.500000
7	4	3	1	0.371429	0.342857	8	4	1	1	1.000000	0.500000
7	4	3	2	0.885714	0.514286	8	4	2	0	0.214286	0.214286
7	4	3	3	1.000000	0.114286	8	4	2	1	0.785714	0.571429
7	4	4	1	0.114286	0.114286	8	4	2	2	1.000000	0.214286
7	4	4	2	0.628571	0.514286	8	4	3	0	0.071429	0.071429
7	4	4	3	0.971428	0.342857	8	4	3	1	0.500000	0.428571
7	4	4	4	1.000000	0.028571	8	4	3	2	0.928571	0.428571
7	5	1	0	0.285714	0.285714	8	4	3	3	1.000000	0.071429
7	5	1	1	1.000000	0.714286	8	4	4	0	0.014286	0.014286
7	5	2	0	0.047619	0.047619	8	4	4	1	0.242857	0.228571
7	5	2	1	0.523809	0.476190	8	4	4	2	0.757143	0.514286
7	5	2	2	1.000000	0.476190	8	4	4	3	0.985714	0.228571
7	5	3	1	0.142857	0.142857	8	4	4	4	1.000000	0.014286
7	5	3	2	0.714286	0.571429	8	5	1	0	0.375000	0.375000
7	5	3	3	1.000000	0.285714	8	5	1	1	1.000000	0.625000
7	5	4	2	0.285714	0.285714	8	5	2	0	0.107143	0.107143
7	5	4	3	0.857143	0.571429	8	5	2	1	0.642857	0.535714
7	5	4	4	1.000000	0.142857	8	5	2	2	1.000000	0.357143
7	5	5	3	0.476190	0.476190	8	5	3	0	0.017857	0.017857
7	5	5	4	0.952381	0.476190	8	5	3	1	0.285714	0.267857
7	5	5	5	1.000000	0.047619	8	5	3	2	0.821429	0.535714
7	6	1	0	0.142857	0.142857	8	5	3	3	1.000000	0.178571
7	6	1	1	1.000000	0.857143	8	5	4	1	0.071429	0.071429
7	6	2	1	0.285714	0.285714	8	5	4	2	0.500000	0.428571
7	6	2	2	1.000000	0.714286	8	5	4	3	0.928571	0.428571
7	6	3	2	0.428571	0.428571	8	5	4	4	1.000000	0.071429
7	6	3	3	1.000000	0.571429	8	5	5	2	0.178571	0.178571
7	6	4	3	0.571429	0.571429	8	5	5	3	0.714286	0.535714
7	6	4	4	1.000000	0.428571	8	5	5	4	0.982143	0.267857
7	6	5	4	0.714286	0.714286	8	5	5	5	1.000000	0.017857
7	6	5	5	1.000000	0.285714	8	6	1	0	0.250000	0.250000
7	6	6	5	0.857143	0.857143	8	6	1	1	1.000000	0.750000
7	6	6	6	1.000000	0.142857	8	6	2	0	0.035714	0.035714
8	1	1	0	0.875000	0.875000	8	6	2	1	0.464286	0.428571
8	1	1	1	1.000000	0.125000	8	6	2	2	1.000000	0.535714
8	2	1	0	0.750000	0.750000	8	6	3	1	0.107143	0.107143
8	2	1	1	1.000000	0.250000	8	6	3	2	0.642857	0.535714
8	2	2	0	0.535714	0.535714	8	6	3	3	1.000000	0.357143
8	2	2	1	0.964286	0.428571	8	6	4	2	0.214286	0.214286
8	2	2	2	1.000000	0.035714	8	6	4	3	0.785714	0.571429
8	3	1	0	0.625000	0.625000	8	6	4	4	1.000000	0.214286
8	3	1	1	1.000000	0.375000	8	6	5	3	0.357143	0.357143
8	3	2	0	0.357143	0.357143	8	6	5	4	0.892857	0.535714
8	3	2	1	0.892857	0.535714	8	6	5	5	1.000000	0.107143
8	3	2	2	1.000000	0.107143	8	6	6	4	0.535714	0.535714
8	3	3	0	0.178571	0.178571	8	6	6	5	0.964286	0.428571
8	3	3	1	0.714286	0.535714	8	6	6	6	1.000000	0.035714

HYPERGEOMETRIC PROBABILITY AND DISTRIBUTION FUNCTIONS

N	n	k	x	$F(x)$	$f(x)$	N	n	k	x	$F(x)$	$f(x)$
8	7	1	0	0.125000	0.125000	9	5	3	1	0.404762	0.357143
8	7	1	1	1.000000	0.875000	9	5	3	2	0.880952	0.476190
8	7	2	1	0.250000	0.250000	9	5	3	3	1.000000	0.119048
8	7	2	2	1.000000	0.750000	9	5	4	0	0.007936	0.007936
8	7	3	2	0.375000	0.375000	9	5	4	1	0.166667	0.158730
8	7	3	3	1.000000	0.625000	9	5	4	2	0.642857	0.476190
8	7	4	3	0.500000	0.500000	9	5	4	3	0.960317	0.317460
8	7	4	4	1.000000	0.500000	9	5	4	4	1.000000	0.039683
8	7	5	4	0.625000	0.625000	9	5	5	1	0.039683	0.039683
8	7	5	5	1.000000	0.375000	9	5	5	2	0.357143	0.317460
8	7	6	5	0.750000	0.750000	9	5	5	3	0.833333	0.476190
8	7	6	6	1.000000	0.250000	9	5	5	4	0.992063	0.158730
8	7	7	6	0.875000	0.875000	9	5	5	5	1.000000	0.007936
8	7	7	7	1.000000	0.125000	9	6	1	0	0.333333	0.333333
9	1	1	0	0.888889	0.888889	9	6	1	1	1.000000	0.666667
9	1	1	1	1.000000	0.111111	9	6	2	0	0.083333	0.083333
9	2	1	0	0.777778	0.777778	9	6	2	1	0.583333	0.500000
9	2	1	1	1.000000	0.222222	9	6	2	2	1.000000	0.416667
9	2	2	0	0.583333	0.583333	9	6	3	0	0.011905	0.011905
9	2	2	1	0.972222	0.388889	9	6	3	1	0.226190	0.214286
9	2	2	2	1.000000	0.027778	9	6	3	2	0.761905	0.535714
9	3	1	0	0.666667	0.666667	9	6	3	3	1.000000	0.238095
9	3	1	1	1.000000	0.333333	9	6	4	1	0.047619	0.047619
9	3	2	0	0.416667	0.416667	9	6	4	2	0.404762	0.357143
9	3	2	1	0.916667	0.500000	9	6	4	3	0.880952	0.476190
9	3	2	2	1.000000	0.083333	9	6	4	4	1.000000	0.119048
9	3	3	0	0.238095	0.238095	9	6	5	2	0.119048	0.119048
9	3	3	1	0.773809	0.535714	9	6	5	3	0.595238	0.476190
9	3	3	2	0.988095	0.214286	9	6	5	4	0.952381	0.357143
9	3	3	3	1.000000	0.011905	9	6	5	5	1.000000	0.047619
9	4	1	0	0.555556	0.555556	9	6	6	3	0.238095	0.238095
9	4	1	1	1.000000	0.444444	9	6	6	4	0.773809	0.535714
9	4	2	0	0.277778	0.277778	9	6	6	5	0.988095	0.214286
9	4	2	1	0.833333	0.555556	9	6	6	6	1.000000	0.011905
9	4	2	2	1.000000	0.166667	9	7	1	0	0.222222	0.222222
9	4	3	0	0.119048	0.119048	9	7	1	1	1.000000	0.777778
9	4	3	1	0.595238	0.476190	9	7	2	0	0.027778	0.027778
9	4	3	2	0.952381	0.357143	9	7	2	1	0.416667	0.388889
9	4	3	3	1.000000	0.047619	9	7	2	2	1.000000	0.583333
9	4	4	0	0.039683	0.039683	9	7	3	1	0.083333	0.083333
9	4	4	1	0.357143	0.317460	9	7	3	2	0.583333	0.500000
9	4	4	2	0.833333	0.476190	9	7	3	3	1.000000	0.416667
9	4	4	3	0.992063	0.158730	9	7	4	2	0.166667	0.166667
9	4	4	4	1.000000	0.007936	9	7	4	3	0.722222	0.555556
9	5	1	0	0.444444	0.444444	9	7	4	4	1.000000	0.277778
9	5	1	1	1.000000	0.555556	9	7	5	3	0.277778	0.277778
9	5	2	0	0.166667	0.166667	9	7	5	4	0.833333	0.555556
9	5	2	1	0.722222	0.555556	9	7	5	5	1.000000	0.166667
9	5	2	2	1.000000	0.277778	9	7	6	4	0.416667	0.416667
9	5	3	0	0.047619	0.047619	9	7	6	5	0.916667	0.500000

HYPERGEOMETRIC PROBABILITY AND DISTRIBUTION FUNCTIONS

N	n	k	x	F(x)	f(x)	N	n	k	x	F(x)	f(x)
9	7	6	6	1.000000	0.083333	10	5	1	0	0.500000	0.500000
9	7	7	5	0.583333	0.583333	10	5	1	1	1.000000	0.500000
9	7	7	6	0.972222	0.388889	10	5	2	0	0.222222	0.222222
9	7	7	7	1.000000	0.027778	10	5	2	1	0.777778	0.555556
9	8	1	0	0.111111	0.111111	10	5	2	2	1.000000	0.222222
9	8	1	1	1.000000	0.888889	10	5	3	0	0.083333	0.083333
9	8	2	1	0.222222	0.222222	10	5	3	1	0.500000	0.416667
9	8	2	2	1.000000	0.777778	10	5	3	2	0.916667	0.416667
9	8	3	2	0.333333	0.333333	10	5	3	3	1.000000	0.083333
9	8	3	3	1.000000	0.666667	10	5	4	0	0.023810	0.023810
9	8	4	3	0.444444	0.444444	10	5	4	1	0.261905	0.238095
9	8	4	4	1.000000	0.555556	10	5	4	2	0.738095	0.476190
9	8	5	4	0.555556	0.555556	10	5	4	3	0.976190	0.238095
9	8	5	5	1.000000	0.444444	10	5	4	4	1.000000	0.023810
9	8	6	5	0.666667	0.666667	10	5	5	0	0.003968	0.003968
9	8	6	6	1.000000	0.333333	10	5	5	1	0.103175	0.099206
9	8	7	6	0.777778	0.777778	10	5	5	2	0.500000	0.396825
9	8	7	7	1.000000	0.222222	10	5	5	3	0.896825	0.396825
9	8	8	7	0.888889	0.888889	10	5	5	4	0.996032	0.099206
9	8	8	8	1.000000	0.111111	10	5	5	5	1.000000	0.003968
10	1	1	0	0.900000	0.900000	10	6	1	0	0.400000	0.400000
10	1	1	1	1.000000	0.100000	10	6	1	1	1.000000	0.600000
10	2	1	0	0.800000	0.800000	10	6	2	0	0.133333	0.133333
10	2	1	1	1.000000	0.200000	10	6	2	1	0.666667	0.533333
10	2	2	0	0.622222	0.622222	10	6	2	2	1.000000	0.333333
10	2	2	1	0.977778	0.355556	10	6	3	0	0.033333	0.033333
10	2	2	2	1.000000	0.022222	10	6	3	1	0.333333	0.300000
10	3	1	0	0.700000	0.700000	10	6	3	2	0.833333	0.500000
10	3	1	1	1.000000	0.300000	10	6	3	3	1.000000	0.166667
10	3	2	0	0.466667	0.466667	10	6	4	0	0.004762	0.004762
10	3	2	1	0.933333	0.466667	10	6	4	1	0.119048	0.114286
10	3	2	2	1.000000	0.066667	10	6	4	2	0.547619	0.428571
10	3	3	0	0.291667	0.291667	10	6	4	3	0.928571	0.380952
10	3	3	1	0.816667	0.525000	10	6	4	4	1.000000	0.071429
10	3	3	2	0.991667	0.175000	10	6	5	1	0.023810	0.023810
10	3	3	3	1.000000	0.008333	10	6	5	2	0.261905	0.238095
10	4	1	0	0.600000	0.600000	10	6	5	3	0.738095	0.476190
10	4	1	1	1.000000	0.400000	10	6	5	4	0.976190	0.238095
10	4	2	0	0.333333	0.333333	10	6	5	5	1.000000	0.023810
10	4	2	1	0.866667	0.533333	10	6	6	2	0.071429	0.071429
10	4	2	2	1.000000	0.133333	10	6	6	3	0.452381	0.380952
10	4	3	0	0.166667	0.166667	10	6	6	4	0.880952	0.428571
10	4	3	1	0.666667	0.500000	10	6	6	5	0.995238	0.114286
10	4	3	2	0.966667	0.300000	10	6	6	6	1.000000	0.004762
10	4	3	3	1.000000	0.033333	10	7	1	0	0.300000	0.300000
10	4	4	0	0.071429	0.071429	10	7	1	1	1.000000	0.700000
10	4	4	1	0.452381	0.380952	10	7	2	0	0.066667	0.066667
10	4	4	2	0.880952	0.428571	10	7	2	1	0.533333	0.466667
10	4	4	3	0.995238	0.114286	10	7	2	2	1.000000	0.466667
10	4	4	4	1.000000	0.004762	10	7	3	0	0.008333	0.008333

III.9   NEGATIVE BINOMIAL DISTRIBUTION

The negative binomial probability function is given by

$$f(x) = \binom{x + r - 1}{r - 1} p^r q^x, \qquad x = 0, 1, 2, \ldots ; p + q = 1,$$

where p is the probability of success and q the probability of failure of a given event. $f(x)$ is the probability that exactly $x + r$ trials will be required to produce r successes. The cumulative distribution is given by

$$f(x) = \sum_{x'=0}^{x} \binom{x' + r - 1}{r - 1} p^r q^{x'} .$$

The cumulative negative binomial distribution is related to the cumulative binomial distribution as follows:

$$\sum_{x'=0}^{x} \binom{x' + r - 1}{r - 1} p^r q^{x'} = \sum_{x'=r}^{x+r} \binom{x + r}{x'} p^{x'} q^{x+r-x'} .$$

NEGATIVE BINOMIAL PROBABILITY AND DISTRIBUTION FUNCTIONS

$$f(x) = \binom{x + r - 1}{r - 1} p^r q^x, \qquad F(x) = \sum_{x'=0}^{x} \binom{x' + r - 1}{r - 1} p^r q^{x'}$$

$p = 0.900, r = 1$			$p = 0.900, r = 4$		
$x + r$	$f(x)$	$F(x)$	$x + r$	$f(x)$	$F(x)$
1	0.90000	0.9000	4	0.65610	0.6561
2	0.09000	0.9900	5	0.26244	0.9185
			6	0.06561	0.9841
			7	0.01312	0.9973

$p = 0.900, r = 2$			$p = 0.900, r = 5$		
$x + r$	$f(x)$	$F(x)$	$x + r$	$f(x)$	$F(x)$
2	0.81000	0.8100	5	0.59049	0.5905
3	0.16200	0.9720	6	0.29524	0.8857
4	0.02430	0.9963	7	0.08857	0.9743
			8	0.02067	0.9950

$p = 0.900, r = 3$			$p = 0.900, r = 6$		
$x + r$	$f(x)$	$F(x)$	$x + r$	$f(x)$	$F(x)$
3	0.72900	0.7290	6	0.53144	0.5314
4	0.21870	0.9477	7	0.31886	0.8503
5	0.04374	0.9914	8	0.11160	0.9619
			9	0.02976	0.9917

IV. Student's *t*-Distribution

IV.1 PERCENTAGE POINTS, STUDENT'S *t*-DISTRIBUTION

This table gives values of t such that

$$F(t) = \int_{-\infty}^{t} \frac{\Gamma\left(\dfrac{n+1}{2}\right)}{\sqrt{n\pi}\,\Gamma\left(\dfrac{n}{2}\right)} \left(1 + \frac{x^2}{n}\right)^{-\frac{n+1}{2}} dx$$

for n, the number of degrees of freedom, equal to 1, 2, . . . , 30, 40, 60, 120, ∞ ; and for $F(t) = 0.60, 0.75, 0.90, 0.95, 0.975, 0.99, 0.995,$ and 0.9995. The *t*-distribution is symmetrical, so that $F(-t) = 1 - F(t)$.

Student's t-Distribution

PERCENTAGE POINTS, STUDENTS t-DISTRIBUTION

$$F(t) = \int_{-\infty}^{t} \frac{\Gamma\left(\frac{n+1}{2}\right)}{\sqrt{n\pi}\ \Gamma\left(\frac{n}{2}\right)}\left(1 + \frac{x^2}{n}\right)^{-\frac{n+1}{2}} dx$$

n \ F	.60	.75	.90	.95	.975	.99	.995	.9995
1	.325	1.000	3.078	6.314	12.706	31.821	63.657	636.61
2	.289	.816	1.886	2.920	4.303	6.965	9.925	31.59
3	.277	.765	1.638	2.353	3.182	4.541	5.841	12.92
4	.271	.741	1.533	2.132	2.776	3.747	4.604	8.61
5	.267	.727	1.476	2.015	2.571	3.365	4.032	6.86
6	.265	.718	1.440	1.943	2.447	3.143	3.707	5.95
7	.263	.711	1.415	1.895	2.365	2.998	3.499	5.40
8	.262	.706	1.397	1.860	2.306	2.896	3.355	5.04
9	.261	.703	1.383	1.833	2.262	2.821	3.250	4.78
10	.260	.700	1.372	1.812	2.228	2.764	3.169	4.58
11	.260	.697	1.363	1.796	2.201	2.718	3.106	4.43
12	.259	.695	1.356	1.782	2.179	2.681	3.055	4.31
13	.259	.694	1.350	1.771	2.160	2.650	3.012	4.22
14	.258	.692	1.345	1.761	2.145	2.624	2.977	4.14
15	.258	.691	1.341	1.753	2.131	2.602	2.947	4.07
16	.258	.690	1.337	1.746	2.120	2.583	2.921	4.01
17	.257	.689	1.333	1.740	2.110	2.567	2.898	3.96
18	.257	.688	1.330	1.734	2.101	2.552	2.878	3.92
19	.257	.688	1.328	1.729	2.093	2.539	2.861	3.88
20	.257	.687	1.325	1.725	2.086	2.528	2.845	3.85
21	.257	.686	1.323	1.721	2.080	2.518	2.831	3.81
22	.256	.686	1.321	1.717	2.074	2.508	2.819	3.79
23	.256	.685	1.319	1.714	2.069	2.500	2.807	3.76
24	.256	.685	1.318	1.711	2.064	2.492	2.797	3.74
25	.256	.684	1.316	1.708	2.060	2.485	2.787	3.72
26	.256	.684	1.315	1.706	2.056	2.479	2.779	3.707
27	.256	.684	1.314	1.703	2.052	2.473	2.771	3.69
28	.256	.683	1.313	1.701	2.048	2.467	2.763	3.674
29	.256	.683	1.311	1.699	2.045	2.462	2.756	3.659
30	.256	.683	1.310	1.697	2.042	2.457	2.750	3.64
40	.255	.681	1.303	1.684	2.021	2.423	2.704	3.55
60	.254	.679	1.296	1.671	2.000	2.390	2.660	3.46
120	.254	.677	1.289	1.658	1.980	2.358	2.617	3.37
∞	.253	.674	1.282	1.645	1.960	2.326	2.576	3.29

IV.2 POWER FUNCTION OF THE *t*-TEST

Any statistic of the form

$$t' = \frac{z + \delta}{s_z} = \frac{z'}{s_z},$$

ere z is normally distributed with expectation zero and standard deviation σ_z, s_z^2 is an
.ependent estimate of σ_z^2 based on ν degrees of freedom, and δ is the noncentrality
:ameter, is distributed as the non-central t-distribution, denoted by $f(t')$.

The power function of the t-test is the value of the following integrals considered as a
.ction of $\dfrac{\delta}{\sigma_z}$

$$\int_{-\infty}^{-t_{\alpha/2}} f(t') \, dt' + \int_{t_{\alpha/2}}^{\infty} f(t') \, dt' \qquad \text{for the double-tail } t\text{-test}$$

$$\int_{t_\alpha}^{\infty} f(t') \, dt' \qquad \text{for the single-tail } t\text{-test } (\delta \geq 0) ,$$

ere t_α denotes the α-level significance point for t.

The graph in this table shows the integral (*a*) for $\alpha = 0.05$ and 0.01, and for various
lues of ν, the degrees of freedom associated with s_z. The horizontal scale is in terms of
$= \dfrac{\delta}{\sigma_z \sqrt{2}}$. The graph can also be used to give the integral (*b*) for significance levels
$= 0.025$ and 0.005.

Student's t-Distribution

POWER FUNCTION OF THE t-TEST

IV.3 NUMBER OF OBSERVATIONS FOR *t*-TEST OF MEAN

This table gives the sample size needed for given values of $\alpha = P$ (Type I error) and $\beta = P$ (Type II error) for a test on a single mean with unknown standard deviation.

To test the hypothesis $H_0: \mu = \mu_0$ against the alternative $H_a: \mu < \mu_0$, the statistic $t = \dfrac{(\bar{x} - \mu_0)\sqrt{n}}{s}$ is used. The distribution of t when H_0 is true is the t-distribution with $n - 1$ degrees of freedom and the critical region $t > t_{n-1,1-\alpha}$ would have a significance level α. The distribution of $t(\delta) = \dfrac{\sqrt{n}\,(\bar{x} - \mu) + \delta\sigma}{s}$ is noncentral t with $n - 1$ degrees of freedom and noncentrality parameter δ. Here $\delta = \dfrac{(\mu - \mu_0)\sqrt{n}}{\sigma}$. This table is used to obtain sample sizes needed to control the values of α and β for various values of $d = \dfrac{\mu - \mu_0}{\sigma}$ for both one-sided tests and two-sided tests.

NUMBER OF OBSERVATIONS FOR t-TEST OF MEAN

Level of t-test

	α = 0.005 (single) / α = 0.01 (double)					α = 0.01 (single) / α = 0.02 (double)					α = 0.025 (single) / α = 0.05 (double)					α = 0.05 (single) / α = 0.1 (double)				
β =	0.01	0.05	0.1	0.2	0.5	0.01	0.05	0.1	0.2	0.5	0.01	0.05	0.1	0.2	0.5	0.01	0.05	0.1	0.2	0.5
0.05																				
0.10																				
0.15																				122
0.20										139					99					70
0.25					110					90				128	64			139	101	45
0.30				134	78				115	63			119	90	45		122	97	71	32
0.35			125	99	58			109	85	47		109	88	67	34		90	72	52	24
0.40		115	97	77	45		101	85	66	37	117	84	68	51	26	101	70	55	40	19
0.45		92	77	62	37	110	81	68	53	30	93	67	54	41	21	80	55	44	33	15
0.50	100	75	63	51	30	90	66	55	43	25	76	54	44	34	18	65	45	36	27	13
0.55	83	63	53	42	26	75	55	46	36	21	63	45	37	28	15	54	38	30	22	11
0.60	71	53	45	36	22	63	47	39	31	18	53	38	32	24	13	46	32	26	19	9
0.65	61	46	39	31	20	55	41	34	27	16	46	33	27	21	12	39	28	22	17	8
0.70	53	40	34	28	17	47	35	30	24	14	40	29	24	19	10	34	24	19	15	8
0.75	47	36	30	25	16	42	31	27	21	13	35	26	21	16	9	30	21	17	13	7
0.80	41	32	27	22	14	37	28	24	19	12	31	22	19	15	9	27	19	15	12	6
0.85	37	29	24	20	13	33	25	21	17	11	28	21	17	13	8	24	17	14	11	6
0.90	34	26	22	18	12	29	23	19	16	10	25	19	16	12	7	21	15	13	10	5
0.95	31	24	20	17	11	27	21	18	14	9	23	17	14	11	7	19	14	11	9	5
1.00	28	22	19	16	10	25	19	16	13	9	21	16	13	10	6	18	13	11	8	5
1.1	24	19	16	14	9	21	16	14	12	8	18	13	11	9	6	15	11	9	7	
1.2	21	16	14	12	8	18	14	12	10	7	15	12	10	8	5	13	10	8	6	
1.3	18	15	13	11	8	16	13	11	9	6	14	10	9	7		11	8	7	6	
1.4	16	13	12	10	7	14	11	10	9	6	12	9	8	7		10	8	7	5	
1.5	15	12	11	9	7	13	10	9	8	6	11	8	7	6		9	7	6		
1.6	13	11	10	8	6	12	10	9	7	5	10	8	7	6		8	6	6		
1.7	12	10	9	8	6	11	9	8	7		9	7	6	5		8	6	5		
1.8	12	10	9	8	6	10	8	7	7		8	7	6			7	6			
1.9	11	9	8	7	6	10	8	7	6		8	6	6			7	5			
2.0	10	8	8	7	5	9	7	7	6		7	6	5			6				
2.1	10	8	7	7		8	7	6	6		7	6				6				
2.2	9	8	7	6		8	7	6	5		7	6				6				
2.3	9	7	7	6		8	6	6			6	5				5				
2.4	8	7	7	6		7	6	6			6									
2.5	8	7	6	6		7	6	6			6									
3.0	7	6	6	5		6	5	5			5									
3.5	6	5	5			5														
4.0	6																			

Value of $\Delta = \dfrac{\mu - \mu_0}{\sigma}$

IV.4 NUMBER OF OBSERVATIONS FOR *t*-TEST OF DIFFERENCE BETWEEN TWO MEANS

This table gives the sample size needed for given values of $\alpha = P$ (Type I error) and $\beta = P$ (Type II error) for a test of the hypothesis of the equality of two means $\mu_1 = \mu_2$, where there is a common but unknown variance. The statistic used is

$$t = \frac{\bar{x}_1 - \bar{x}_2}{s\sqrt{\dfrac{1}{n_1} + \dfrac{1}{n_2}}}$$

which is distributed as Students *t*-distribution with $n_1 + n_2 - 2$ degrees of freedom. Here

$$s = \left[\frac{(n_1 - 1)s_1^2 + (n_2 - 1)s_2^2}{n_1 + n_2 - 2}\right]^{1/2}$$

The noncentrality parameter in this case is

$$\sigma = \frac{\mu_1 - \mu_2}{\sigma\sqrt{\dfrac{1}{n_1} + \dfrac{1}{n_2}}} \cdot$$

This table is used to obtain sample sizes needed to control the values of α and β for various values of $\Delta = \dfrac{\mu_1 - \mu_2}{\sigma}$ for both one-sided and two-sided tests for the case $n_1 = n_2 = n$.

Student's t-Distribution

NUMBER OF OBSERVATIONS FOR t-TEST OF DIFFERENCE BETWEEN TWO MEA

	Level of t-test																		
Single-sided test Double-sided test	α = 0.005 α = 0.01					α = 0.01 α = 0.02					α = 0.025 α = 0.05					α = 0.05 α = 0.1			
β =	0.01	0.05	0.1	0.2	0.5	0.01	0.05	0.1	0.2	0.5	0.01	0.05	0.1	0.2	0.5	0.01	0.05	0.1	0.2
0.05																			
0.10																			
0.15																			
0.20																			
0.25															124				
0.30										123					87				
0.35					110					90					64				102
0.40					85					70				100	50			108	78
0.45				118	68				101	55			105	79	39		108	86	62
0.50				96	55			106	82	45		106	86	64	32		88	70	51
0.55			101	79	46		106	88	68	38		87	71	53	27	112	73	58	42
0.60		101	85	67	39		90	74	58	32	104	74	60	45	23	89	61	49	36
0.65		87	73	57	34	104	77	64	49	27	88	63	51	39	20	76	52	42	30
0.70	100	75	63	50	29	90	66	55	43	24	76	55	44	34	17	66	45	36	26
0.75	88	66	55	44	26	79	58	48	38	21	67	48	39	29	15	57	40	32	23
0.80	77	58	49	39	23	70	51	43	33	19	59	42	34	26	14	50	35	28	21
0.85	69	51	43	35	21	62	46	38	30	17	52	37	31	23	12	45	31	25	18
0.90	62	46	39	31	19	55	41	34	27	15	47	34	27	21	11	40	28	22	16
0.95	55	42	35	28	17	50	37	31	24	14	42	30	25	19	10	36	25	20	15
1.00	50	38	32	26	15	45	33	28	22	13	38	27	23	17	9	33	23	18	14
1.1	42	32	27	22	13	38	28	23	19	11	32	23	19	14	8	27	19	15	12
1.2	36	27	23	18	11	32	24	20	16	9	27	20	16	12	7	23	16	13	10
1.3	31	23	20	16	10	28	21	17	14	8	23	17	14	11	6	20	14	11	9
1.4	27	20	17	14	9	24	18	15	12	8	20	15	12	10	6	17	12	10	8
1.5	24	18	15	13	8	21	16	14	11	7	18	13	11	9	5	15	11	9	7
1.6	21	16	14	11	7	19	14	12	10	6	16	12	10	8	5	14	10	8	6
1.7	19	15	13	10	7	17	13	11	9	6	14	11	9	7	4	12	9	7	6
1.8	17	13	11	10	6	15	12	10	8	5	13	10	8	6	4	11	8	7	5
1.9	16	12	11	9	6	14	11	9	8	5	12	9	7	6	4	10	7	6	5
2.0	14	11	10	8	6	13	10	9	7	5	11	8	7	6	4	9	7	6	4
2.1	13	10	9	8	5	12	9	8	7	5	10	8	6	5	3	8	6	5	4
2.2	12	10	8	7	5	11	9	7	6	4	9	7	6	5		8	6	5	4
2.3	11	9	8	7	5	10	8	7	6	4	9	7	6	5		7	5	5	4
2.4	11	9	8	6	5	10	8	7	6	4	8	6	5	4		7	5	4	4
2.5	10	8	7	6	4	9	7	6	5	4	8	6	5	4		6	5	4	3
3.0	8	6	6	5	4	7	6	5	4	3	6	5	4	4		5	4	3	
3.5	6	5	5	4	3	6	5	4	4		5	4	4	3		4	3		
4.0	6	5	4	4		5	4	4	3		4	4	3			4			

Value of $\Delta = \dfrac{\mu_1 - \mu_2}{\sigma}$

V. Chi-Square Distribution

V.1 PERCENTAGE POINTS, CHI-SQUARE DISTRIBUTION

This table gives values of χ^2 such that

$$F(\chi^2) = \int_0^{\chi^2} \frac{1}{2^{\frac{n}{2}} \, \Gamma\left(\frac{n}{2}\right)} x^{\frac{n-2}{2}} e^{-\frac{x}{2}} \, dx$$

n, the number of degrees of freedom, equal to 1, 2, . . . , 30. For $n > 30$, a normal approximation is quite accurate. The expression $\sqrt{2\chi^2} - \sqrt{2n-1}$ is approximately normally distributed as the standard normal distribution. Thus χ_α^2, the α-point of the distribution, may be computed by the formula

$$\chi_\alpha^2 = \tfrac{1}{2}[x_\alpha + \sqrt{2n-1}]^2,$$

where x_α is the α-point of the cumulative normal distribution. For even values of n, $F(\chi^2)$ may be written as

$$1 - F(\chi^2) = \sum_{x=0}^{x'-1} \frac{e^{-m}m^x}{x!}$$

with $m = \tfrac{1}{2}\chi^2$ and $x' = \tfrac{1}{2}n$. Thus the cumulative Chi-Square distribution is related to the cumulative Poisson distribution.

89

PERCENTAGE POINTS, CHI-SQUARE DISTRIBUTION

$$F(x^2) = \int_0^{x^2} \frac{1}{2^{\frac{n}{2}} \Gamma\left(\frac{n}{2}\right)} x^{\frac{n-2}{2}} e^{-\frac{x}{2}} \, dx$$

F / n	.005	.010	.025	.050	.100	.250	.500	.750	.900	.950	.975	.990	.995
1	.0000393	.000157	.000982	.00393	.0158	.102	.455	1.32	2.71	3.84	5.02	6.63	7.88
2	.0100	.0201	.0506	.103	.211	.575	1.39	2.77	4.61	5.99	7.38	9.21	10.6
3	.0717	.115	.216	.352	.584	1.21	2.37	4.11	6.25	7.81	9.35	11.3	12.8
4	.207	.297	.484	.711	1.06	1.92	3.36	5.39	7.78	9.49	11.1	13.3	14.9
5	.412	.554	.831	1.15	1.61	2.67	4.35	6.63	9.24	11.1	12.8	15.1	16.7
6	.676	.872	1.24	1.64	2.20	3.45	5.35	7.84	10.6	12.6	14.4	16.8	18.5
7	.989	1.24	1.69	2.17	2.83	4.25	6.35	9.04	12.0	14.1	16.0	18.5	20.3
8	1.34	1.65	2.18	2.73	3.49	5.07	7.34	10.2	13.4	15.5	17.5	20.1	22.0
9	1.73	2.09	2.70	3.33	4.17	5.90	8.34	11.4	14.7	16.9	19.0	21.7	23.6
10	2.16	2.56	3.25	3.94	4.87	6.74	9.34	12.5	16.0	18.3	20.5	23.2	25.2
11	2.60	3.05	3.82	4.57	5.58	7.58	10.3	13.7	17.3	19.7	21.9	24.7	26.8
12	3.07	3.57	4.40	5.23	6.30	8.44	11.3	14.8	18.5	21.0	23.3	26.2	28.3
13	3.57	4.11	5.01	5.89	7.04	9.30	12.3	16.0	19.8	22.4	24.7	27.7	29.8
14	4.07	4.66	5.63	6.57	7.79	10.2	13.3	17.1	21.1	23.7	26.1	29.1	31.3
15	4.60	5.23	6.26	7.26	8.55	11.0	14.3	18.2	22.3	25.0	27.5	30.6	32.8
16	5.14	5.81	6.91	7.96	9.31	11.9	15.3	19.4	23.5	26.3	28.8	32.0	34.3
17	5.70	6.41	7.56	8.67	10.1	12.8	16.3	20.5	24.8	27.6	30.2	33.4	35.7
18	6.26	7.01	8.23	9.39	10.9	13.7	17.3	21.6	26.0	28.9	31.5	34.8	37.2
19	6.84	7.63	8.91	10.1	11.7	14.6	18.3	22.7	27.2	30.1	32.9	36.2	38.6
20	7.43	8.26	9.59	10.9	12.4	15.5	19.3	23.8	28.4	31.4	34.2	37.6	40.0
21	8.03	8.90	10.3	11.6	13.2	16.3	20.3	24.9	29.6	32.7	35.5	38.9	41.4
22	8.64	9.54	11.0	12.3	14.0	17.2	21.3	26.0	30.8	33.9	36.8	40.3	42.8
23	9.26	10.2	11.7	13.1	14.8	18.1	22.3	27.1	32.0	35.2	38.1	41.6	44.2
24	9.89	10.9	12.4	13.8	15.7	19.0	23.3	28.2	33.2	36.4	39.4	43.0	45.6
25	10.5	11.5	13.1	14.6	16.5	19.9	24.3	29.3	34.4	37.7	40.6	44.3	46.9
26	11.2	12.2	13.8	15.4	17.3	20.8	25.3	30.4	35.6	38.9	41.9	45.6	48.3
27	11.8	12.9	14.6	16.2	18.1	21.7	26.3	31.5	36.7	40.1	43.2	47.0	49.6
28	12.5	13.6	15.3	16.9	18.9	22.7	27.3	32.6	37.9	41.3	44.5	48.3	51.0
29	13.1	14.3	16.0	17.7	19.8	23.6	28.3	33.7	39.1	42.6	45.7	49.6	52.3

V.2 PERCENTAGE POINTS, CHI-SQUARE OVER DEGREES OF FREEDOM DISTRIBUTION

This table gives the percentage points of the sampling distribution of $\frac{s^2}{\sigma^2}$, referred to he percentage points of the $\frac{\chi^2}{\text{d.f.}}$ distribution (read "chi-square over degrees of freedom"). The percentage points are different for different sample sizes.

Chi-Square Distribution

PERCENTAGE POINTS, CHI-SQUARE OVER DEGREES OF FREEDOM DISTRIBUTIO

F \ n	Probability in per cent						Probability in per cent					
	0.05	0.1	0.5	1.0	2.5	5.0	95.0	97.5	99.0	99.5	99.9	99.95
1	.0000	.0000	.0000	.0002	.0010	.0039	3.8410	5.0240	6.6350	7.8790	10.8280	12.116
2	.0005	.0010	.0050	.0100	.0253	.0515	2.9955	3.6890	4.6050	5.2985	6.9080	7.601
3	.0051	.0081	.0239	.0383	.0720	.1173	2.6050	3.1160	3.7817	4.2793	5.4220	5.910
4	.0160	.0227	.0518	.0742	.1210	.1778	2.3720	2.7858	3.3192	3.7150	4.6168	4.99
5	.0316	.0420	.0824	.1108	.1662	.2290	2.2140	2.5664	3.0172	3.3500	4.1030	4.42
6	.0499	.0635	.1127	.1453	.2062	.2725	2.0987	2.4082	2.8020	3.0913	3.7430	4.01
7	.0693	.0854	.1413	.1770	.2414	.3096	2.0096	2.2876	2.6393	2.8969	3.4746	3.71
8	.0888	.1071	.1680	.2058	.2725	.3416	1.9384	2.1919	2.5112	2.7444	3.2656	3.48
9	.1080	.1281	.1928	.2320	.3000	.3694	1.8799	2.1137	2.4073	2.6210	3.0974	3.29
10	.1265	.1479	.2156	.2558	.3247	.3940	1.8307	2.0483	2.3209	2.5188	2.9588	3.14
11	.1443	.1667	.2366	.2775	.3469	.4159	1.7886	1.9927	2.2477	2.4325	2.8422	3.01
12	.1612	.1845	.2562	.2976	.3670	.4355	1.7522	1.9447	2.1848	2.3583	2.7424	2.90
13	.1773	.2013	.2742	.3159	.3853	.4532	1.7202	1.9028	2.1298	2.2938	2.6560	2.80
14	.1926	.2172	.2911	.3329	.4021	.4694	1.6918	1.8656	2.0815	2.2371	2.5802	2.72
15	.2072	.2322	.3067	.3486	.4175	.4841	1.6664	1.8325	2.0385	2.1867	2.5131	2.64
16	.2210	.2464	.3214	.3632	.4318	.4976	1.6435	1.8028	2.0000	2.1417	2.4532	2.58
17	.2341	.2598	.3351	.3769	.4449	.5101	1.6228	1.7759	1.9652	2.1011	2.3994	2.52
18	.2466	.2725	.3481	.3897	.4573	.5217	1.6038	1.7514	1.9336	2.0642	2.3507	2.46
19	.2585	.2846	.3602	.4017	.4688	.5325	1.5865	1.7291	1.9048	2.0306	2.3063	2.41
20	.2699	.2961	.3717	.4130	.4796	.5426	1.5705	1.7085	1.8783	1.9998	2.2658	2.37
21	.2808	.3070	.3826	.4237	.4897	.5520	1.5558	1.6895	1.8539	1.9715	2.2284	2.33
22	.2911	.3174	.3929	.4337	.4992	.5608	1.5420	1.6719	1.8313	1.9453	2.1940	2.29
23	.3010	.3273	.4026	.4433	.5082	.5692	1.5292	1.6555	1.8103	1.9209	2.1621	2.26
24	.3105	.3369	.4119	.4523	.5167	.5770	1.5173	1.6402	1.7908	1.8982	2.1325	2.22
25	.3196	.3460	.4208	.4610	.5248	.5844	1.5061	1.6258	1.7726	1.8771	2.1048	2.19
26	.3284	.3547	.4292	.4692	.5325	.5915	1.4956	1.6124	1.7555	1.8573	2.0789	2.16
27	.3368	.3631	.4373	.4770	.5397	.5982	1.4857	1.5998	1.7394	1.8387	2.0547	2.14
28	.3449	.3711	.4450	.4845	.5467	.6046	1.4763	1.5879	1.7242	1.8212	2.0319	2.11
29	.3527	.3788	.4524	.4916	.5533	.6106	1.4675	1.5766	1.7099	1.8047	2.0104	2.09
30	.3601	.3863	.4596	.4984	.5597	.6164	1.4591	1.5660	1.6964	1.7891	1.9901	2.07
31	.3674	.3934	.4664	.5050	.5658	.6220	1.4511	1.5559	1.6836	1.7743	1.9709	2.05
32	.3743	.4003	.4729	.5113	.5716	.6272	1.4436	1.5462	1.6714	1.7602	1.9527	2.03
33	.3811	.4070	.4792	.5174	.5772	.6323	1.4364	1.5371	1.6599	1.7469	1.9355	2.01
34	.3876	.4134	.4853	.5232	.5825	.6372	1.4295	1.5284	1.6489	1.7342	1.9190	1.99
35	.3939	.4197	.4912	.5288	.5877	.6419	1.4229	1.5201	1.6383	1.7221	1.9034	1.97
36	.4000	.4257	.4969	.5342	.5927	.6464	1.4166	1.5121	1.6283	1.7106	1.8885	1.96
37	.4059	.4315	.5023	.5395	.5975	.6507	1.4106	1.5045	1.6187	1.6995	1.8742	1.94
38	.4117	.4371	.5076	.5445	.6021	.6548	1.4048	1.4972	1.6095	1.6890	1.8606	1.93
39	.4173	.4426	.5127	.5494	.6065	.6588	1.3993	1.4903	1.6007	1.6789	1.8476	1.91
40	.4226	.4479	.5177	.5541	.6108	.6627	1.3940	1.4836	1.5923	1.6692	1.8350	1.90
41	.4279	.4530	.5225	.5587	.6150	.6665	1.3888	1.4771	1.5841	1.6598	1.8230	1.88
42	.4330	.4580	.5271	.5631	.6190	.6701	1.3839	1.4709	1.5763	1.6509	1.8115	1.87
43	.4380	.4629	.5316	.5674	.6229	.6736	1.3792	1.4649	1.5688	1.6422	1.8004	1.85
44	.4428	.4676	.5360	.5715	.6267	.6770	1.3746	1.4591	1.5616	1.6339	1.7898	1.85
45	.4475	.4722	.5402	.5756	.6304	.6803	1.3701	1.4536	1.5546	1.6259	1.7795	1.84
46	.4520	.4767	.5444	.5795	.6339	.6835	1.3659	1.4482	1.5478	1.6182	1.7696	1.83
47	.4565	.4811	.5484	.5833	.6374	.6866	1.3617	1.4430	1.5413	1.6107	1.7600	1.82
48	.4609	.4853	.5523	.5870	.6407	.6895	1.3577	1.4380	1.5351	1.6035	1.7508	1.81
49	.4651	.4894	.5561	.5906	.6440	.6924	1.3539	1.4331	1.5290	1.5966	1.7418	1.80
50	.4692	.4935	.5598	.5941	.6471	.6953	1.3501	1.4284	1.5231	1.5898	1.7332	1.79

RCENTAGE POINTS, CHI-SQUARE OVER DEGREES OF FREEDOM DISTRIBUTION

F	Probability in per cent						Probability in per cent					
	0.05	0.1	0.5	1.0	2.5	5.0	95.0	97.5	99.0	99.5	99.9	99.95
51	.4733	.4974	.5634	.5975	.6502	.6980	1.3465	1.4238	1.5174	1.5833	1.7249	1.7821
52	.4772	.5012	.5669	.6009	.6532	.7007	1.3429	1.4194	1.5118	1.5769	1.7168	1.7733
53	.4810	.5050	.5704	.6041	.6562	.7033	1.3395	1.4151	1.5065	1.5708	1.7089	1.7648
54	.4848	.5087	.5737	.6073	.6590	.7059	1.3362	1.4110	1.5013	1.5649	1.7013	1.7565
55	.4885	.5122	.5770	.6104	.6618	.7083	1.3329	1.4069	1.4962	1.5591	1.6939	1.7484
56	.4921	.5157	.5802	.6134	.6645	.7107	1.3298	1.4030	1.4913	1.5535	1.6868	1.7406
57	.4956	.5191	.5833	.6163	.6671	.7131	1.3267	1.3992	1.4865	1.5480	1.6798	1.7331
58	.4990	.5225	.5863	.6192	.6697	.7154	1.3238	1.3954	1.4819	1.5427	1.6731	1.7257
59	.5024	.5258	.5893	.6220	.6722	.7176	1.3209	1.3918	1.4774	1.5375	1.6665	1.7185
60	.5057	.5290	.5922	.6248	.6747	.7198	1.3180	1.3883	1.4730	1.5325	1.6601	1.7116
61	.5089	.5321	.5951	.6274	.6771	.7219	1.3153	1.3849	1.4687	1.5276	1.6539	1.7048
62	.5121	.5352	.5979	.6300	.6795	.7240	1.3126	1.3815	1.4645	1.5229	1.6478	1.6982
63	.5152	.5382	.6006	.6326	.6817	.7260	1.3100	1.3783	1.4605	1.5182	1.6419	1.6918
64	.5182	.5411	.6033	.6351	.6840	.7280	1.3074	1.3751	1.4565	1.5137	1.6362	1.6855
65	.5212	.5440	.6059	.6376	.6862	.7300	1.3049	1.3720	1.4526	1.5093	1.6306	1.6794
66	.5241	.5469	.6085	.6400	.6883	.7319	1.3025	1.3689	1.4489	1.5050	1.6251	1.6735
67	.5270	.5496	.6110	.6424	.6905	.7338	1.3001	1.3660	1.4452	1.5008	1.6198	1.6677
68	.5298	.5524	.6134	.6447	.6925	.7356	1.2978	1.3631	1.4416	1.4967	1.6146	1.6620
69	.5325	.5550	.6159	.6469	.6946	.7374	1.2955	1.3602	1.4381	1.4927	1.6095	1.6565
70	.5352	.5577	.6182	.6492	.6965	.7391	1.2933	1.3575	1.4346	1.4888	1.6045	1.6511
71	.5379	.5602	.6205	.6514	.6985	.7408	1.2911	1.3548	1.4313	1.4850	1.5997	1.6458
72	.5405	.5628	.6228	.6535	.7004	.7425	1.2890	1.3521	1.4280	1.4812	1.5949	1.6407
73	.5431	.5653	.6251	.6556	.7023	.7442	1.2869	1.3495	1.4248	1.4776	1.5903	1.6356
74	.5456	.5677	.6273	.6576	.7041	.7458	1.2849	1.3470	1.4216	1.4740	`1.5858	1.6307
75	.5481	.5701	.6294	.6597	.7059	.7474	1.2829	1.3445	1.4186	1.4705	1.5813	1.6259
76	.5505	.5724	.6316	.6617	.7077	.7489	1.2809	1.3421	1.4156	1.4670	1.5770	1.6212
77	.5529	.5748	.6336	.6636	.7094	.7505	1.2790	1.3397	1.4126	1.4637	1.5727	1.6166
78	.5553	.5771	.6357	.6655	.7111	.7520	1.2771	1.3374	1.4097	1.4604	1.5686	1.6120
79	.5576	.5793	.6377	.6674	.7128	.7534	1.2753	1.3351	1.4069	1.4572	1.5645	1.6076
80	.5599	.5815	.6396	.6692	.7144	.7549	1.2735	1.3329	1.4041	1.4540	1.5605	1.6033
81	.5621	.5837	.6416	.6711	.7160	.7563	1.2717	1.3307	1.4014	1.4509	1.5566	1.5990
82	.5643	.5858	.6435	.6729	.7176	.7577	1.2700	1.3285	1.3987	1.4479	1.5527	1.5948
83	.5665	.5879	.6454	.6746	.7192	.7591	1.2683	1.3264	1.3961	1.4449	1.5490	1.5908
84	.5687	.5900	.6472	.6763	.7207	.7604	1.2666	1.3243	1.3935	1.4420	1.5453	1.5868
85	.5708	.5920	.6491	.6780	.7222	.7618	1.2650	1.3223	1.3910	1.4391	1.5417	1.5828
86	.5728	.5940	.6508	.6797	.7237	.7631	1.2633	1.3203	1.3885	1.4363	1.5381	1.5790
87	.5749	.5960	.6526	.6814	.7252	.7643	1.2618	1.3183	1.3861	1.4335	1.5346	1.5752
88	.5769	.5979	.6543	.6830	.7266	.7656	1.2602	1.3164	1.3837	1.4308	1.5312	1.5715
89	.5789	.5998	.6561	.6846	.7280	.7668	1.2587	1.3145	1.3814	1.4282	1.5278	1.5678
90	.5808	.6017	.6577	.6862	.7294	.7681	1.2572	1.3126	1.3791	1.4255	1.5245	1.5643
91	.5828	.6036	.6594	.6877	.7308	.7693	1.2557	1.3108	1.3768	1.4230	1.5213	1.5607
92	.5847	.6054	.6610	.6892	.7321	.7705	1.2542	1.3090	1.3746	1.4204	1.5181	1.5573
93	.5865	.6072	.6626	.6907	.7335	.7716	1.2528	1.3072	1.3724	1.4180	1.5150	1.5539
94	.5884	.6090	.6642	.6922	.7348	.7728	1.2514	1.3055	1.3702	1.4155	1.5119	1.5505
95	.5902	.6108	.6658	.6937	.7361	.7739	1.2500	1.3038	1.3681	1.4131	1.5089	1.5473
96	.5920	.6125	.6673	.6951	.7373	.7750	1.2487	1.3021	1.3661	1.4108	1.5059	1.5440
97	.5938	.6142	.6688	.6965	.7386	.7761	1.2473	1.3004	1.3640	1.4084	1.5030	1.5409
98	.5955	.6159	.6703	.6979	.7398	.7772	1.2460	1.2988	1.3620	1.4062	1.5001	1.5377
99	.5973	.6175	.6718	.6993	.7410	.7782	1.2447	1.2972	1.3600	1.4039	1.4973	1.5347
100	.5990	.6192	.6733	.7007	.7422	.7793	1.2434	1.2956	1.3581	1.4017	1.4945	1.5317

Chi-Square Distribution

PERCENTAGE POINTS, CHI-SQUARE OVER DEGREES OF FREEDOM DISTRIBUTION

F n	Probability in per cent						Probability in per cent					
	0.05	0.1	0.5	1.0	2.5	5.0	95.0	97.5	99.0	99.5	99.9	99.95
100	.5990	.6192	.6733	.7007	.7422	.7793	1.2434	1.2956	1.3581	1.4017	1.4945	1.531
105	.6072	.6271	.6802	.7071	.7480	.7843	1.2373	1.2881	1.3488	1.3911	1.4812	1.517
110	.6148	.6344	.6868	.7132	.7534	.7890	1.2316	1.2811	1.3401	1.3813	1.4689	1.504
115	.6221	.6414	.6930	.7190	.7584	.7934	1.2263	1.2746	1.3321	1.3722	1.4575	1.491
120	.6289	.6480	.6988	.7243	.7632	.7975	1.2214	1.2685	1.3246	1.3637	1.4468	1.480
125	.6353	.6542	.7042	.7294	.7676	.8014	1.2167	1.2627	1.3175	1.3557	1.4368	1.469
130	.6414	.6600	.7094	.7342	.7718	.8051	1.2124	1.2574	1.3109	1.3484	1.4275	1.459
135	.6473	.6656	.7143	.7388	.7757	.8085	1.2083	1.2523	1.3047	1.3413	1.4187	1.449
140	.6528	.6709	.7190	.7431	.7795	.8119	1.2043	1.2475	1.2988	1.3346	1.4104	1.440
145	.6581	.6760	.7234	.7472	.7831	.8150	1.2007	1.2430	1.2933	1.3284	1.4026	1.432
150	.6631	.6808	.7276	.7511	.7865	.8180	1.1972	1.2387	1.2880	1.3224	1.3951	1.424
155	.6679	.6854	.7316	.7549	.7898	.8208	1.1939	1.2346	1.2830	1.3168	1.3881	1.416
160	.6725	.6898	.7355	.7584	.7930	.8235	1.1907	1.2308	1.2783	1.3114	1.3813	1.409
165	.6769	.6939	.7392	.7618	.7959	.8260	1.1877	1.2270	1.2737	1.3063	1.3751	1.402
170	.6811	.6980	.7427	.7651	.7987	.8285	1.1848	1.2235	1.2694	1.3014	1.3690	1.395
175	.6852	.7019	.7461	.7682	.8015	.8309	1.1821	1.2201	1.2653	1.2968	1.3632	1.389
180	.6891	.7056	.7494	.7712	.8041	.8332	1.1795	1.2170	1.2614	1.2924	1.3577	1.383
185	.6929	.7092	.7525	.7741	.8066	.8353	1.1769	1.2138	1.2576	1.2881	1.3523	1.377
190	.6964	.7127	.7555	.7768	.8090	.8374	1.1745	1.2109	1.2541	1.2840	1.3472	1.372
195	.6999	.7160	.7584	.7795	.8114	.8394	1.1722	1.2081	1.2506	1.2801	1.3424	1.367
200	.7033	.7192	.7612	.7821	.8136	.8414	1.1700	1.2053	1.2473	1.2763	1.3377	1.362
210	.7097	.7254	.7665	.7870	.8179	.8451	1.1657	1.2001	1.2409	1.2692	1.3288	1.352
220	.7157	.7311	.7715	.7916	.8219	.8485	1.1618	1.1953	1.2351	1.2626	1.3207	1.343
230	.7213	.7365	.7762	.7959	.8256	.8517	1.1582	1.1908	1.2297	1.2564	1.3131	1.335
240	.7266	.7415	.7805	.7999	.8291	.8547	1.1547	1.1867	1.2246	1.2507	1.3060	1.327
250	.7317	.7463	.7847	.8037	.8324	.8576	1.1515	1.1828	1.2198	1.2453	1.2994	1.320
260	.7364	.7507	.7886	.8073	.8355	.8602	1.1485	1.1791	1.2153	1.2403	1.2931	1.314
270	.7408	.7550	.7923	.8107	.8384	.8628	1.1457	1.1756	1.2111	1.2356	1.2872	1.307
280	.7450	.7590	.7958	.8139	.8412	.8652	1.1430	1.1723	1.2071	1.2312	1.2817	1.301
290	.7491	.7629	.7991	.8170	.8438	.8674	1.1404	1.1692	1.2033	1.2269	1.2764	1.296
300	.7529	.7665	.8023	.8199	.8463	.8696	1.1380	1.1663	1.1997	1.2229	1.2714	1.296
350	.7698	.7826	.8160	.8326	.8573	.8790	1.1275	1.1535	1.1843	1.2055	1.2500	1.267
400	.7836	.7957	.8272	.8429	.8662	.8866	1.1191	1.1433	1.1718	1.1915	1.2378	1.249
450	.7951	.8066	.8366	.8515	.8736	.8929	1.1121	1.1349	1.1616	1.1801	1.2187	1.234
500	.8050	.8160	.8446	.8588	.8799	.8983	1.1063	1.1277	1.1530	1.1704	1.2070	1.221
550	.8135	.8239	.8515	.8651	.8853	.9029	1.1012	1.1216	1.1456	1.1622	1.1968	1.21
600	.8208	.8310	.8575	.8706	.8900	.9070	1.0968	1.1163	1.1392	1.1550	1.1880	1.20
650	.8275	.8373	.8629	.8755	.8942	.9106	1.0929	1.1116	1.1335	1.1487	1.1803	1.19
700	.8334	.8429	.8677	.8799	.8980	.9137	1.0895	1.1074	1.1285	1.1430	1.1734	1.18
750	.8387	.8480	.8720	.8838	.9013	.9166	1.0864	1.1037	1.1240	1.1380	1.1672	1.17
800	.8436	.8526	.8759	.8874	.9044	.9192	1.0836	1.1004	1.1200	1.1335	1.1617	1.17
850	.8480	.8568	.8795	.8906	.9072	.9216	1.0811	1.0973	1.1163	1.1294	1.1567	1.16
900	.8521	.8606	.8827	.8936	.9097	.9237	1.0788	1.0945	1.1129	1.1256	1.1520	1.16
950	.8559	.8642	.8858	.8964	.9121	.9257	1.0767	1.0919	1.1098	1.1221	1.1478	1.15
1000	.8594	.8675	.8886	.8989	.9143	.9276	1.0747	1.0895	1.1070	1.1190	1.1440	1.15
2000	.8992	.9051	.9204	.9279	.9390	.9486	1.0526	1.0629	1.0750	1.0833	1.1006	1.10
3000	.9172	.9221	.9348	.9409	.9500	.9579	1.0429	1.0513	1.0611	1.0678	1.0817	1.08
4000	.9280	.9323	.9433	.9487	.9566	.9635	1.0370	1.0443	1.0527	1.0585	1.0705	1.07
5000	.9355	.9393	.9493	.9541	.9612	.9673	1.0331	1.0396	1.0471	1.0523	1.0630	1.06
10000	.9541	.9569	.9640	.9674	.9725	.9769	1.0234	1.0279	1.0332	1.0368	1.0443	1.04

V.3 NUMBER OF OBSERVATIONS REQUIRED FOR THE COMPARISON OF A POPULATION VARIANCE WITH A STANDARD VALUE USING THE CHI-SQUARE TEST

The tabular entries show the value of the ratio R of the population variance σ_1^2 to a standard variance σ_0^2 which is undetected with probability β in a χ^2 test at significance level α of an estimate s_1^2 of σ_1^2 based on n degrees of freedom.

NUMBER OF OBSERVATIONS REQUIRED FOR THE COMPARISON OF A POPULATION VARIANCE WITH A STANDARD VALUE USING THE CHI-SQUARE TEST

n	$\alpha = 0.01$				$\alpha = 0.05$			
	$\beta = 0.01$	$\beta = 0.05$	$\beta = 0.1$	$\beta = 0.5$	$\beta = 0.01$	$\beta = 0.05$	$\beta = 0.1$	$\beta = 0.5$
1	42,240	1,687	420.2	14.58	25,450	977.0	243.3	8.444
2	458.2	89.78	43.71	6.644	298.1	58.40	28.43	4.322
3	98.79	32.24	19.41	4.795	68.05	22.21	13.37	3.303
4	44.69	18.68	12.48	3.955	31.93	13.35	8.920	2.826
5	27.22	13.17	9.369	3.467	19.97	9.665	6.875	2.544
6	19.28	10.28	7.628	3.144	14.44	7.699	5.713	2.354
7	14.91	8.524	6.521	2.911	11.35	6.491	4.965	2.217
8	12.20	7.352	5.757	2.736	9.418	5.675	4.444	2.112
9	10.38	6.516	5.198	2.597	8.103	5.088	4.059	2.028
10	9.072	5.890	4.770	2.484	7.156	4.646	3.763	1.960
12	7.343	5.017	4.159	2.312	5.889	4.023	3.335	1.854
15	5.847	4.211	3.578	2.132	4.780	3.442	2.925	1.743
20	4.548	3.462	3.019	1.943	3.802	2.895	2.524	1.624
24	3.959	3.104	2.745	1.842	3.354	2.630	2.326	1.560
30	3.403	2.752	2.471	1.735	2.927	2.367	2.125	1.492
40	2.874	2.403	2.192	1.619	2.516	2.103	1.919	1.418
60	2.358	2.046	1.902	1.490	2.110	1.831	1.702	1.333
120	1.829	1.661	1.580	1.332	1.686	1.532	1.457	1.228
∞	1.000	1.000	1.000	1.000	1.000	1.000	1.000	1.000

Examples

Testing for an increase in variance. Let $\alpha = 0.05$, $\beta = 0.01$, and $R = 4$. Entering the table with these values it is found that the value 4 occurs between the rows corresponding to $n = 15$ and $n = 20$. Using rough interpolation it is indicated that the estimate of variance should be based on 19 degrees of freedom.

Testing for a decrease in variance. Let $\alpha = 0.05$, $\beta = 0.01$, and $R = 0.33$. The table is entered with $\alpha' = \beta = 0.01$, $\beta' = \alpha = 0.05$, and $R' = 1/R = 3$. It is found that the value 3 occurs between the rows corresponding to $n = 24$ and $n = 30$. Using rough interpolation it is indicated that the estimate of variance should be based on 26 degrees of freedom.

VI. F-Distribution

This table gives values of F such that

$$F(F) = \int_0^F \frac{\Gamma\left(\dfrac{m+n}{2}\right)}{\Gamma\left(\dfrac{m}{2}\right)\Gamma\left(\dfrac{n}{2}\right)} m^{\frac{m}{2}} n^{\frac{n}{2}} x^{\frac{m-2}{2}} (n + mx)^{-\frac{m+n}{2}} dx$$

for selected values of m, the number of degrees of freedom of the numerator of F; and for selected values of n, the number of degrees of freedom of the denominator of F. The table also provides values corresponding to $F(F) = .10, .05, .025, .01, .005, .001$ since $F_{1-\alpha}$ for m and n degrees of freedom is the reciprocal of F_α for n and m degrees of freedom. Thus

$$F_{.05}(4, 7) = \frac{1}{F_{.95}(7, 4)} = \frac{1}{6.09} = .164 \ .$$

PERCENTAGE POINTS, F-DISTRIBUTION

$$F(F) = \int_0^F \frac{\Gamma\left(\frac{m+n}{2}\right)}{\Gamma\left(\frac{m}{2}\right)\Gamma\left(\frac{n}{2}\right)}\, m^{\frac{m}{2}} n^{\frac{n}{2}} x^{\frac{m}{2}-1}\,(n+mx)^{-\frac{m+n}{2}}\,dx = .90$$

m \ n	1	2	3	4	5	6	7	8	9	10	12	15	20	24	30	40	60	120	∞
1	39.86	49.50	53.59	55.83	57.24	58.20	58.91	59.44	59.86	60.19	60.71	61.22	61.74	62.00	62.26	62.53	62.79	63.06	63.33
2	8.53	9.00	9.16	9.24	9.29	9.33	9.35	9.37	9.38	9.39	9.41	9.42	9.44	9.45	9.46	9.47	9.47	9.48	9.49
3	5.54	5.46	5.39	5.34	5.31	5.28	5.27	5.25	5.24	5.23	5.22	5.20	5.18	5.18	5.17	5.16	5.15	5.14	5.13
4	4.54	4.32	4.19	4.11	4.05	4.01	3.98	3.95	3.94	3.92	3.90	3.87	3.84	3.83	3.82	3.80	3.79	3.78	3.76
5	4.06	3.78	3.62	3.52	3.45	3.40	3.37	3.34	3.32	3.30	3.27	3.24	3.21	3.19	3.17	3.16	3.14	3.12	3.10
6	3.78	3.46	3.29	3.18	3.11	3.05	3.01	2.98	2.96	2.94	2.90	2.87	2.84	2.82	2.80	2.78	2.76	2.74	2.72
7	3.59	3.26	3.07	2.96	2.88	2.83	2.78	2.75	2.72	2.70	2.67	2.63	2.59	2.58	2.56	2.54	2.51	2.49	2.47
8	3.46	3.11	2.92	2.81	2.73	2.67	2.62	2.59	2.56	2.54	2.50	2.46	2.42	2.40	2.38	2.36	2.34	2.32	2.29
9	3.36	3.01	2.81	2.69	2.61	2.55	2.51	2.47	2.44	2.42	2.38	2.34	2.30	2.28	2.25	2.23	2.21	2.18	2.16
10	3.29	2.92	2.73	2.61	2.52	2.46	2.41	2.38	2.35	2.32	2.28	2.24	2.20	2.18	2.16	2.13	2.11	2.08	2.06
11	3.23	2.86	2.66	2.54	2.45	2.39	2.34	2.30	2.27	2.25	2.21	2.17	2.12	2.10	2.08	2.05	2.03	2.00	1.97
12	3.18	2.81	2.61	2.48	2.39	2.33	2.28	2.24	2.21	2.19	2.15	2.10	2.06	2.04	2.01	1.99	1.96	1.93	1.90
13	3.14	2.76	2.56	2.43	2.35	2.28	2.23	2.20	2.16	2.14	2.10	2.05	2.01	1.98	1.96	1.93	1.90	1.88	1.85
14	3.10	2.73	2.52	2.39	2.31	2.24	2.19	2.15	2.12	2.10	2.05	2.01	1.96	1.94	1.91	1.89	1.86	1.83	1.80
15	3.07	2.70	2.49	2.36	2.27	2.21	2.16	2.12	2.09	2.06	2.02	1.97	1.92	1.90	1.87	1.85	1.82	1.79	1.76
16	3.05	2.67	2.46	2.33	2.24	2.18	2.13	2.09	2.06	2.03	1.99	1.94	1.89	1.87	1.84	1.81	1.78	1.75	1.72
17	3.03	2.64	2.44	2.31	2.22	2.15	2.10	2.06	2.03	2.00	1.96	1.91	1.86	1.84	1.81	1.78	1.75	1.72	1.69
18	3.01	2.62	2.42	2.29	2.20	2.13	2.08	2.04	2.00	1.98	1.93	1.89	1.84	1.81	1.78	1.75	1.72	1.69	1.66
19	2.99	2.61	2.40	2.27	2.18	2.11	2.06	2.02	1.98	1.96	1.91	1.86	1.81	1.79	1.76	1.73	1.70	1.67	1.63
20	2.97	2.59	2.38	2.25	2.16	2.09	2.04	2.00	1.96	1.94	1.89	1.84	1.79	1.77	1.74	1.71	1.68	1.64	1.61
21	2.96	2.57	2.36	2.23	2.14	2.08	2.02	1.98	1.95	1.92	1.87	1.83	1.78	1.75	1.72	1.69	1.66	1.62	1.59
22	2.95	2.56	2.35	2.22	2.13	2.06	2.01	1.97	1.93	1.90	1.86	1.81	1.76	1.73	1.70	1.67	1.64	1.60	1.57
23	2.94	2.55	2.34	2.21	2.11	2.05	1.99	1.95	1.92	1.89	1.84	1.80	1.74	1.72	1.69	1.66	1.62	1.59	1.55
24	2.93	2.54	2.33	2.19	2.10	2.04	1.98	1.94	1.91	1.88	1.83	1.78	1.73	1.70	1.67	1.64	1.61	1.57	1.53
25	2.92	2.53	2.32	2.18	2.09	2.02	1.97	1.93	1.89	1.87	1.82	1.77	1.72	1.69	1.66	1.63	1.59	1.56	1.52
26	2.91	2.52	2.31	2.17	2.08	2.01	1.96	1.92	1.88	1.86	1.81	1.76	1.71	1.68	1.65	1.61	1.58	1.54	1.50
27	2.90	2.51	2.30	2.17	2.07	2.00	1.95	1.91	1.87	1.85	1.80	1.75	1.70	1.67	1.64	1.60	1.57	1.53	1.49
28	2.89	2.50	2.29	2.16	2.06	2.00	1.94	1.90	1.87	1.84	1.79	1.74	1.69	1.66	1.63	1.59	1.56	1.52	1.48
29	2.89	2.50	2.28	2.15	2.06	1.99	1.93	1.89	1.86	1.83	1.78	1.73	1.68	1.65	1.62	1.58	1.55	1.51	1.47
30	2.88	2.49	2.28	2.14	2.05	1.98	1.93	1.88	1.85	1.82	1.77	1.72	1.67	1.64	1.61	1.57	1.54	1.50	1.46
40	2.84	2.44	2.23	2.09	2.00	1.93	1.87	1.83	1.79	1.76	1.71	1.66	1.61	1.57	1.54	1.51	1.47	1.42	1.38
60	2.79	2.39	2.18	2.04	1.95	1.87	1.82	1.77	1.74	1.71	1.66	1.60	1.54	1.51	1.48	1.44	1.40	1.35	1.29
120	2.75	2.35	2.13	1.99	1.90	1.82	1.77	1.72	1.68	1.65	1.60	1.55	1.48	1.45	1.41	1.37	1.32	1.26	1.19
∞	2.71	2.30	2.08	1.94	1.85	1.77	1.72	1.67	1.63	1.60	1.55	1.49	1.42	1.38	1.34	1.30	1.24	1.17	1.00

$F = \dfrac{s_1^2}{s_2^2} = \dfrac{S_1}{m}\Big/\dfrac{S_2}{n}$, where $s_1^2 = S_1/m$ and $s_2^2 = S_2/n$ are independent mean squares estimating a common variance σ^2 and based on m and n degrees of freedom, respectively.

PERCENTAGE POINTS, F-DISTRIBUTION

$$F(F) = \int_0^F \frac{\Gamma\left(\dfrac{m+n}{2}\right)}{\Gamma\left(\dfrac{m}{2}\right)\Gamma\left(\dfrac{n}{2}\right)} m^{\frac{m}{2}} n^{\frac{n}{2}} x^{\frac{m}{2}-1} (n+mx)^{-\frac{m+n}{2}}\, dx = .95$$

$\frac{m}{n}$	1	2	3	4	5	6	7	8	9	10	12	15	20	24	30	40	60	120	∞
1	161.4	199.5	215.7	224.6	230.2	234.0	236.8	238.9	240.5	241.9	243.9	245.9	248.0	249.1	250.1	251.1	252.2	253.3	254.3
2	18.51	19.00	19.16	19.25	19.30	19.33	19.35	19.37	19.38	19.40	19.41	19.43	19.45	19.45	19.46	19.47	19.48	19.49	19.50
3	10.13	9.55	9.28	9.12	9.01	8.94	8.89	8.85	8.81	8.79	8.74	8.70	8.66	8.64	8.62	8.59	8.57	8.55	8.53
4	7.71	6.94	6.59	6.39	6.26	6.16	6.09	6.04	6.00	5.96	5.91	5.86	5.80	5.77	5.75	5.72	5.69	5.66	5.63
5	6.61	5.79	5.41	5.19	5.05	4.95	4.88	4.82	4.77	4.74	4.68	4.62	4.56	4.53	4.50	4.46	4.43	4.40	4.36
6	5.99	5.14	4.76	4.53	4.39	4.28	4.21	4.15	4.10	4.06	4.00	3.94	3.87	3.84	3.81	3.77	3.74	3.70	3.67
7	5.59	4.74	4.35	4.12	3.97	3.87	3.79	3.73	3.68	3.64	3.57	3.51	3.44	3.41	3.38	3.34	3.30	3.27	3.23
8	5.32	4.46	4.07	3.84	3.69	3.58	3.50	3.44	3.39	3.35	3.28	3.22	3.15	3.12	3.08	3.04	3.01	2.97	2.93
9	5.12	4.26	3.86	3.63	3.48	3.37	3.29	3.23	3.18	3.14	3.07	3.01	2.94	2.90	2.86	2.83	2.79	2.75	2.71
10	4.96	4.10	3.71	3.48	3.33	3.22	3.14	3.07	3.02	2.98	2.91	2.85	2.77	2.74	2.70	2.66	2.62	2.58	2.54
11	4.84	3.98	3.59	3.36	3.20	3.09	3.01	2.95	2.90	2.85	2.79	2.72	2.65	2.61	2.57	2.53	2.49	2.45	2.40
12	4.75	3.89	3.49	3.26	3.11	3.00	2.91	2.85	2.80	2.75	2.69	2.62	2.54	2.51	2.47	2.43	2.38	2.34	2.30
13	4.67	3.81	3.41	3.18	3.03	2.92	2.83	2.77	2.71	2.67	2.60	2.53	2.46	2.42	2.38	2.34	2.30	2.25	2.21
14	4.60	3.74	3.34	3.11	2.96	2.85	2.76	2.70	2.65	2.60	2.53	2.46	2.39	2.35	2.31	2.27	2.22	2.18	2.13
15	4.54	3.68	3.29	3.06	2.90	2.79	2.71	2.64	2.59	2.54	2.48	2.40	2.33	2.29	2.25	2.20	2.16	2.11	2.07
16	4.49	3.63	3.24	3.01	2.85	2.74	2.66	2.59	2.54	2.49	2.42	2.35	2.28	2.24	2.19	2.15	2.11	2.06	2.01
17	4.45	3.59	3.20	2.96	2.81	2.70	2.61	2.55	2.49	2.45	2.38	2.31	2.23	2.19	2.15	2.10	2.06	2.01	1.96
18	4.41	3.55	3.16	2.93	2.77	2.66	2.58	2.51	2.46	2.41	2.34	2.27	2.19	2.15	2.11	2.06	2.02	1.97	1.92
19	4.38	3.52	3.13	2.90	2.74	2.63	2.54	2.48	2.42	2.38	2.31	2.23	2.16	2.11	2.07	2.03	1.98	1.93	1.88
20	4.35	3.49	3.10	2.87	2.71	2.60	2.51	2.45	2.39	2.35	2.28	2.20	2.12	2.08	2.04	1.99	1.95	1.90	1.84
21	4.32	3.47	3.07	2.84	2.68	2.57	2.49	2.42	2.37	2.32	2.25	2.18	2.10	2.05	2.01	1.96	1.92	1.87	1.81
22	4.30	3.44	3.05	2.82	2.66	2.55	2.46	2.40	2.34	2.30	2.23	2.15	2.07	2.03	1.98	1.94	1.89	1.84	1.78
23	4.28	3.42	3.03	2.80	2.64	2.53	2.44	2.37	2.32	2.27	2.20	2.13	2.05	2.01	1.96	1.91	1.86	1.81	1.76
24	4.26	3.40	3.01	2.78	2.62	2.51	2.42	2.36	2.30	2.25	2.18	2.11	2.03	1.98	1.94	1.89	1.84	1.79	1.73
25	4.24	3.39	2.99	2.76	2.60	2.49	2.40	2.34	2.28	2.24	2.16	2.09	2.01	1.96	1.92	1.87	1.82	1.77	1.71
26	4.23	3.37	2.98	2.74	2.59	2.47	2.39	2.32	2.27	2.22	2.15	2.07	1.99	1.95	1.90	1.85	1.80	1.75	1.69
27	4.21	3.35	2.96	2.73	2.57	2.46	2.37	2.31	2.25	2.20	2.13	2.06	1.97	1.93	1.88	1.84	1.79	1.73	1.67
28	4.20	3.34	2.95	2.71	2.56	2.45	2.36	2.29	2.24	2.19	2.12	2.04	1.96	1.91	1.87	1.82	1.77	1.71	1.65
29	4.18	3.33	2.93	2.70	2.55	2.43	2.35	2.28	2.22	2.18	2.10	2.03	1.94	1.90	1.85	1.81	1.75	1.70	1.64
30	4.17	3.32	2.92	2.69	2.53	2.42	2.33	2.27	2.21	2.16	2.09	2.01	1.93	1.89	1.84	1.79	1.74	1.68	1.62
40	4.08	3.23	2.84	2.61	2.45	2.34	2.25	2.18	2.12	2.08	2.00	1.92	1.84	1.79	1.74	1.69	1.64	1.58	1.51
60	4.00	3.15	2.76	2.53	2.37	2.25	2.17	2.10	2.04	1.99	1.92	1.84	1.75	1.70	1.65	1.59	1.53	1.47	1.39
120	3.92	3.07	2.68	2.45	2.29	2.17	2.09	2.02	1.96	1.91	1.83	1.75	1.66	1.61	1.55	1.50	1.43	1.35	1.25
∞	3.84	3.00	2.60	2.37	2.21	2.10	2.01	1.94	1.88	1.83	1.75	1.67	1.57	1.52	1.46	1.39	1.32	1.22	1.00

$F = \dfrac{s_1^2}{s_2^2} = \dfrac{S_1}{m} / \dfrac{S_2}{n}$, where $s_1^2 = S_1/m$ and $s_2^2 = S_2/n$ are independent mean squares estimating a common variance σ^2 and based on m and n degrees of freedom, respectively.

$$F(F) = \int_0^F \frac{\Gamma\left(\frac{m+n}{2}\right)}{\Gamma\left(\frac{m}{2}\right)\Gamma\left(\frac{n}{2}\right)} m^{\frac{m}{2}} n^{\frac{n}{2}} x^{\frac{m}{2}-1} (n+mx)^{-\frac{m+n}{2}} \, dx = .975$$

n \ m	1	2	3	4	5	6	7	8	9	10	12	15	20	24	30	40	60	120	∞
1	647.8	799.5	864.2	899.6	921.8	937.1	948.2	956.7	963.3	968.6	976.7	984.9	993.1	997.2	1001	1006	1010	1014	1018
2	38.51	39.00	39.17	39.25	39.30	39.33	39.36	39.37	39.39	39.40	39.41	39.43	39.45	39.46	39.46	39.47	39.48	39.49	39.50
3	17.44	16.04	15.44	15.10	14.88	14.73	14.62	14.54	14.47	14.42	14.34	14.25	14.17	14.12	14.08	14.04	13.99	13.95	13.90
4	12.22	10.65	9.98	9.60	9.36	9.20	9.07	8.98	8.90	8.84	8.75	8.66	8.56	8.51	8.46	8.41	8.36	8.31	8.26
5	10.01	8.43	7.76	7.39	7.15	6.98	6.85	6.76	6.68	6.62	6.52	6.43	6.33	6.28	6.23	6.18	6.12	6.07	6.02
6	8.81	7.26	6.60	6.23	5.99	5.82	5.70	5.60	5.52	5.46	5.37	5.27	5.17	5.12	5.07	5.01	4.96	4.90	4.85
7	8.07	6.54	5.89	5.52	5.29	5.12	4.99	4.90	4.82	4.76	4.67	4.57	4.47	4.42	4.36	4.31	4.25	4.20	4.14
8	7.57	6.06	5.42	5.05	4.82	4.65	4.53	4.43	4.36	4.30	4.20	4.10	4.00	3.95	3.89	3.84	3.78	3.73	3.67
9	7.21	5.71	5.08	4.72	4.48	4.32	4.20	4.10	4.03	3.96	3.87	3.77	3.67	3.61	3.56	3.51	3.45	3.39	3.33
10	6.94	5.46	4.83	4.47	4.24	4.07	3.95	3.85	3.78	3.72	3.62	3.52	3.42	3.37	3.31	3.26	3.20	3.14	3.08
11	6.72	5.26	4.63	4.28	4.04	3.88	3.76	3.66	3.59	3.53	3.43	3.33	3.23	3.17	3.12	3.06	3.00	2.94	2.88
12	6.55	5.10	4.47	4.12	3.89	3.73	3.61	3.51	3.44	3.37	3.28	3.18	3.07	3.02	2.96	2.91	2.85	2.79	2.72
13	6.41	4.97	4.35	4.00	3.77	3.60	3.48	3.39	3.31	3.25	3.15	3.05	2.95	2.89	2.84	2.78	2.72	2.66	2.60
14	6.30	4.86	4.24	3.89	3.66	3.50	3.38	3.29	3.21	3.15	3.05	2.95	2.84	2.79	2.73	2.67	2.61	2.55	2.49
15	6.20	4.77	4.15	3.80	3.58	3.41	3.29	3.20	3.12	3.06	2.96	2.86	2.76	2.70	2.64	2.59	2.52	2.46	2.40
16	6.12	4.69	4.08	3.73	3.50	3.34	3.22	3.12	3.05	2.99	2.89	2.79	2.68	2.63	2.57	2.51	2.45	2.38	2.32
17	6.04	4.62	4.01	3.66	3.44	3.28	3.16	3.06	2.98	2.92	2.82	2.72	2.62	2.56	2.50	2.44	2.38	2.32	2.25
18	5.98	4.56	3.95	3.61	3.38	3.22	3.10	3.01	2.93	2.87	2.77	2.67	2.56	2.50	2.44	2.38	2.32	2.26	2.19
19	5.92	4.51	3.90	3.56	3.33	3.17	3.05	2.96	2.88	2.82	2.72	2.62	2.51	2.45	2.39	2.33	2.27	2.20	2.13
20	5.87	4.46	3.86	3.51	3.29	3.13	3.01	2.91	2.84	2.77	2.68	2.57	2.46	2.41	2.35	2.29	2.22	2.16	2.09
21	5.83	4.42	3.82	3.48	3.25	3.09	2.97	2.87	2.80	2.73	2.64	2.53	2.42	2.37	2.31	2.25	2.18	2.11	2.04
22	5.79	4.38	3.78	3.44	3.22	3.05	2.93	2.84	2.76	2.70	2.60	2.50	2.39	2.33	2.27	2.21	2.14	2.08	2.00
23	5.75	4.35	3.75	3.41	3.18	3.02	2.90	2.81	2.73	2.67	2.57	2.47	2.36	2.30	2.24	2.18	2.11	2.04	1.97
24	5.72	4.32	3.72	3.38	3.15	2.99	2.87	2.78	2.70	2.64	2.54	2.44	2.33	2.27	2.21	2.15	2.08	2.01	1.94
25	5.69	4.29	3.69	3.35	3.13	2.97	2.85	2.75	2.68	2.61	2.51	2.41	2.30	2.24	2.18	2.12	2.05	1.98	1.91
26	5.66	4.27	3.67	3.33	3.10	2.94	2.82	2.73	2.65	2.59	2.49	2.39	2.28	2.22	2.16	2.09	2.03	1.95	1.88
27	5.63	4.24	3.65	3.31	3.08	2.92	2.80	2.71	2.63	2.57	2.47	2.36	2.25	2.19	2.13	2.07	2.00	1.93	1.85
28	5.61	4.22	3.63	3.29	3.06	2.90	2.78	2.69	2.61	2.55	2.45	2.34	2.23	2.17	2.11	2.05	1.98	1.91	1.83
29	5.59	4.20	3.61	3.27	3.04	2.88	2.76	2.67	2.59	2.53	2.43	2.32	2.21	2.15	2.09	2.03	1.96	1.89	1.81
30	5.57	4.18	3.59	3.25	3.03	2.87	2.75	2.65	2.57	2.51	2.41	2.31	2.20	2.14	2.07	2.01	1.94	1.87	1.79
40	5.42	4.05	3.46	3.13	2.90	2.74	2.62	2.53	2.45	2.39	2.29	2.18	2.07	2.01	1.94	1.88	1.80	1.72	1.64
60	5.29	3.93	3.34	3.01	2.79	2.63	2.51	2.41	2.33	2.27	2.17	2.06	1.94	1.88	1.82	1.74	1.67	1.58	1.48
120	5.15	3.80	3.23	2.89	2.67	2.52	2.39	2.30	2.22	2.16	2.05	1.94	1.82	1.76	1.69	1.61	1.53	1.43	1.31
∞	5.02	3.69	3.12	2.79	2.57	2.41	2.29	2.19	2.11	2.05	1.94	1.83	1.71	1.64	1.57	1.48	1.39	1.27	1.00

$F = \dfrac{s_1^2}{s_2^2} = \dfrac{S_1/m}{S_2/n}$, where $s_1^2 = S_1/m$ and $s_2^2 = S_2/n$ are independent mean squares estimating a common variance σ^2 and based on m and n degrees of freedom, respectively.

F-Distribution

PERCENTAGE POINTS, F-DISTRIBUTION

$$F(F) = \int_0^F \frac{\Gamma\left(\frac{m+n}{2}\right)}{\Gamma\left(\frac{m}{2}\right)\Gamma\left(\frac{n}{2}\right)}\, m^{\frac{m}{2}} n^{\frac{n}{2}} x^{\frac{m}{2}-1} (n+mx)^{-\frac{m+n}{2}}\, dx = .99$$

m \ n	1	2	3	4	5	6	7	8	9	10	12	15	20	24	30	40	60	120	∞
1	4052	4999.5	5403	5625	5764	5859	5928	5982	6022	6056	6106	6157	6209	6235	6261	6287	6313	6339	6366
2	98.50	99.00	99.17	99.25	99.30	99.33	99.36	99.37	99.39	99.40	99.42	99.43	99.45	99.46	99.47	99.47	99.48	99.49	99.50
3	34.12	30.82	29.46	28.71	28.24	27.91	27.67	27.49	27.35	27.23	27.05	26.87	26.69	26.60	26.50	26.41	26.32	26.22	26.13
4	21.20	18.00	16.69	15.98	15.52	15.21	14.98	14.80	14.66	14.55	14.37	14.20	14.02	13.93	13.84	13.75	13.65	13.56	13.46
5	16.26	13.27	12.06	11.39	10.97	10.67	10.46	10.29	10.16	10.05	9.89	9.72	9.55	9.47	9.38	9.29	9.20	9.11	9.02
6	13.75	10.92	9.78	9.15	8.75	8.47	8.26	8.10	7.98	7.87	7.72	7.56	7.40	7.31	7.23	7.14	7.06	6.97	6.88
7	12.25	9.55	8.45	7.85	7.46	7.19	6.99	6.84	6.72	6.62	6.47	6.31	6.16	6.07	5.99	5.91	5.82	5.74	5.65
8	11.26	8.65	7.59	7.01	6.63	6.37	6.18	6.03	5.91	5.81	5.67	5.52	5.36	5.28	5.20	5.12	5.03	4.95	4.86
9	10.56	8.02	6.99	6.42	6.06	5.80	5.61	5.47	5.35	5.26	5.11	4.96	4.81	4.73	4.65	4.57	4.48	4.40	4.31
10	10.04	7.56	6.55	5.99	5.64	5.39	5.20	5.06	4.94	4.85	4.71	4.56	4.41	4.33	4.25	4.17	4.08	4.00	3.91
11	9.65	7.21	6.22	5.67	5.32	5.07	4.89	4.74	4.63	4.54	4.40	4.25	4.10	4.02	3.94	3.86	3.78	3.69	3.60
12	9.33	6.93	5.95	5.41	5.06	4.82	4.64	4.50	4.39	4.30	4.16	4.01	3.86	3.78	3.70	3.62	3.54	3.45	3.36
13	9.07	6.70	5.74	5.21	4.86	4.62	4.44	4.30	4.19	4.10	3.96	3.82	3.66	3.59	3.51	3.43	3.34	3.25	3.17
14	8.86	6.51	5.56	5.04	4.69	4.46	4.28	4.14	4.03	3.94	3.80	3.66	3.51	3.43	3.35	3.27	3.18	3.09	3.00
15	8.68	6.36	5.42	4.89	4.56	4.32	4.14	4.00	3.89	3.80	3.67	3.52	3.37	3.29	3.21	3.13	3.05	2.96	2.87
16	8.53	6.23	5.29	4.77	4.44	4.20	4.03	3.89	3.78	3.69	3.55	3.41	3.26	3.18	3.10	3.02	2.93	2.84	2.75
17	8.40	6.11	5.18	4.67	4.34	4.10	3.93	3.79	3.68	3.59	3.46	3.31	3.16	3.08	3.00	2.92	2.83	2.75	2.65
18	8.29	6.01	5.09	4.58	4.25	4.01	3.84	3.71	3.60	3.51	3.37	3.23	3.08	3.00	2.92	2.84	2.75	2.66	2.57
19	8.18	5.93	5.01	4.50	4.17	3.94	3.77	3.63	3.52	3.43	3.30	3.15	3.00	2.92	2.84	2.76	2.67	2.58	2.49
20	8.10	5.85	4.94	4.43	4.10	3.87	3.70	3.56	3.46	3.37	3.23	3.09	2.94	2.86	2.78	2.69	2.61	2.52	2.42
21	8.02	5.78	4.87	4.37	4.04	3.81	3.64	3.51	3.40	3.31	3.17	3.03	2.88	2.80	2.72	2.64	2.55	2.46	2.36
22	7.95	5.72	4.82	4.31	3.99	3.76	3.59	3.45	3.35	3.26	3.12	2.98	2.83	2.75	2.67	2.58	2.50	2.40	2.31
23	7.88	5.66	4.76	4.26	3.94	3.71	3.54	3.41	3.30	3.21	3.07	2.93	2.78	2.70	2.62	2.54	2.45	2.35	2.26
24	7.82	5.61	4.72	4.22	3.90	3.67	3.50	3.36	3.26	3.17	3.03	2.89	2.74	2.66	2.58	2.49	2.40	2.31	2.21
25	7.77	5.57	4.68	4.18	3.85	3.63	3.46	3.32	3.22	3.13	2.99	2.85	2.70	2.62	2.54	2.45	2.36	2.27	2.17
26	7.72	5.53	4.64	4.14	3.82	3.59	3.42	3.29	3.18	3.09	2.96	2.81	2.66	2.58	2.50	2.42	2.33	2.23	2.13
27	7.68	5.49	4.60	4.11	3.78	3.56	3.39	3.26	3.15	3.06	2.93	2.78	2.63	2.55	2.47	2.38	2.29	2.20	2.10
28	7.64	5.45	4.57	4.07	3.75	3.53	3.36	3.23	3.12	3.03	2.90	2.75	2.60	2.52	2.44	2.35	2.26	2.17	2.06
29	7.60	5.42	4.54	4.04	3.73	3.50	3.33	3.20	3.09	3.00	2.87	2.73	2.57	2.49	2.41	2.33	2.23	2.14	2.03
30	7.56	5.39	4.51	4.02	3.70	3.47	3.30	3.17	3.07	2.98	2.84	2.70	2.55	2.47	2.39	2.30	2.21	2.11	2.01
40	7.31	5.18	4.31	3.83	3.51	3.29	3.12	2.99	2.89	2.80	2.66	2.52	2.37	2.29	2.20	2.11	2.02	1.92	1.80
60	7.08	4.98	4.13	3.65	3.34	3.12	2.95	2.82	2.72	2.63	2.50	2.35	2.20	2.12	2.03	1.94	1.84	1.73	1.60
120	6.85	4.79	3.95	3.48	3.17	2.96	2.79	2.66	2.56	2.47	2.34	2.19	2.03	1.95	1.86	1.76	1.66	1.53	1.38
∞	6.63	4.61	3.78	3.32	3.02	2.80	2.64	2.51	2.41	2.32	2.18	2.04	1.88	1.79	1.70	1.59	1.47	1.32	1.00

$F = \dfrac{s_1^2}{s_2^2} = \dfrac{S_1/m}{S_2/n}$, where $s_1^2 = S_1/m$ and $s_2^2 = S_2/n$ are independent mean squares estimating a common variance σ^2 and based on m and n degrees of freedom, respectively.

PERCENTAGE POINTS, F-DISTRIBUTION

$$F(F) = \int_0^F \frac{\Gamma\left(\dfrac{m+n}{2}\right)}{\Gamma\left(\dfrac{m}{2}\right)\Gamma\left(\dfrac{n}{2}\right)}\, m^{\frac{m}{2}} n^{\frac{n}{2}} x^{\frac{m}{2}-1} (n+mx)^{-\frac{m+n}{2}}\, dx = .995$$

$n \backslash m$	1	2	3	4	5	6	7	8	9	10	12	15	20	24	30	40	60	120	∞
1	16211	20000	21615	22500	23056	23437	23715	23925	24091	24224	24426	24630	24836	24940	25044	25148	25253	25359	25465
2	198.5	199.0	199.2	199.2	199.3	199.3	199.4	199.4	199.4	199.4	199.4	199.4	199.4	199.5	199.5	199.5	199.5	199.5	199.5
3	55.55	49.80	47.47	46.19	45.39	44.84	44.43	44.13	43.88	43.69	43.39	43.08	42.78	42.62	42.47	42.31	42.15	41.99	41.83
4	31.33	26.28	24.26	23.15	22.46	21.97	21.62	21.35	21.14	20.97	20.70	20.44	20.17	20.03	19.89	19.75	19.61	19.47	19.32
5	22.78	18.31	16.53	15.56	14.94	14.51	14.20	13.96	13.77	13.62	13.38	13.15	12.90	12.78	12.66	12.53	12.40	12.27	12.14
6	18.63	14.54	12.92	12.03	11.46	11.07	10.79	10.57	10.39	10.25	10.03	9.81	9.59	9.47	9.36	9.24	9.12	9.00	8.88
7	16.24	12.40	10.88	10.05	9.52	9.16	8.89	8.68	8.51	8.38	8.18	7.97	7.75	7.65	7.53	7.42	7.31	7.19	7.08
8	14.69	11.04	9.60	8.81	8.30	7.95	7.69	7.50	7.34	7.21	7.01	6.81	6.61	6.50	6.40	6.29	6.18	6.06	5.95
9	13.61	10.11	8.72	7.96	7.47	7.13	6.88	6.69	6.54	6.42	6.23	6.03	5.83	5.73	5.62	5.52	5.41	5.30	5.19
10	12.83	9.43	8.08	7.34	6.87	6.54	6.30	6.12	5.97	5.85	5.66	5.47	5.27	5.17	5.07	4.97	4.86	4.75	4.64
11	12.23	8.91	7.60	6.88	6.42	6.10	5.86	5.68	5.54	5.42	5.24	5.05	4.86	4.76	4.65	4.55	4.44	4.34	4.23
12	11.75	8.51	7.23	6.52	6.07	5.76	5.52	5.35	5.20	5.09	4.91	4.72	4.53	4.43	4.33	4.23	4.12	4.01	3.90
13	11.37	8.19	6.93	6.23	5.79	5.48	5.25	5.08	4.94	4.82	4.64	4.46	4.27	4.17	4.07	3.97	3.87	3.76	3.65
14	11.06	7.92	6.68	6.00	5.56	5.26	5.03	4.86	4.72	4.60	4.43	4.25	4.06	3.96	3.86	3.76	3.66	3.55	3.44
15	10.80	7.70	6.48	5.80	5.37	5.07	4.85	4.67	4.54	4.42	4.25	4.07	3.88	3.79	3.69	3.58	3.48	3.37	3.26
16	10.58	7.51	6.30	5.64	5.21	4.91	4.69	4.52	4.38	4.27	4.10	3.92	3.73	3.64	3.54	3.44	3.33	3.22	3.11
17	10.38	7.35	6.16	5.50	5.07	4.78	4.56	4.39	4.25	4.14	3.97	3.79	3.61	3.51	3.41	3.31	3.21	3.10	2.98
18	10.22	7.21	6.03	5.37	4.96	4.66	4.44	4.28	4.14	4.03	3.86	3.68	3.50	3.40	3.30	3.20	3.10	2.99	2.87
19	10.07	7.09	5.92	5.27	4.85	4.56	4.34	4.18	4.04	3.93	3.76	3.59	3.40	3.31	3.21	3.11	3.00	2.89	2.78
20	9.94	6.99	5.82	5.17	4.76	4.47	4.26	4.09	3.96	3.85	3.68	3.50	3.32	3.22	3.12	3.02	2.92	2.81	2.69
21	9.83	6.89	5.73	5.09	4.68	4.39	4.18	4.01	3.88	3.77	3.60	3.43	3.24	3.15	3.05	2.95	2.84	2.73	2.61
22	9.73	6.81	5.65	5.02	4.61	4.32	4.11	3.94	3.81	3.70	3.54	3.36	3.18	3.08	2.98	2.88	2.77	2.66	2.55
23	9.63	6.73	5.58	4.95	4.54	4.26	4.05	3.88	3.75	3.64	3.47	3.30	3.12	3.02	2.92	2.82	2.71	2.60	2.48
24	9.55	6.66	5.52	4.89	4.49	4.20	3.99	3.83	3.69	3.59	3.42	3.25	3.06	2.97	2.87	2.77	2.66	2.55	2.43
25	9.48	6.60	5.46	4.84	4.43	4.15	3.94	3.78	3.64	3.54	3.37	3.20	3.01	2.92	2.82	2.72	2.61	2.50	2.38
26	9.41	6.54	5.41	4.79	4.38	4.10	3.89	3.73	3.60	3.49	3.33	3.15	2.97	2.87	2.77	2.67	2.56	2.45	2.33
27	9.34	6.49	5.36	4.74	4.34	4.06	3.85	3.69	3.56	3.45	3.28	3.11	2.93	2.83	2.73	2.63	2.52	2.41	2.29
28	9.28	6.44	5.32	4.70	4.30	4.02	3.81	3.65	3.52	3.41	3.25	3.07	2.89	2.79	2.69	2.59	2.48	2.37	2.25
29	9.23	6.40	5.28	4.66	4.26	3.98	3.77	3.61	3.48	3.38	3.21	3.04	2.86	2.76	2.66	2.56	2.45	2.33	2.24
30	9.18	6.35	5.24	4.62	4.23	3.95	3.74	3.58	3.45	3.34	3.18	3.01	2.82	2.73	2.63	2.52	2.42	2.30	2.18
40	8.83	6.07	4.98	4.37	3.99	3.71	3.51	3.35	3.22	3.12	2.95	2.78	2.60	2.50	2.40	2.30	2.18	2.06	1.93
60	8.49	5.79	4.73	4.14	3.76	3.49	3.29	3.13	3.01	2.90	2.74	2.57	2.39	2.29	2.19	2.08	1.96	1.83	1.69
120	8.18	5.54	4.50	3.92	3.55	3.28	3.09	2.93	2.81	2.71	2.54	2.37	2.19	2.09	1.98	1.87	1.75	1.61	1.43
∞	7.88	5.30	4.28	3.72	3.35	3.09	2.90	2.74	2.62	2.52	2.36	2.19	2.00	1.90	1.79	1.67	1.53	1.36	1.00

$F = \dfrac{s_1^2}{s_2^2} = \dfrac{S_1/m}{S_2/n}$, where $s_1^2 = S_1/m$ and $s_2^2 = S_2/n$ are independent mean squares estimating a common variance σ^2 and based on m and n degrees of freedom, respectively.

F-Distribution

PERCENTAGE POINTS, F-DISTRIBUTION

$$F(F) = \int_0^F \frac{\Gamma\left(\dfrac{m+n}{2}\right)}{\Gamma\left(\dfrac{m}{2}\right)\Gamma\left(\dfrac{n}{2}\right)} m^{\frac{m}{2}} n^{\frac{n}{2}} x^{\frac{m}{2}-1}(n+mx)^{-\frac{m+n}{2}}\,dx = .999$$

n \ m	1	2	3	4	5	6	7	8	9	10	12	15	20	24	30	40	60	120	∞
1	4053*	5000*	5404*	5625*	5764*	5859*	5929*	5981*	6023*	6056*	6107*	6158*	6209*	6235*	6261*	6287*	6313*	6340*	6366*
2	998.5	999.0	999.2	999.2	999.3	999.3	999.4	999.4	999.4	999.4	999.4	999.4	999.4	999.5	999.5	999.5	999.5	999.5	999.5
3	167.0	148.5	141.1	137.1	134.6	132.8	131.6	130.6	129.9	129.2	128.3	127.4	126.4	125.9	125.4	125.0	124.5	124.0	123.5
4	74.14	61.25	56.18	53.44	51.71	50.53	49.66	49.00	48.47	48.05	47.41	46.76	46.10	45.77	45.43	45.09	44.75	44.40	44.05
5	47.18	37.12	33.20	31.09	29.75	28.84	28.16	27.64	27.24	26.92	26.42	25.91	25.39	25.14	24.87	24.60	24.33	24.06	23.79
6	35.51	27.00	23.70	21.92	20.81	20.03	19.46	19.03	18.69	18.41	17.99	17.56	17.12	16.89	16.67	16.44	16.21	15.99	15.75
7	29.25	21.69	18.77	17.19	16.21	15.52	15.02	14.63	14.33	14.08	13.71	13.32	12.93	12.73	12.53	12.33	12.12	11.91	11.70
8	25.42	18.49	15.83	14.39	13.49	12.86	12.40	12.04	11.77	11.54	11.19	10.84	10.48	10.30	10.11	9.92	9.73	9.53	9.33
9	22.86	16.39	13.90	12.56	11.71	11.13	10.70	10.37	10.11	9.89	9.57	9.24	8.90	8.72	8.55	8.37	8.19	8.00	7.81
10	21.04	14.91	12.55	11.28	10.48	9.92	9.52	9.20	8.96	8.75	8.45	8.13	7.80	7.64	7.47	7.30	7.12	6.94	6.76
11	19.69	13.81	11.56	10.35	9.58	9.05	8.66	8.35	8.12	7.92	7.63	7.32	7.01	6.85	6.68	6.52	6.35	6.17	6.00
12	18.64	12.97	10.80	9.63	8.89	8.38	8.00	7.71	7.48	7.29	7.00	6.71	6.40	6.25	6.09	5.93	5.76	5.59	5.42
13	17.81	12.31	10.21	9.07	8.35	7.86	7.49	7.21	6.98	6.80	6.52	6.23	5.93	5.78	5.63	5.47	5.30	5.14	4.97
14	17.14	11.78	9.73	8.62	7.92	7.43	7.08	6.80	6.58	6.40	6.13	5.85	5.56	5.41	5.25	5.10	4.94	4.77	4.60
15	16.59	11.34	9.34	8.25	7.57	7.09	6.74	6.47	6.26	6.08	5.81	5.54	5.25	5.10	4.95	4.80	4.64	4.47	4.31
16	16.12	10.97	9.00	7.94	7.27	6.81	6.46	6.19	5.98	5.81	5.55	5.27	4.99	4.85	4.70	4.54	4.39	4.23	4.06
17	15.72	10.66	8.73	7.68	7.02	6.56	6.22	5.96	5.75	5.58	5.32	5.05	4.78	4.63	4.48	4.33	4.18	4.02	3.85
18	15.38	10.39	8.49	7.46	6.81	6.35	6.02	5.76	5.56	5.39	5.13	4.87	4.59	4.45	4.30	4.15	4.00	3.84	3.67
19	15.08	10.16	8.28	7.26	6.62	6.18	5.85	5.59	5.39	5.22	4.97	4.70	4.43	4.29	4.14	3.99	3.84	3.68	3.51
20	14.82	9.95	8.10	7.10	6.46	6.02	5.69	5.44	5.24	5.08	4.82	4.56	4.29	4.15	4.00	3.86	3.70	3.54	3.38
21	14.59	9.77	7.94	6.95	6.32	5.88	5.56	5.31	5.11	4.95	4.70	4.44	4.17	4.03	3.88	3.74	3.58	3.42	3.26
22	14.38	9.61	7.80	6.81	6.19	5.76	5.44	5.19	4.99	4.83	4.58	4.33	4.06	3.92	3.78	3.63	3.48	3.32	3.15
23	14.19	9.47	7.67	6.69	6.08	5.65	5.33	5.09	4.89	4.73	4.48	4.23	3.96	3.82	3.68	3.53	3.38	3.22	3.05
24	14.03	9.34	7.55	6.59	5.98	5.55	5.23	4.99	4.80	4.64	4.39	4.14	3.87	3.74	3.59	3.45	3.29	3.14	2.97
25	13.88	9.22	7.45	6.49	5.88	5.46	5.15	4.91	4.71	4.56	4.31	4.06	3.79	3.66	3.52	3.37	3.22	3.06	2.89
26	13.74	9.12	7.36	6.41	5.80	5.38	5.07	4.83	4.64	4.48	4.24	3.99	3.72	3.59	3.44	3.30	3.15	2.99	2.82
27	13.61	9.02	7.27	6.33	5.73	5.31	5.00	4.76	4.57	4.41	4.17	3.92	3.66	3.52	3.38	3.23	3.08	2.92	2.75
28	13.50	8.93	7.19	6.25	5.66	5.24	4.93	4.69	4.50	4.35	4.11	3.86	3.60	3.46	3.32	3.18	3.02	2.86	2.69
29	13.39	8.85	7.12	6.19	5.59	5.18	4.87	4.64	4.45	4.29	4.05	3.80	3.54	3.41	3.27	3.12	2.97	2.81	2.64
30	13.29	8.77	7.05	6.12	5.53	5.12	4.82	4.58	4.39	4.24	4.00	3.75	3.49	3.36	3.22	3.07	2.92	2.76	2.59
40	12.61	8.25	6.60	5.70	5.13	4.73	4.44	4.21	4.02	3.87	3.64	3.40	3.15	3.01	2.87	2.73	2.57	2.41	2.23
60	11.97	7.76	6.17	5.31	4.76	4.37	4.09	3.87	3.69	3.54	3.31	3.08	2.83	2.69	2.55	2.41	2.25	2.08	1.89
120	11.38	7.32	5.79	4.95	4.42	4.04	3.77	3.55	3.38	3.24	3.02	2.78	2.53	2.40	2.26	2.11	1.95	1.76	1.54
∞	10.83	6.91	5.42	4.62	4.10	3.74	3.47	3.27	3.10	2.96	2.74	2.51	2.27	2.13	1.99	1.84	1.66	1.45	1.00

* Multiply these entries by 100.

VI.2 POWER FUNCTIONS OF THE ANALYSIS-OF-VARIANCE TESTS

The noncentral F-distribution, $f(F')$, arises in the ratio of a non-central chi-square with m degrees of freedom and noncentrality parameter λ to an independent chi-square with n degrees of freedom.

P. C. Tang has compiled tables of

$$\int_0^{F_\alpha} f(F') \, dF'$$

for certain values of F_α. These tables are given in terms of E^2, where

$$E^2 = \frac{mF'}{mF' + n} .$$

If the frequency function of E^2 is denoted by $g(E^2;m,n,\lambda)$, $g(E^2)$ is a beta distribution if $\lambda = 0$ and a noncentral beta distribution if $\lambda \neq 0$.

The integral

$$\int_0^{E_\alpha^2} g(E^2;m,n,\lambda) \, dE^2$$

equals $1 - \beta(\lambda)$, or unity minus the power of the test, which is the probability of a type II error. Here E_α^2 is obtained from the integral

$$\int_{E_\alpha^2}^1 g(E^2;m,n, \lambda = 0) \, dE^2 = \lambda.$$

P. C. Tang evaluated the integral

$$P(\text{II}) = 1 - \beta(\phi) = \int_0^{E_\alpha^2} g(E^2;f_1,f_2,\phi) \, dE^2$$

for various values of f_1, f_2, ϕ, and E_α^2 for $\alpha = 0.05$ and 0.01, where

$$\phi = \sqrt{\frac{2\lambda}{f_1 + 1}}$$

and where f_1 is the degrees of freedom in the numerator of the F statistic.

In this table, graphs are shown with $1 - \beta$ on the vertical scale corresponding to ϕ on the horizontal. The graphs are for two levels of significance, $\alpha = 0.01$ and 0.05, for eight values of ν_1, the number of degrees of freedom for the numerator, and several values of ν_2, the number of degrees of freedom for the denominator of the F ratio. There is a different curve for each set of values α, ν_1, and ν_2.

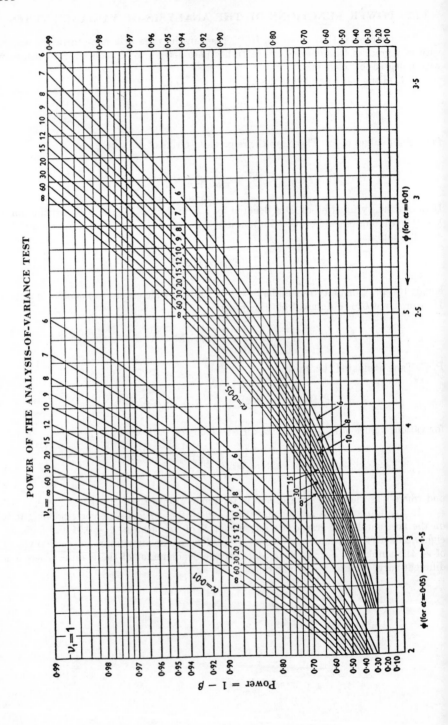

POWER OF THE ANALYSIS-OF-VARIANCE TEST

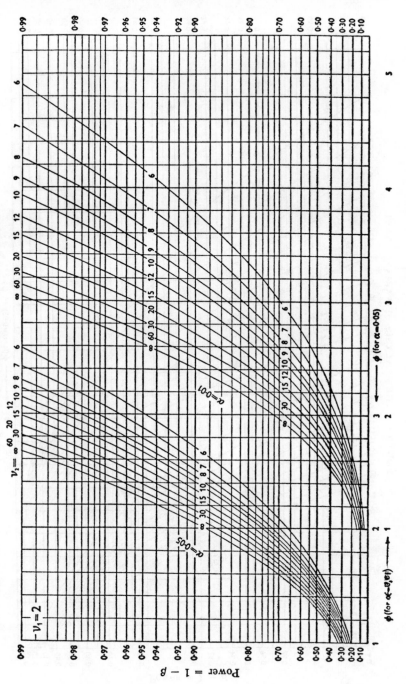

POWER OF THE ANALYSIS-OF-VARIANCE TEST

F-Distribution

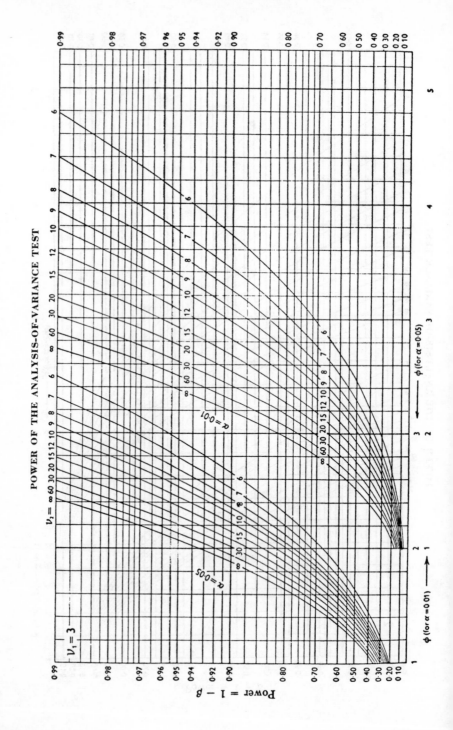

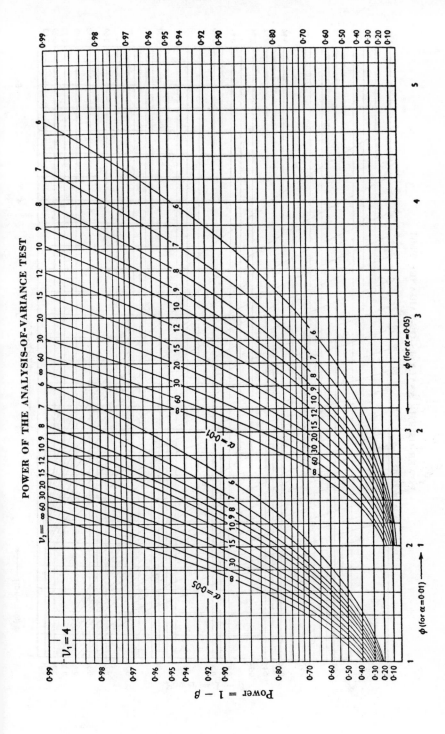

POWER OF THE ANALYSIS-OF-VARIANCE TEST

F-Distribution

POWER OF THE ANALYSIS-OF-VARIANCE TEST

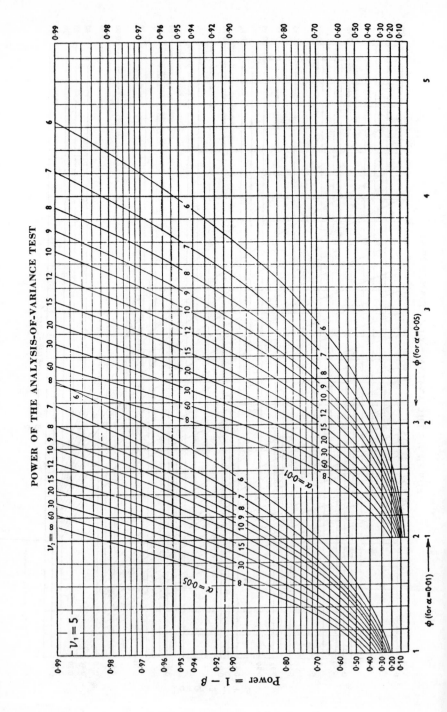

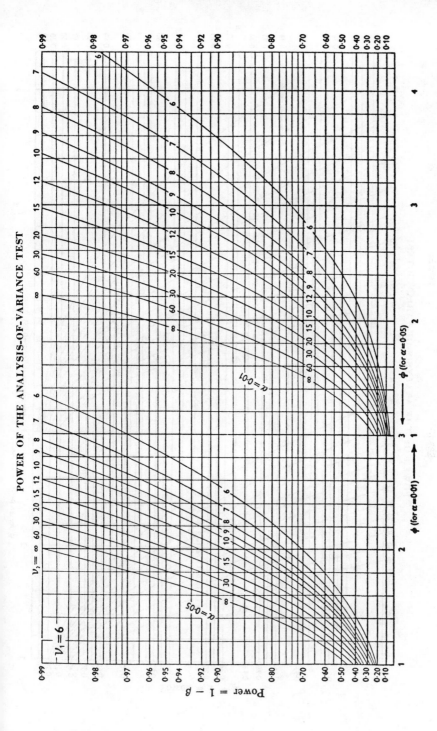

POWER OF THE ANALYSIS-OF-VARIANCE TEST

POWER OF THE ANALYSIS-OF-VARIANCE TEST

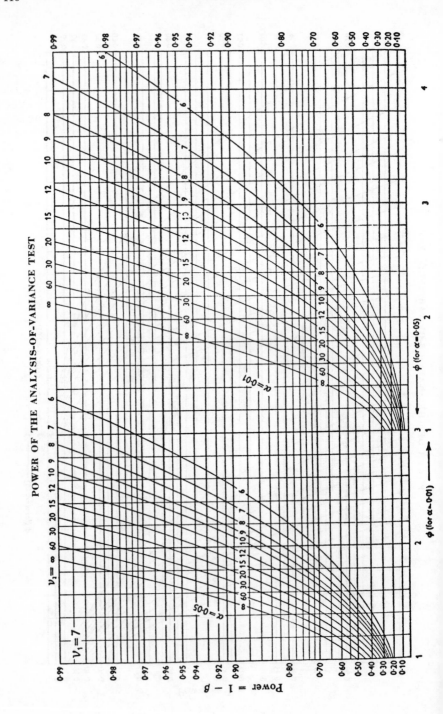

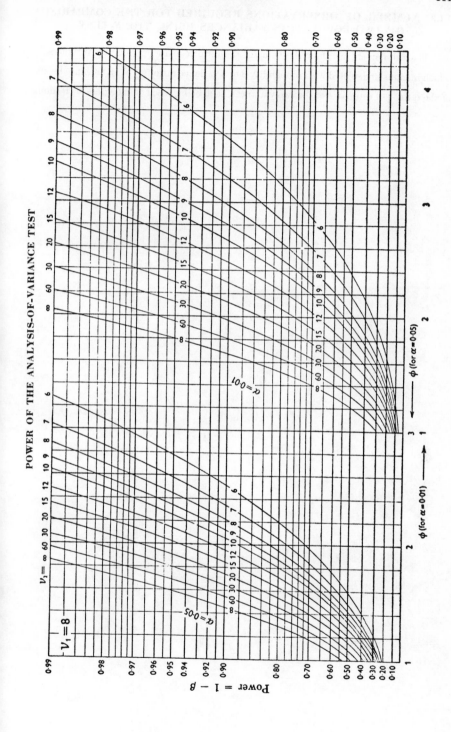

POWER OF THE ANALYSIS-OF-VARIANCE TEST

$V_1 = 8$

VI.3 NUMBER OF OBSERVATIONS REQUIRED FOR THE COMPARISON OF TWO POPULATION VARIANCES USING THE F-TEST

The tabular entries show the value of the ratio R of two population variances $\dfrac{\sigma_2^2}{\sigma_1^2}$, which remains undetected with probability β in a variance ratio test at significance level α of the ratio $\dfrac{s_2^2}{s_1^2}$ of estimates of the two variances, each being based on n degrees of freedom.

NUMBER OF OBSERVATIONS REQUIRED FOR THE COMPARISON OF TWO POPULATION VARIANCES USING THE F-TEST

η	α = 0.01				α = 0.05				α = 0.5		
	β = 0.01	β = 0.05	β = 0.1	β = 0.5	β = 0.01	β = 0.05	β = 0.1	β = 0.5	β = 0.05	β = 0.1	β = 0.5
1	16,420,000	654,200	161,500	4052	654,200	26,070	6,436	161.5	161.5	39.85	1.000
2	9,000	1,881	891.0	99.00	1,881	361.0	171.0	19.00	19.00	9.000	1.000
3	867.7	273.3	158.8	29.46	273.3	86.06	50.01	9.277	9.277	5.391	1.000
4	255.3	102.1	65.62	15.98	102.1	40.81	26.24	6.388	6.388	4.108	1.000
5	120.3	55.39	37.87	10.97	55.39	25.51	17.44	5.050	5.050	3.453	1.000
6	71.67	36.27	25.86	8.466	36.27	18.35	13.09	4.284	4.284	3.056	1.000
7	48.90	26.48	19.47	6.993	26.48	14.34	10.55	3.787	3.787	2.786	1.000
8	36.35	20.73	15.61	6.029	20.73	11.82	8.902	3.438	3.438	2.589	1.000
9	28.63	17.01	13.06	5.351	17.01	10.11	7.757	3.179	3.179	2.440	1.000
10	23.51	14.44	11.26	4.849	14.44	8.870	6.917	2.978	2.978	2.323	1.000
12	17.27	11.16	8.923	4.155	11.16	7.218	5.769	2.687	2.687	2.147	1.000
15	12.41	8.466	6.946	3.522	8.466	5.777	4.740	2.404	2.404	1.972	1.000
20	8.630	6.240	5.270	2.938	6.240	4.512	3.810	2.124	2.124	1.794	1.000
24	7.071	5.275	4.526	2.659	5.275	3.935	3.376	1.984	1.984	1.702	1.000
30	5.693	4.392	3.833	2.386	4.392	3.389	2.957	1.841	1.841	1.606	1.000
40	4.470	3.579	3.183	2.114	3.579	2.866	2.549	1.693	1.693	1.506	1.000
60	3.372	2.817	2.562	1.836	2.817	2.354	2.141	1.534	1.534	1.396	1.000
120	2.350	2.072	1.939	1.533	2.072	1.828	1.710	1.352	1.352	1.265	1.000
∞	1.000	1.000	1.000	1.000	1.000	1.000	1.000	1.000	1.000	1.000	1.000

VII. Order Statistics

VII.1 EXPECTED VALUES OF ORDER STATISTICS FROM A STANDARD NORMAL POPULATION

If a sample of n observations $x_1, x_2, \ldots, x_n$ is drawn from a standard normal distribution, and the observations are arranged in ascending order of magnitude $x_{(1)}, \ldots, x_{(n)}$, the ith value of the set $\{x_{(i)}\}$ is called the ith normal order statistic and its expectation is given by

$$E[x_{(i)}] = \frac{n!}{(i-1)!(n-i)!} \int_{-\infty}^{\infty} xf(x)F^{i-1}(x)[1 - F(x)]^{n-i} \, dx \ ,$$

where $f(x) = \dfrac{1}{\sqrt{2\pi}} e^{-\frac{1}{2}x^2}$,

$$F(x) = \int_{-\infty}^{x} \frac{1}{\sqrt{2\pi}} e^{-t^2/2} \, dt.$$

This table gives value of $E[x_{(i)}]$ for various values of n. Missing values may be obtained by noting that

$$E[x_{(i)}] = -E[x_{(n-i+1)}] \ .$$

114

EXPECTED VALUES OF NORMAL ORDER STATISTICS

$i \backslash n$	2	3	4	5	6	7	8	9
1		0.56419	0.84628	1.02938	1.16296	1.26721	1.35218	1.42360	1.48501
2		—	.00000	0.29701	0.49502	0.64176	0.75737	0.85222	0.93230
3		—	—	—	.00000	.20155	.35271	.47282	.57197
4		—	—	—	—	—	.00000	.15251	.27453
5		—	—	—	—	—	—	—	.00000

$i \backslash n$	10	11	12	13	14	15	16	17	18	19
1	1.53875	1.58644	1.62923	1.66799	1.70338	1.73591	1.76599	1.79394	1.82003	1.84448
2	1.00136	1.06192	1.11573	1.16408	1.20790	1.24794	1.28474	1.31878	1.35041	1.37994
3	0.65606	0.72884	0.79284	0.84983	0.90113	0.94769	0.99027	1.02946	1.06473	1.09945
4	.37576	.46198	.53684	.60285	.66176	.71488	.76317	0.80738	0.84812	0.88586
5	.12267	.22489	.31225	.38833	.45557	.51570	.57001	.61946	.66479	.70661
6	—	0.00000	0.10259	0.19052	0.26730	0.33530	0.39622	0.45133	0.50158	0.54771
7	—	—	—	.00000	0.08816	.16530	.23375	.29519	.35084	.40164
8	—	—	—	—	—	.00000	.07729	.14599	.20774	.26374
9	—	—	—	—	—	—	—	.00000	.06880	.13072
10	—	—	—	—	—	—	—	—	—	.00000

$i \backslash n$	20	21	22	23	24	25	26	27	28	29
1	1.86748	1.88917	1.90969	1.92916	1.94767	1.96531	1.98216	1.99827	2.01371	2.02852
2	1.40760	1.43362	1.45816	1.48137	1.50338	1.52430	1.54423	1.56326	1.58145	1.59888
3	1.13095	1.16047	1.18824	1.21445	1.23924	1.26275	1.28511	1.30641	1.32674	1.34619
4	0.92098	0.95380	0.98459	1.01356	1.04091	1.06679	1.09135	1.11471	1.13697	1.15822
5	.74538	.78150	.81527	.84697	.87682	.90501	.93171	.95705	.98115	1.00414
6	0.59030	0.62982	0.66667	0.70115	0.73354	0.76405	0.79289	0.82021	0.84615	0.87084
7	.44833	.49148	.53157	.56896	.60399	.63690	.66794	.69727	.72508	.75150
8	.31493	.36203	.40559	.44609	.48391	.51935	.55267	.58411	.61385	.64205
9	.18696	.23841	.28579	.32965	.37047	.40860	.44436	.47801	.50977	.53982
10	.06200	.11836	.16997	.21755	.26163	.30268	.34105	.37706	.41096	.44298
11	—	0.00000	0.05642	0.10813	0.15583	0.20006	0.24128	0.27983	0.31603	0.35013
12	—	—	—	.00000	.05176	.09953	.14387	.18520	.22389	.26023
13	—	—	—	—	—	.00000	.04781	.09220	.13361	.17240
14	—	—	—	—	—	—	—	.00000	.04442	.08588
15	—	—	—	—	—	—	—	—	—	.00000

$i \backslash n$	30	31	32	33	34	35	36	37	38	39
1	2.04276	2.05646	2.06967	2.08241	2.09471	2.10661	2.11812	2.12928	2.14009	2.15059
2	1.61560	1.63166	1.64712	1.66200	1.67636	1.69023	1.70362	1.71659	1.72914	1.74131
3	1.36481	1.38268	1.39985	1.41637	1.43228	1.44762	1.46244	1.47676	1.49061	1.50402
4	1.17855	1.19803	1.21672	1.23468	1.25196	1.26860	1.28466	1.30016	1.31514	1.32964
5	1.02609	1.04709	1.06721	1.08652	1.10509	1.12295	1.14016	1.15677	1.17280	1.18830
6	0.89439	0.91688	0.93841	0.95905	0.97886	0.99790	1.01624	1.03390	1.05095	1.06741
7	.77666	.80066	.82359	.84555	.86660	.88681	0.90625	0.92496	0.94300	0.96041
8	.66885	.69438	.71875	.74204	.76435	.78574	.80629	.82605	.84508	.86343
9	.56834	.59545	.62129	.64596	.66954	.69214	.71382	.73465	.75468	.77398
10	.47329	.50206	.52943	.55552	.58043	.60427	.62710	.64902	.67009	.69035
11	0.38235	0.41287	0.44185	0.46942	0.49572	0.52084	0.54488	0.56793	0.59005	0.61131
12	.29449	.32686	.35755	.38669	.41444	.44091	.46620	.49042	.51363	.53592
13	.20885	.24322	.27573	.30654	.33582	.36371	.39032	.41576	.44012	.46348
14	.12473	.16126	.19572	.22832	.25924	.28863	.31663	.34336	.36892	.39340
15	.04148	.08037	.11695	.15147	.18415	.21515	.24463	.27272	.29954	.32520
16	—	0.00000	0.03890	0.07552	0.11009	0.14282	0.17388	0.20342	0.23159	0.25849
17	—	—	—	.00000	.03663	.07123	.10399	.13509	.16469	.19292
18	—	—	—	—	—	.00000	.03461	.06739	.09853	.12817
19	—	—	—	—	—	—	—	.00000	.03280	.06395
20	—	—	—	—	—	—	—	—	—	.00000

VII.2   VARIANCES AND COVARIANCES OF ORDER STATISTICS

If a sample of n observations $x_1, x_2, \ldots, x_n$ is drawn from a standard normal distribution, and the observations are arranged in ascending order of magnitude $x_{(1)}, x_{(2)}, \ldots, x_{(n)}$, then the variances and covariances of these order statistics may be obtained from the following expressions for expected values and product moments:

$$E[x_{(i)}] = \frac{n!}{(i-1)!(n-i)!} \int_{-\infty}^{\infty} x f(x) [F(x)]^{i-1} [1 - F(x)]^{n-i}\, dx,$$

$$E[x_{(i)}^2] = \frac{n!}{(i-1)!(n-i)!} \int_{-\infty}^{\infty} x^2 f(x) [F(x)]^{i-1} [1 - F(x)]^{n-i}\, dx,$$

$$E[x_{(i)}x_{(j)}] = \frac{n!}{(i-1)!(j-i-1)!(n-j)!}$$
$$\int_{-\infty}^{\infty} \int_{-\infty}^{y} xy f(x) f(y) [F(x)]^{i-1} [1 - F(y)]^{n-j} [F(y) - F(x)]^{j-i-1}\, dx\, dy$$

where $f(x) = \dfrac{1}{\sqrt{2\pi}}\, e^{-\frac{1}{2}x^2}$,

$$F(x) = \int_{-\infty}^{x} \frac{1}{\sqrt{2\pi}}\, e^{-\frac{1}{2}t^2}\, dt.$$

This table gives the variances and covariances of order statistics in samples of sizes up to 20 from a standard normal distribution. Missing values may be supplied from $E[x_{(i)}] = -E[x_{(n-i+1)}]$; $E[x_{(i)}x_{(j)}] = E[x_{(j)}x_{(i)}] = E[x_{(n-i+1)}x_{(n-j+1)}]$.

VARIANCES AND COVARIANCES OF ORDER STATISTICS

n	i	j	Value	n	i	j	Value	n	i	j	Value
2	1	1	.6816901139	8				10	3		
		2	.3183098861		2	2	.2394010458			5	.1077445336
	2	2	.6816901139			3	.1631958727			6	.0892254012
3	1	1	.5594672038			4	.1232633317			7	.0749183943
		2	.2756644477			5	.0975647193			8	.0630332449
		3	.1648683485			6	.0787224662		4	4	.1579389144
	2	2	.4486711046			7	.0632466118			5	.1275089295
4	1	1	.4917152369		3	3	.2007687900			6	.1057858169
		2	.2455926930			4	.1523584312			7	.0889462026
		3	.1580080701			5	.1209637555		5	5	.1510539039
		4	.1046840000			6	.0978171355			6	.1255989678
	2	2	.3604553434		4	4	.1871862195	11	1	1	.3332474428
		3	.2359438935			5	.1491754908			2	.1653647712
5	1	1	.4475340691	9	1	1	.3573533264			3	.1123584351
		2	.2243309596			2	.1781434240			4	.0855170596
		3	.1481477252			3	.1207454442			5	.0688483064
		4	.1057719776			4	.0913071400			6	.0572007586
		5	.0742152685			5	.0727422354			7	.0483754063
	2	2	.3115189521			6	.0594831125			8	.0412423472
		3	.2084354440			7	.0490764061			9	.0351103357
		4	.1499426668			8	.0400936927			10	.0294198503
	3	3	.2868336616			9	.0310552188			11	.0233152868
6	1	1	.4159271090		2	2	.2256968778		2	2	.2051975798
		2	.2085030023			3	.1541163526			3	.1403096511
		3	.1394352565			4	.1170056918			4	.1071492595
		4	.1024293940			5	.0934477394			5	.0864430257
		5	.0773637839			6	.0765461431			6	.0719305024
		6	.0563414544			7	.0632354695			7	.0608869662
	2	2	.2795777392			8	.0517146091			8	.0519504506
		3	.1889859560		3	3	.1863826133			9	.0442549455
		4	.1396640604			4	.1420779776			10	.0371029977
		5	.1059054582			5	.1137680176		3	3	.1657242880
	3	3	.2462125354			6	.0933625386			4	.1269672925
		4	.1832727978			7	.0772351806			5	.1026407291
7	1	1	.3919177761		4	4	.1705588454			6	.0855178832
		2	.1961990246			5	.1369913669			7	.0724741050
		3	.1321155811			6	.1126671842			8	.0618873278
		4	.0984868607		5	5	.1661012814			9	.0527550069
		5	.0765598346	10	1	1	.3443348233		4	4	.1479546565
		6	.0599187124			2	.1712629030			5	.1198752861
		7	.0448022105			3	.1162590989			6	.1000346585
	2	2	.2567328862			4	.0882494247			7	.0848765182
		3	.1744833274			5	.0707413677			8	.0725451434
		4	.1307298656			6	.0583987134		5	5	.1396410804
		5	.1019550089			7	.0489206279			6	.1167449805
		6	.0799811748			8	.0410844589			7	.0991935960
	3	3	.2197215626			9	.0340406470		6	6	.1371624335
		4	.1655598429			10	.0266989351	12	1	1	.3236363870
		5	.1296048425		2	2	.2145241430			2	.1602373762
	4	4	.2104468615			3	.1466226180			3	.1089309641
8	1	1	.3728971434			4	.1117015961			4	.0830686767
		2	.1863073997			5	.0897428245			5	.0670884464
		3	.1259660300			6	.0741995414			6	.0559933694
		4	.0947230277			7	.0622278486			7	.0476620974
		5	.0747650242			8	.0523067222			8	.0410208554
		6	.0602075169			9	.0433711561			9	.0354439060
		7	.0482985508		3	3	.1750032834			10	.0305012591
		8	.0368353073			4	.1338022448			11	.0257945392

Order Statistics

VARIANCES AND COVARIANCES OF ORDER STATISTICS

n	i	j	Value
12	1		
		12	.0206221233
	2	2	.1972646039
		3	.1349020328
		4	.1031959206
		5	.0835045822
		6	.0697859658
		7	.0594590652
		8	.0512113198
		9	.0442747124
		10	.0381191478
		11	.0322507340
	3	3	.1579786877
		4	.1212063211
		5	.0982605602
		6	.0822228461
		7	.0701213964
		8	.0604384621
		9	.0522825611
		10	.0450357615
	4	4	.1398109405
		5	.1135687821
		6	.0951645279
		7	.0812419810
		8	.0700795832
		9	.0606620874
	5	5	.1306137359
		6	.1096212247
		7	.0936951520
		8	.0808972960
	6	6	.1266377911
		7	.1083945831
13	1	1	.3152053842
		2	.1557272904
		3	.1058908842
		4	.0808649736
		5	.0654634499
		6	.0548221797
		7	.0468833088
		8	.0406132548
		9	.0354226462
		10	.0309322744
		11	.0268537250
		12	.0228858068
		13	.0184348220
	2	2	.1904130721
		3	.1302055829
		4	.0997262696
		5	.0808785938
		6	.0678145832
		7	.0580457285
		8	.0503167946
		9	.0439095087
		10	.0383601798
		11	.0333147765
		12	.0284018130
	3	3	.1513917013
		4	.1162698131

n	i	j	Value
13	3	5	.0944566603
		6	.0792922993
		7	.0679282354
		8	.0589221432
		9	.0514460445
		10	.0449637542
		11	.0390643799
	4	4	.1330111820
		5	.1082512667
		6	.0909855605
		7	.0780173339
		8	.0677217143
		9	.0591628729
		10	.0517328050
	5	5	.1232503256
		6	.1037367701
		7	.0890434754
		8	.0773552864
		9	.0676230994
	6	6	.1183175325
		7	.1016824204
		8	.0884194610
	7	7	.1167989950
14	1	1	.3077301026
		2	.1517203662
		3	.1031719531
		4	.0788715916
		5	.0639657428
		6	.0537064714
		7	.0460899189
		8	.0401141688
		9	.0352141760
		10	.0310371163
		11	.0273362865
		12	.0239061001
		13	.0205080257
		14	.0166279801
	2	2	.1844200252
		3	.1260791989
		4	.0966524633
		5	.0785202981
		6	.0660028340
		7	.0566896715
		8	.0493708148
		9	.0433617156
		10	.0382337404
		11	.0336863221
		12	.0294681314
		13	.0252863928
	3	3	.1457045665
		4	.1119816877
		5	.0911181271
		6	.0766754957
		7	.0659084825
		8	.0574341188
		9	.0504677802
		10	.0445169192

n	i	j	Value
14	3	11	.0392352316
		12	.0343322071
	4	4	.1272273070
		5	.1036931108
		6	.0873562483
		7	.0751519909
		8	.0655310936
		9	.0576120957
		10	.0508402240
		11	.0448243469
	5	5	.1171012461
		6	.0987747550
		7	.0850536546
		8	.0742181416
		9	.0652867776
		10	.0576401464
	6	6	.1115324579
		7	.0961405595
		8	.0839617110
		9	.0739069221
	7	7	.1090269480
		8	.0953087256
15	1	1	.3010415703
		2	.1481297708
		3	.1007223449
		4	.0770594060
		5	.0625845851
		6	.0526530129
		7	.0453078886
		8	.0395736673
		9	.0349035905
		10	.0309614122
		11	.0275211039
		12	.0244126313
		13	.0214819828
		14	.0185333263
		15	.0151137071
	2	2	.1791215291
		3	.1224176953
		4	.0939067144
		5	.0763912337
		6	.0643390895
		7	.0554074400
		8	.0484238833
		9	.0427294113
		10	.0379177516
		11	.0337151721
		12	.0299152347
		13	.0263303885
		14	.0227213594
	3	3	.1407322502
		4	.1082138452
		5	.0881605755
		6	.0743268436
		7	.0640558183
		8	.0560136122
		9	.0494485109

VARIANCES AND COVARIANCES OF ORDER STATISTICS

i	j	Value	n	i	j	Value	n	i	j	Value
3			16	2			17	1		
	10	.0438960670			14	.0237301562			14	.0199690651
	11	.0390426915			15	.0205785433			15	.0177476891
	12	.0346513382		3	3	.1363385612			16	.0154552071
	13	.0305060359			4	.1048706756			17	.0127264751
4	4	.1222328270			5	.0855189036		2	2	.1701426762
	5	.0997323941			6	.0722075087			3	.1161866734
	6	.0841705696			7	.0623568515			4	.0891982557
	7	.0725946869			8	.0546749107			5	.0726970385
	8	.0635175907			9	.0484366096			6	.0613998459
	9	.0560990511			10	.0431979377			7	.0530761573
	10	.0498187836			11	.0386652995			8	.0466140918
	11	.0443247452			12	.0346277256			9	.0413928192
	12	.0393501820			13	.0309149135			10	.0370349110
5	5	.1118698986			14	.0273595378			11	.0332940892
	6	.0945206004		4	4	.1178657554			12	.0299982825
	7	.0815891122			5	.0962513413			13	.0270170379
	8	.0714331681			6	.0813480448			14	.0242386812
	9	.0631224388			7	.0703000911			15	.0215459396
	10	.0560795065			8	.0616728990			16	.0187658306
	11	.0499127743			9	.0546595026		3	3	.1324207975
6	6	.1058666366			10	.0487647746			4	.1018792434
	7	.0914683204			11	.0436607328			5	.0831421716
	8	.0801407559			12	.0391112669			6	.0702850403
	9	.0708582099			13	.0349253749			7	.0607964413
	10	.0629824402		5	5	.1073517089			8	.0534208202
7	7	.1026916923			6	.0908232622			9	.0474555487
	8	.0900499964			7	.0785480532			10	.0424726884
	9	.0796738323			8	.0689488802			11	.0381925587
8	8	.1016946521			9	.0611364182			12	.0344194567
1	1	.2950098090			10	.0545638941			13	.0310047771
	2	.1448881689			11	.0488684327			14	.0278210708
	3	.0985009764			12	.0437882959			15	.0247342095
	4	.0754040023		6	6	.1010461906		4	4	.1140068197
	5	.0613086724			7	.0874627156			5	.0931620339
	6	.0516624963			8	.0768239668			6	.0788266621
	7	.0445503705			9	.0681545540			7	.0682298909
	8	.0390194716			10	.0608534805			8	.0599826092
	9	.0345378158			11	.0545210724			9	.0533057575
	10	.0307810093		7	7	.0974026613			10	.0477239973
	11	.0275353612			8	.0856181916			11	.0429261816
	12	.0246479007			9	.0760015577			12	.0386942630
	13	.0219956755			10	.0678931922			13	.0348624030
	14	.0194585037		8	8	.0957213007			14	.0312881041
	15	.0168710289			9	.0850291218		5	5	.1034004377
	16	.0138287378	17	1	1	.2895330037			6	.0875729930
2	2	.1743940788			2	.1419424629			7	.0758534534
	3	.1191409287			3	.0964748737			8	.0667204245
	4	.0914359918			4	.0733849615			9	.0593187706
	5	.0744591145			5	.0601272302			10	.0531257771
	6	.0628093909			6	.0507326948			11	.0477987292
	7	.0542033941			7	.0438236491			12	.0430970793
	8	.0475009769			8	.0384672834			13	.0388375657
	9	.0420638230			9	.0341441055		6	6	.0968824669
	10	.0375018250			10	.0305389548			7	.0839811738
	11	.0335574912			11	.0274465527			8	.0739130260
	12	.0300461298			12	.0247237144			9	.0657442736
	13	.0268189579			13	.0222620771			10	.0589030403

VARIANCES AND COVARIANCES OF ORDER STATISTICS

n	i	j	Value	n	i	j	Value	n	i	j	Value
17	6	11	.0530137275	18	3	15	.0252244786	19	1	14	.02040073
		12	.0478122599			16	.0225161109			15	.018543153
	7	7	.0929031780		4	4	.1105660331			16	.01677311
		8	.0818194607			5	.0903973787			17	.0150223
		9	.0728154074			6	.0765579277			18	.01317899
		10	.0652667274			7	.0663522086			19	.01093825
		11	.0587626219			8	.0584310521		2	2	.162785665
	8	8	.0907361650			9	.0520394281			3	.111059014
		9	.0808000267			10	.0467183404			4	.085293108
		10	.0724599963			11	.0421694861			5	.069597075
	9	9	.0900465814			12	.0381869632			6	.058891019
18	1	1	.2845301297			13	.0346192645			7	.051035105
		2	.1392501620			14	.0313452497			8	.044965224
		3	.0946172637			15	.0282548286			9	.040089178
		4	.0724851730		5	5	.0999084321			10	.036049004
		5	.0590304274			6	.0846879168			11	.032613754
		6	.0498600635			7	.0734460811			12	.029625823
		7	.0431302310			8	.0647101858			13	.026971659
		8	.0379260195			9	.0576543520			14	.024564190
		9	.0337388141			10	.0517756675			15	.022330688
		10	.0302610667			11	.0467468133			16	.020201724
		11	.0272938041			12	.0423415563			17	.018095219
		12	.0247002471			13	.0383932046			18	.015876728
		13	.0223801573			14	.0347682770		3	3	.125713890
		14	.0202537421		6	6	.0932407331			4	.096736709
		15	.0182488619			7	.0809202644			5	.079029879
		16	.0162850441			8	.0713338046			6	.066927369
		17	.0142368875			9	.0635829688			7	.058033612
		18	.0117719054			10	.0571197288			8	.051154141
	2	2	.1662929294			11	.0515868552			9	.045622881
		3	.1135058132			12	.0467370896			10	.041036562
		4	.0871597604			13	.0423879846			11	.037134642
		5	.0710825990		7	7	.0890167025			12	.033739117
		6	.0600975754			8	.0785179677			13	.030721591
		7	.0520217423			9	.0700199026			14	.027983502
		8	.0457683625			10	.0629269074			15	.025442410
		9	.0407317967			11	.0568501034			16	.023019506
		10	.0365451034			12	.0515199092			17	.020621464
		11	.0329704894		8	8	.0864960639		4	4	.107474083
		12	.0298442464			9	.0771762286			5	.087905196
		13	.0270462261			10	.0693891332			6	.074503387
		14	.0244806359			11	.0627116906			7	.064640618
		15	.0220607111		9	9	.0853127880			8	.057003228
		16	.0196894667			10	.0767442321			9	.050857260
		17	.0172154925	19	1	1	.2799358050			10	.045757659
	3	3	.1288998943			2	.1367768168			11	.041416509
		4	.0991828539			3	.0929061763			12	.037636875
		5	.0809899792			4	.0711902425			13	.034276554
		6	.0685324700			5	.0580094835			14	.031226254
		7	.0593598602			6	.0490405678			15	.028394452
		8	.0522488413			7	.0424705246			16	.025693514
		9	.0465162123			8	.0374006329		5	5	.096794474
		10	.0417473296			9	.0333319395			6	.082105569
		11	.0376730987			10	.0299634144			7	.071279674
		12	.0341080171			11	.0271011338			8	.062887009
		13	.0309157650			12	.0246129452			9	.056127202
		14	.0279875014			13	.0224037540			10	.050514163

VARIANCES AND COVARIANCES OF ORDER STATISTICS

n	i	j	Value	n	i	j	Value	n	i	j	Value
19	5			20	1			20	4		
		11	.0457330144			18	.0139227072			14	.0310045146
		12	.0415681234			19	.0122530117			15	.0283650517
		13	.0378636088			20	.0102047204			16	.0258897454
		14	.0344995261		2	2	.1595731636			17	.0235070343
		15	.0313752928			3	.1088143707		5	5	.0939960007
	6	6	.0900218693			4	.0835758044			6	.0797773755
		7	.0782029063			5	.0682247554			7	.0693175756
		8	.0690294360			6	.0577699656			8	.0612251429
		9	.0616336896			7	.0501109523			9	.0547222526
		10	.0554877905			8	.0442041191			10	.0493374275
		11	.0502493169			9	.0394693443			11	.0447662310
		12	.0456834841			10	.0355565554			12	.0408014074
		13	.0416203596			11	.0322405467			13	.0372948400
		14	.0379290224			12	.0293684960			14	.0341351571
	7	7	.0856172981			13	.0268315105			15	.0312332040
		8	.0756153413			14	.0245479493			16	.0285109200
		9	.0675433161			15	.0224526609		6	6	.0871511254
		10	.0608297030			16	.0204888032			7	.0757703360
		11	.0551032224			17	.0185994024			8	.0669555789
		12	.0501089625			18	.0167136502			9	.0598659769
		13	.0456621835			19	.0147107671			10	.0539910639
	8	8	.0828339961		3	3	.1228134687			11	.0490008080
		9	.0740273546			4	.0945049010			12	.0446702771
		10	.0666958229			5	.0772355098			13	.0408385549
		11	.0604372723			6	.0654510179			14	.0373845194
		12	.0549752083			7	.0568056677			15	.0342111024
	9	9	.0812876330			8	.0501310269		7	7	.0826123955
		10	.0732703911			9	.0447763202			8	.0730383676
		11	.0664202898			10	.0403482354			9	.0653307665
	10	10	.0807909751			11	.0365934287			10	.0589387428
20	1	1	.2756966156			12	.0333397949			11	.0535056766
		2	.1344941714			13	.0304645792			12	.0487882257
		3	.0913234064			14	.0278756579			13	.0446121090
		4	.0699879991			15	.0254994381			14	.0408459989
		5	.0570566384			16	.0232716371		8	8	.0796309757
		6	.0482701093			17	.0211277373			9	.0712591607
		7	.0418437826			18	.0189874448			10	.0643103375
		8	.0368937058		4	4	.1046766243			11	.0583997310
		9	.0329296302			5	.0856442356			12	.0532644495
		10	.0296562523			6	.0726321560			13	.0487159834
		11	.0268838808			7	.0630731775		9	9	.0778118317
		12	.0244839567			8	.0556855081			10	.0702526464
		13	.0223649803			9	.0497539273			11	.0638176734
		14	.0204584277			10	.0448455403			12	.0582229133
		15	.0187096782			11	.0406811669		10	10	.0769474356
		16	.0170711408			12	.0370709493			11	.0699266198
		17	.0154951854			13	.0338793392				

VII.3   CONFIDENCE INTERVALS FOR MEDIANS

If the observations $x_1, x_2, \ldots, x_n$ are arranged in ascending order $x_{(1)}, x_{(2)}, \ldots$ $x_{(n)}$, a $100(1 - \alpha)\%$ confidence interval on the median of the population can be foun This table gives values of k and α such that one can be $100(1 - \alpha)\%$ confident that t population median is between $x_{(k)}$ and $x_{(n-k+1)}$.

CONFIDENCE INTERVALS FOR THE MEDIAN

n	Largest k	Actual $\alpha \le 0.05$	Largest k	Actual $\alpha \le 0.01$	N	Largest k	Actual $\alpha \le 0.05$	Largest k	Actual $\alpha \le 0.$
6	1	0.031			36	12	0.029	10	0.00
7	1	0.016			37	13	0.047	11	0.008
8	1	0.008	1	0.008	38	13	0.034	11	0.00
9	2	0.039	1	0.004	39	13	0.024	12	0.009
10	2	0.021	1	0.002	40	14	0.038	12	0.00
11	2	0.012	1	0.001	41	14	0.028	12	0.004
12	3	0.039	2	0.006	42	15	0.044	13	0.008
13	3	0.022	2	0.003	43	15	0.032	13	0.00
14	3	0.013	2	0.002	44	16	0.049	14	0.010
15	4	0.035	3	0.007	45	16	0.036	14	0.007
16	4	0.021	3	0.004	46	16	0.026	14	0.00
17	5	0.049	3	0.002	47	17	0.040	15	0.008
18	5	0.031	4	0.008	48	17	0.029	15	0.006
19	5	0.019	4	0.004	49	18	0.044	16	0.009
20	6	0.041	4	0.003	50	18	0.033	16	0.007
21	6	0.027	5	0.007	51	19	0.049	16	0.005
22	6	0.017	5	0.004	52	19	0.036	17	0.008
23	7	0.035	5	0.003	53	19	0.027	17	0.005
24	7	0.023	6	0.007	54	20	0.040	18	0.009
25	8	0.043	6	0.004	55	20	0.030	18	0.006
26	8	0.029	7	0.009	56	21	0.044	18	0.005
27	8	0.019	7	0.006	57	21	0.033	19	0.008
28	9	0.036	7	0.004	58	22	0.048	19	0.005
29	9	0.024	8	0.008	59	22	0.036	20	0.009
30	10	0.043	8	0.005	60	22	0.027	20	0.006
31	10	0.029	8	0.003	61	23	0.040	21	0.010
32	10	0.020	9	0.007	62	23	0.030	21	0.007
33	11	0.035	9	0.005	63	24	0.043	21	0.005
34	11	0.024	10	0.009	64	24	0.033	22	0.008
35	12	0.041	10	0.006	65	25	0.046	22	0.006

VII.4 CRITICAL VALUES FOR TESTING OUTLIERS

Tests for outliers may be based on the largest deviation $\max\limits_{i=1,2,\ldots,n} (x_i - \bar{x})$ of the

servations from their mean or on the range w, where these statistics have to be divided
the standard deviation σ or by an estimate of σ, depending on whether σ is known or not.
alternate set of test statistics is considered in this table. The ratios r_{10} and r_{11} are
table for the detection of a single outlier on the left, r_{21} and r_{22} for the detection of two
liers on the left. For outliers on the right replace $x_{(i)}$ by $x_{(n-i+1)}$ and use the same

ical values; e.g., r_{11} becomes $\dfrac{x_{(n)} - x_{(n-1)}}{x_{(n)} - x_{(2)}}$.

CRITICAL VALUES FOR TESTING OUTLIERS

Statistic	Number of obs., n	Critical values						
		$\alpha = .30$	$\alpha = .20$	$\alpha = .10$	$\alpha = .05$	$\alpha = .02$	$\alpha = .01$	$\alpha = .005$
$= \dfrac{x_{(2)} - x_{(1)}}{x_{(n)} - x_{(1)}}$	3	.684	.781	.886	.941	.976	.988	.994
	4	.471	.560	.679	.765	.846	.889	.926
	5	.373	.451	.557	.642	.729	.780	.821
	6	.318	.386	.482	.560	.644	.698	.740
	7	.281	.344	.434	.507	.586	.637	.680
$= \dfrac{x_{(2)} - x_{(1)}}{x_{(n-1)} - x_{(1)}}$	8	.318	.385	.479	.554	.631	.683	.725
	9	.288	.352	.441	.512	.587	.635	.677
	10	.265	.325	.409	.477	.551	.597	.639
$= \dfrac{x_{(3)} - x_{(1)}}{x_{(n-1)} - x_{(1)}}$	11	.391	.442	.517	.576	.638	.679	.713
	12	.370	.419	.490	.546	.605	.642	.675
	13	.351	.399	.467	.521	.578	.615	.649
$= \dfrac{x_{(3)} - x_{(1)}}{x_{(n-2)} - x_{(1)}}$	14	.370	.421	.492	.546	.602	.641	.674
	15	.353	.402	.472	.525	.579	.616	.647
	16	.338	.386	.454	.507	.559	.595	.624
	17	.325	.373	.438	.490	.542	.577	.605
	18	.314	.361	.424	.475	.527	.561	.589
	19	.304	.350	.412	.462	.514	.547	.575
	20	.295	.340	.401	.450	.502	.535	.562
	21	.287	.331	.391	.440	.491	.524	.551
	22	.280	.323	.382	.430	.481	.514	.541
	23	.274	.316	.374	.421	.472	.505	.532
	24	.268	.310	.367	.413	.464	.497	.524
	25	.262	.304	.360	.406	.457	.489	.516

VII.5 PERCENTILE ESTIMATES IN LARGE SAMPLES

A. Estimates of Population Mean

In sampling from a normal population with variance σ^2, the sampling distribution of the median has variance $\dfrac{1.57\sigma^2}{n}$ for large sample size n. For estimating the mean μ the efficiency of the median P_{50} is .637, i.e.

$$E_{P_{50}} = \frac{\sigma^2/n}{1.57\sigma^2/n} = .637 .$$

Higher efficiencies can be obtained from the mean of several percentile values. The efficiencies in this table are for the percentile estimates obtained from the mean of the indicated percentile.

B. Estimates of the Population Standard Deviation

This table gives the efficiencies for estimating the population standard deviation obtained from percentile estimates.

C. Estimates of Mean and Standard Deviation

This table gives percentile values for estimating both the mean and standard deviation and the efficiencies for the estimation of each. The values of K are the multipliers for the estimate of σ.

PERCENTILE ESTIMATES IN LARGE SAMPLES

A. Mean.

	Percentile estimate	Eff.
1	P_{50}	.64
2	$.5(P_{25} + P_{75})$	.81
3	$.3333(P_{17} + P_{50} + P_{83})$	.88
4	$.25(P_{12.5} + P_{37.5} + P_{62.5} + P_{87.5})$	.91
5	$.20(P_{10} + P_{30} + P_{50} + P_{70} + P_{90})$	.93
...	....	...
10	$.10(P_{05} + P_{15} + P_{25} + P_{35} + P_{45} + P_{55} + P_{65} + P_{75} + P_{85} + P_{95})$	.97

B. Standard deviation.

	Percentile estimate	Eff.
2	$.3388(P_{93} - P_{07})$	.65
4	$.1714(P_{97} + P_{85} - P_{15} - P_{03})$	.80
6	$.1180(P_{98} + P_{91} + P_{80} - P_{20} - P_{09} - P_{02})$	.87
8	$.0935(P_{98} + P_{93} + P_{86} + P_{77} - P_{23} - P_{14} - P_{07} - P_{02})$	.90
10	$.0739(P_{98.5} + P_{95} + P_{90} + P_{84} + P_{75} - P_{25} - P_{16} - P_{10} - P_{05} - P_{01.5})$	.92

C. Mean and standard d v ation.

	Percentile	Efficiency Mean	Efficiency Standard deviation	K
2	15, 85	.73	.56	.4824
4	05, 30, 70, 95	.80	.74	.2305
6	05, 15, 40, 60, 85, 95	.89	.80	.1704
8	03, 10, 25, 45, 55, 75, 90, 97	.90	.86	.1262
10	03, 10, 20, 30, 50, 50, 70, 80, 90, 97	.94	.87	.1104

VII.6   SIMPLE ESTIMATES IN SMALL SAMPLES

The observations $x_1, x_2, \ldots, x_n$ are arranged in ascending order $x_{(1)}, x_{(2)}, \ldots, x_{(n)}$

A. Estimates of the Population Mean

This tables gives variance and efficiency for several alternate estimates of the population mean. The midrange is defined as $(x_{(1)} + x_{(n)})/2$.

B. Estimates of the Population Standard Deviation

(i) The range is a biased estimate of σ. By multiplying the range by a factor, an unbiased estimate is obtained. This table gives values of this factor, along with variance and efficiency of the range.

(ii) The mean deviation can also be used to estimate σ. The mean deviation is given by

$$\text{M.D.} = \frac{1}{n} \sum_{i=1}^{n} |x_i - \bar{x}| \ .$$

It is easy to compute for small samples if the deviations are taken from the median. This table indicates the computation of the sum of these deviations. The multiplier to convert this sum to an unbiased estimate of σ and the efficiencies of this estimate are also included

(iii) The efficiencies of the range and mean deviation are less than for estimates which are easier to compute than the mean deviation. This table indicates the values to use in computing an estimate of σ which will give the highest efficiency for an estimate of this type. The coefficient is such that this statistic will give an unbiased estimate of σ.

(iv) This table indicates the values to use in computing the best linear estimate of σ. The efficiency of each of these estimates is also given.

SIMPLE ESTIMATES IN SMALL SAMPLES

Several estimates of the mean. (Variance to be multiplied by σ^2.)

n	Median Var.	Median Eff.	Midrange Var.	Midrange Eff.	Av. of best two Statistic	Av. of best two Var.	Av. of best two Eff.	$(x_2 + x_3 + \cdots + x_{n-1})/(n-2)$ Var.	$(x_2 + x_3 + \cdots + x_{n-1})/(n-2)$ Eff.
2	.500	1.000	.500	1.000	$\frac{1}{2}(x_1 + x_2)$	.500	1.000		
3	.449	.743	.362	.920	$\frac{1}{2}(x_1 + x_3)$	.362	.920	.449	.743
4	.298	.838	.298	.838	$\frac{1}{2}(x_2 + x_3)$	.298	.838	.298	.838
5	.287	.697	.261	.767	$\frac{1}{2}(x_2 + x_4)$	.231	.867	.227	.881
6	.215	.776	.236	.706	$\frac{1}{2}(x_2 + x_5)$	.193	.865	.184	.906
7	.210	.679	.218	.654	$\frac{1}{2}(x_2 + x_6)$	.168	.849	.155	.922
8	.168	.743	.205	.610	$\frac{1}{2}(x_3 + x_6)$	.149	.837	.134	.934
9	.166	.669	.194	.572	$\frac{1}{2}(x_3 + x_7)$	.132	.843	.118	.942
10	.138	.723	.186	.539	$\frac{1}{2}(x_3 + x_8)$	.119	.840	.105	.949
11	.137	.663	.178	.510	$\frac{1}{2}(x_3 + x_9)$	.109	.832	.0952	.955
12	.118	.709	.172	.484	$\frac{1}{2}(x_4 + x_9)$	.100	.831	.0869	.959
13	.117	.659	.167	.461	$\frac{1}{2}(x_4 + x_{10})$	.0924	.833	.0799	.963
14	.102	.699	.162	.440	$\frac{1}{2}(x_4 + x_{11})$	.0860	.830	.0739	.966
15	.102	.656	.158	.422	$\frac{1}{2}(x_4 + x_{12})$	.0808	.825	.0688	.969
16	.0904	.692	.154	.392	$\frac{1}{2}(x_5 + x_{12})$	.0756	.827	.0644	.971
17	.0901	.653	.151	.389	$\frac{1}{2}(x_5 + x_{13})$	.0711	.827	.0605	.973
18	.0810	.686	.148	.375	$\frac{1}{2}(x_5 + x_{14})$	.0673	.825	.0570	.975
19	.0808	.651	.145	.362	$\frac{1}{2}(x_6 + x_{14})$	.0640	.823	.0539	.976
20	.0734	.681	.143	.350	$\frac{1}{2}(x_6 + x_{15})$	.0607	.824	.0511	.978
∞	$\dfrac{1.57}{n}$	.637		.000	$\frac{1}{2}(P_{25} + P_{75})$	$\dfrac{1.24}{n}$	.808		1.000

B. Estimates of mean and dispersion in small samples.
(i) Unbiased estimate of σ using w. (Variance to be multiplied by σ^2.)

Sample size	Estimate	Variance	Eff.	Sample size	Estimate	Variance	Eff.
2	.886w	.571	1.000	11	.315w	.0616	.831
3	.591w	.275	.992	12	.307w	.0571	.814
4	.486w	.183	.975	13	.300w	.0533	.797
5	.430w	.138	.955	14	.294w	.0502	.781
6	.395w	.112	.933	15	.288w	.0474	.766
7	.370w	.0949	.911	16	.283w	.0451	.751
8	.351w	.0829	.890	17	.279w	.0430	.738
9	.337w	.0740	.869	18	.275w	.0412	.725
10	.325w	.0671	.850	19	.271w	.0395	.712
				20	.268w	.0381	.700

(ii) Mean deviation estimate of σ.

Sample size	Estimate	Eff.
2	$.8862(x_2 - x_1)$	1.00
3	$.5908(x_3 - x_1)$	.99
4	$.3770(x_4 + x_3 - x_2 - x_1)$	.91
5	$.3016(x_5 + x_4 - x_2 - x_1)$	.94
6	$.2369(x_6 + x_5 + x_4 - x_3 - x_2 - x_1)$	.90
7	$.2031(x_7 + x_6 + x_5 - x_3 - x_2 - x_1)$	.92
8	$.1723(x_8 + x_7 + x_6 + x_5 - x_4 - x_3 - x_2 - x_1)$	.90
9	$.1532(x_9 + x_8 + x_7 + x_6 - x_4 - x_3 - x_2 - x_1)$	.91
10	$.1353(x_{10} + x_9 + x_8 + x_7 + x_6 - x_5 - x_4 - x_3 - x_2 - x_1)$	.89

(iii) Modified linear estimate of σ. (Variance to be multiplied by σ^2.)

Sample size	Estimate	Variance	Eff.
2	$.8862(x_2 - x_1)$	.571	1.000
3	$.5908(x_3 - x_1)$	.275	.992
4	$.4857(x_4 - x_1)$	.183	.975
5	$.4299(x_5 - x_1)$	.138	.955
6	$.2619(x_6 + x_5 - x_2 - x_1)$	.109	.957
7	$.2370(x_7 + x_6 - x_2 - x_1)$	.0895	.967
8	$.2197(x_8 + x_7 - x_2 - x_1)$	.0761	.970
9	$.2068(x_9 + x_8 - x_2 - x_1)$	.0664	.968
10	$.1968(x_{10} + x_9 - x_2 - x_1)$	.0591	.964
11	$.1608(x_{11} + x_{10} + x_8 - x_4 - x_2 - x_1)$	.0529	.967
12	$.1524(x_{12} + x_{11} + x_9 - x_4 - x_2 - x_1)$	.0478	.972
13	$.1456(x_{13} + x_{12} + x_{10} - x_4 - x_2 - x_1)$	.0436	.975
14	$.1399(x_{14} + x_{13} + x_{11} - x_4 - x_2 - x_1)$	.0401	.977
15	$.1352(x_{15} + x_{14} + x_{12} - x_4 - x_2 - x_1)$	.0372	.977
16	$.1311(x_{16} + x_{15} + x_{13} - x_4 - x_2 - x_1)$	.0347	.975
17	$.1050(x_{17} + x_{16} + x_{15} + x_{13} - x_5 - x_3 - x_2 - x_1)$	.0325	.978
18	$.1020(x_{18} + x_{17} + x_{16} + x_{14} - x_5 - x_3 - x_2 - x_1)$	.0305	.978
19	$.09939(x_{19} + x_{18} + x_{17} + x_{15} - x_5 - x_3 - x_2 - x_1)$	.0288	.979
20	$.09706(x_{20} + x_{19} + x_{18} + x_{16} - x_5 - x_3 - x_2 - x_1)$	.0272	.978

(iv) Best linear estimate of σ.

Sample size	Estimate	Eff.
2	$.8862(x_2 - x_1)$	1.000
3	$.5908(x_3 - x_1)$	.992
4	$.4539(x_4 - x_1) + .1102(x_3 - x_2)$	.989
5	$.3724(x_5 - x_1) + .1352(x_4 - x_2)$	.988
6	$.3175(x_6 - x_1) + .1386(x_5 - x_2) + .0432(x_4 - x_3)$	.988
7	$.2778(x_7 - x_1) + .1351(x_6 - x_2) + .0625(x_5 - x_3)$	.989
8	$.2476(x_8 - x_1) + .1294(x_7 - x_2) + .0713(x_6 - x_3) + .0230(x_5 - x_4)$	.989
9	$.2237(x_9 - x_1) + .1233(x_8 - x_2) + .0751(x_7 - x_3) + .0360(x_6 - x_4)$	.989
10	$.2044(x_{10} - x_1) + .1172(x_9 - x_2) + .0763(x_8 - x_3) + .0436(x_7 - x_4)$ $+ .0142(x_6 - x_5)$	.990

VIII. Range and Studentized Range

VIII.1 PROBABILITY INTEGRAL OF THE RANGE

Let $x_1, x_2, \ldots, x_n$ denote a random sample of size n from a population with density function $f(x)$ and distribution function $F(x)$.

Let $x_{(1)}, x_{(2)}, \ldots, x_{(n)}$ denote the same values in ascending order of magnitude. Then the sample range w is defined by

$$w = x_{(n)} - x_{(1)} .$$

In standardized form

$$W = \frac{x_{(n)} - x_{(1)}}{\sigma} = X_{(n)} - X_{(1)} .$$

The probability integral for W for a sample of size n is given by

$$P(W;n) = n \int_{-\infty}^{\infty} [F(X + w) - F(X)]^{n-1} f(X) \, dX .$$

This table gives values of $P(W;n)$ for the normal density function $f(x) = \dfrac{1}{\sqrt{2\pi}} e^{-\frac{1}{2}x^2}$ and for various values of n and W.

Range and Studentized Range

PROBABILITY INTEGRAL OF THE RANGE

n / W	2	3	4	5	6	7	8	9	10
0.00	0.0000	0.0000							
0.05	.0282	.0007	0.0000						
0.10	.0564	.0028	.0001						
0.15	.0845	.0062	.0004	0.0000					
0.20	.1125	.0110	.0010	.0001					
0.25	0.1403	0.0171	0.0020	0.0002					
0.30	.1680	.0245	.0034	.0004	0.0001				
0.35	.1955	.0332	.0053	.0008	.0001				
0.40	.2227	.0431	.0079	.0014	.0002	0.0000			
0.45	.2497	.0543	.0111	.0022	.0004	.0001			
0.50	0.2763	0.0666	0.0152	0.0033	0.0007	0.0002	0.0000		
0.55	.3027	.0800	.0200	.0048	.0011	.0003	.0001		
0.60	.3286	.0944	.0257	.0068	.0017	.0004	.0001	0.0000	
0.65	.3542	.1099	.0322	.0092	.0026	.0007	.0002	.0001	
0.70	.3794	.1263	.0398	.0121	.0036	.0011	.0003	.0001	
0.75	0.4041	0.1436	0.0483	0.0157	0.0050	0.0016	0.0005	0.0002	0.0000
0.80	.4284	.1616	.0578	.0200	.0068	.0023	.0008	.0002	.0001
0.85	.4522	.1805	.0682	.0250	.0090	.0032	.0011	.0004	.0001
0.90	.4755	.2000	.0797	.0308	.0117	.0044	.0016	.0006	.0002
0.95	.4983	.2201	.0922	.0375	.0150	.0059	.0023	.0009	.0003
1.00	0.5205	0.2407	0.1057	0.0450	0.0188	0.0078	0.0032	0.0013	0.0005
1.05	.5422	.2618	.1201	.0535	.0234	.0101	.0043	.0018	.0008
1.10	.5633	.2833	.1355	.0629	.0287	.0129	.0058	.0025	.0011
1.15	.5839	.3052	.1517	.0733	.0348	.0163	.0075	.0035	.0016
1.20	.6039	.3272	.1688	.0847	.0417	.0203	.0098	.0047	.0022
1.25	0.6232	0.3495	0.1867	0.0970	0.0495	0.0250	0.0125	0.0062	0.0030
1.30	.6420	.3719	.2054	.1104	.0583	.0304	.0157	.0080	.0041
1.35	.6602	.3943	.2248	.1247	.0680	.0366	.0195	.0103	.0054
1.40	.6778	.4168	.2448	.1400	.0787	.0437	.0240	.0131	.0071
1.45	.6948	.4392	.2654	.1562	.0904	0.516	.0292	.0164	.0092
1.50	0.7112	0.4614	0.2865	0.1733	0.1031	0.0606	0.0353	0.0204	0.0117
1.55	.7269	.4835	.3080	.1913	.1168	.0705	.0421	.0250	.0148
1.60	.7421	.5053	.3299	.2101	.1315	.0814	.0499	.0304	.0184
1.65	.7567	.5269	.3521	.2296	.1473	.0934	.0587	.0366	.0227
1.70	.7707	.5481	.3745	.2498	.1639	.1064	.0684	.0437	.0278
1.75	0.7841	0.5690	0.3970	0.2706	0.1815	0.1204	0.0792	0.0517	0.0336
1.80	.7969	.5894	.4197	.2920	.2000	.1355	.0910	.0607	.0403
1.85	.8092	.6094	.4423	.3138	.2193	.1516	.1039	.0707	.0479
1.90	.8209	.6290	.4649	.3361	.2394	.1686	.1178	.0818	.0565
1.95	.8321	.6480	.4874	.3587	.2602	.1867	.1329	.0939	.0661
2.00	0.8427	0.6665	0.5096	0.3816	0.2816	0.2056	0.1489	0.1072	0.0768
2.05	.8528	.6845	.5317	.4046	.3035	.2254	.1661	.1216	.0886
2.10	.8624	.7019	.5534	.4277	.3260	.2460	.1842	.1371	.1015
2.15	.8716	.7187	.5748	.4508	.3489	.2673	.2032	.1536	.1155
2.20	.8802	.7349	.5957	.4739	.3720	.2893	.2232	.1712	.1307
2.25	0.8884	0.7505	0.6163	0.4969	0.3955	0.3118	0.2440	0.1899	0.1470

PROBABILITY INTEGRAL OF THE RANGE

n / W	11	12	13	14	15	16	17	18	19	20
0.85	0.0000									
0.90	.0001									
0.95	.0001	0.0000								
1.00	0.0002	0.0001	0.0000							
1.05	.0003	.0001	.0001	0.0000						
1.10	.0005	.0002	.0001	0.0001						
1.15	.0007	.0003	.0001	.0001	0.0000					
1.20	.0010	.0005	.0002	.0001	0.0001					
1.25	0.0015	0.0007	0.0004	0.0002	0.0001	0.0000				
1.30	.0021	.0010	.0005	.0003	.0001	.0001	0.0000	0.0000		
1.35	.0028	.0015	.0008	.0004	.0002	.0001	.0001	0.0001		
1.40	.0038	.0021	.0011	.0006	.0003	.0002	.0001	.0001		
1.45	.0051	.0028	.0016	.0009	.0005	.0003	.0001	.0001	0.0000	
1.50	0.0067	0.0038	0.0022	0.0012	0.0007	0.0004	0.0002	0.0001	0.0001	0.0000
1.55	.0087	.0051	.0030	.0017	.0010	.0006	.0003	.0002	.0001	.0001
1.60	.0111	.0067	.0040	.0024	.0014	.0008	.0005	.0003	.0002	.0001
1.65	.0140	.0086	.0053	.0032	.0020	.0012	.0007	.0004	.0003	.0002
1.70	.0176	.0111	.0070	.0044	.0027	.0017	.0011	.0007	.0004	.0003
1.75	0.0217	0.0140	0.0090	0.0058	0.0037	0.0023	0.0015	0.0010	0.0006	0.0004
1.80	.0266	.0175	.0115	.0075	.0049	.0032	.0021	.0014	.0009	.0006
1.85	.0323	.0217	.0145	.0097	.0065	.0043	.0029	.0019	.0013	.0008
1.90	.0388	.0266	.0182	.0124	.0084	.0057	.0039	.0026	.0018	.0012
1.95	.0463	.0323	.0070	.0044	.0108	.0075	.0052	.0036	.0024	.0017
2.00	0.0548	0.0389	0.0276	0.0195	0.0137	0.0097	0.0068	0.0048	0.0033	0.0023
2.05	.0643	.0465	.0335	.0241	.0173	.0124	.0088	.0063	.0045	.0032
2.10	.0748	.0550	.0403	.0295	.0215	.0156	.0114	.0082	.0060	.0043
2.15	.0866	.0646	.0481	.0357	.0265	.0196	.0144	.0106	.0078	.0058
2.20	.0994	.0753	.0569	.0429	.0323	.0242	.0182	.0136	.0102	.0076
2.25	0.1134	0.0872	0.0669	0.0511	0.0390	0.0297	0.0226	0.0172	0.0130	0.0099

Range and Studentized Range

PROBABILITY INTEGRAL OF THE RANGE

n W	2	3	4	5	6	7	8	9	10
2.25	0.8884	0.7505	0.6163	0.4969	0.3955	0.3118	0.2440	0.1899	0.1470
2.30	.8961	.7655	.6363	.5196	.4190	.3348	.2656	.2095	.1648
2.35	.9034	.7799	.6559	.5421	.4427	.3582	.2878	.2300	.1829
2.40	.9103	.7937	.6748	.5643	.4663	.3820	.3107	.2514	.2023
2.45	.9168	.8069	.6932	.5861	.4899	.4059	.3341	.2735	.2229
2.50	0.9229	0.8195	0.7110	0.6075	0.5132	0.4300	0.3579	0.2963	0.2443
2.55	.9286	.8315	.7282	.6283	.5364	.4541	.3820	.3198	.2665
2.60	.9340	.8429	.7448	.6487	.5592	.4782	.4064	.3437	.2894
2.65	.9390	.8537	.7607	.6685	.5816	.5022	.4310	.3680	.3130
2.70	.9438	.8640	.7759	.6877	.6036	.5259	.4555	.3927	.3372
2.75	0.9482	0.8737	0.7905	0.7063	0.6252	0.5494	0.4801	0.4175	0.3617
2.80	.9523	.8828	.8045	.7242	.6461	.5725	.5045	.4425	.3867
2.85	.9561	.8915	.8177	.7415	.6665	.5952	.5286	.4675	.4119
2.90	.9597	.8996	.8304	.7581	.6863	.6174	.5525	.4923	.4372
2.95	.9630	.9073	.8424	.7739	.7055	.6391	.5760	.5171	.4625
3.00	0.9661	0.9145	0.8537	0.7891	0.7239	0.6601	0.5991	0.5415	0.4878
3.05	.9690	.9212	.8645	.8036	.7416	.6806	.6216	.5656	.5129
3.10	.9716	.9275	.8746	.8174	.7587	.7003	.6436	.5892	.5378
3.15	.9741	.9334	.8842	.8305	.7750	.7194	.6649	.6124	.5623
3.20	.9763	.9388	.8931	.8429	.7905	.7377	.6856	.6350	.5864
3.25	0.9784	0.9439	0.9016	0.8546	0.8053	0.7553	0.7055	0.6569	0.6099
3.30	.9804	.9487	.9095	.8657	.8194	.7721	.7248	.6782	.6329
3.35	.9822	.9531	.9168	.8761	.8327	.7881	.7432	.6988	.6553
3.40	.9838	.9572	.9237	.8859	.8454	.8034	.7609	.7186	.6769
3.45	.9853	.9610	.9302	.8951	.8573	.8179	.7778	.7376	.6978
3.50	0.9867	0.9644	0.9361	0.9037	0.8685	0.8316	0.7938	0.7558	0.7180
3.55	.9879	.9677	.9417	.9117	.8790	.8446	.8091	.7732	.7373
3.60	.9891	.9706	.9468	.9192	.8889	.8568	.8236	.7898	.7558
3.65	.9901	.9734	.9516	.9261	.8981	.8683	.8372	.8055	.7735
3.70	.9911	.9759	.9560	.9326	.9067	.8790	.8501	.8204	.7903
3.75	0.9920	0.9782	0.9600	0.9386	0.9147	0.8891	0.8622	0.8345	0.8062
3.80	.9928	.9803	.9637	.9441	.9222	.8985	.8736	.8477	.8212
3.85	.9935	.9822	.9672	.9493	.9291	.9073	.8842	.8602	.8355
3.90	.9942	.9840	.9703	.9540	.9355	.9155	.8941	.8718	.8488
3.95	.9948	.9856	.9732	.9583	.9415	.9230	.9034	.8827	.8614
4.00	0.9953	0.9870	0.9758	0.9623	0.9469	0.9300	0.9120	0.8929	0.8731
4.05	.9958	.9883	.9782	.9660	.9520	.9365	.9199	.9024	.8841
4.10	.9963	.9895	.9804	.9693	.9566	.9425	.9273	.9112	.8943
4.15	.9967	.9906	.9824	.9724	.9608	.9480	.9341	.9193	.9038
4.20	.9970	.9916	.9842	.9752	.9647	.9530	.9404	.9268	.9126
4.25	0.9973	0.9925	0.9859	0.9777	0.9682	0.9576	0.9461	0.9338	0.9208
4.30	.9976	.9933	.9874	.9800	.9715	.9520	.9514	.9402	.9283
4.35	.9979	.9941	.9887	.9821	.9744	.9657	.9562	.9460	.9352
4.40	.9981	.9947	.9899	.9840	.9771	.9692	.9607	.9514	.9416
4.45	.9983	.9953	.9910	.9857	.9795	.9724	.9647	.9563	.9474
4.50	0.9985	0.9958	0.9920	0.9873	0.9817	0.9754	0.9684	0.9608	0.9527

PROBABILITY INTEGRAL OF THE RANGE

n / W	11	12	13	14	15	16	17	18	19	20
.25	0.1134	0.0872	0.0669	0.0511	0.0390	0.0297	0.0226	0.0172	0.0130	0.0099
.30	.1286	.1003	.0779	.0604	.0468	.0361	.0279	.0214	.0165	.0127
.35	.1450	.1145	.0902	.0709	.0556	.0435	.0340	.0265	.0207	.0161
.40	.1624	.1299	.1036	.0825	.0655	.0519	.0411	.0325	.0256	.0202
.45	.1810	.1466	.1183	.0953	.0766	.0615	.0493	.0394	.0315	.0251
.50	0.2007	0.1643	0.1342	0.1094	0.0890	0.0722	0.0585	0.0474	0.0383	0.0309
.55	.2213	.1833	.1513	.1247	.1025	.0842	.0690	.0565	.0462	.0377
.60	.2429	.2032	.1696	.1413	.1174	.0974	.0807	.0668	.0552	.0455
.65	.2653	.2243	.1891	.1590	.1335	.1119	.0937	.0783	.0654	.0545
.70	.2885	.2462	.2096	.1780	.1509	.1278	.1080	.0911	.0768	.0647
.75	0.3124	0.2690	0.2311	0.1981	0.1696	0.1449	0.1236	0.1053	0.0896	0.0761
.80	.3368	.2926	.2536	.2194	.1894	.1632	.1405	.1208	.1037	.0889
.85	.3618	.3169	.2770	.2416	.2103	.1828	.1587	.1376	.1191	.1031
.90	.3870	.3417	.3011	.2647	.2323	.2036	.1782	.1557	.1360	.1186
.95	.4125	.3670	.3258	.2887	.2553	.2255	.1989	.1752	.1541	.1355
.00	0.4382	0.3927	0.3511	0.3134	0.2792	0.2484	0.2207	0.1959	0.1736	0.1537
.05	.4639	.4186	.3769	.3387	.3039	.2723	.2436	.2177	.1944	.1733
.10	.4895	.4446	.4029	.3645	.3292	.2969	.2675	.2407	.2163	.1942
.15	.5150	.4706	.4291	.3907	.3551	.3223	.2922	.2646	.2394	.2163
.20	.5401	.4965	.4554	.4171	.3814	.3483	.3177	.2894	.2634	.2395
.25	0.5649	0.5222	0.4817	0.4437	0.4080	0.3748	0.3438	0.3151	0.2884	0.2638
.30	.5893	.5475	.5078	.4703	.4348	.4016	.3704	.3413	.3142	.2890
.35	.6131	.5725	.5337	.4967	.4617	.4286	.3974	.3681	.3407	.3150
.40	.6363	.5970	.5592	.5230	.4885	.4557	.4246	.3953	.3676	.3416
.45	.6589	.6209	.5842	.5489	.5150	.4827	.4519	.4227	.3950	.3688
.50	0.6807	0.6442	0.6087	0.5744	0.5413	0.5096	0.4792	0.4502	0.4226	0.3964
.55	.7017	.6668	.6326	.5994	.5672	.5362	.5063	.4777	.4504	.4242
.60	.7220	.6886	.6558	.6237	.5926	.5624	.5332	.5051	.4781	.4522
.65	.7414	.7096	.6782	.6474	.6173	.5881	.5597	.5322	.5056	.4801
.70	.7600	.7298	.6999	.6704	.6414	.6132	.5856	.5588	.5329	.5078
.75	0.7776	0.7491	0.7206	0.6925	0.6648	0.6376	0.6110	0.5850	0.5598	0.5352
.80	.7944	.7675	.7406	.7138	.6874	.6613	.6357	.6106	.5861	.5622
.85	.8103	.7850	.7596	.7342	.7090	.6842	.6596	.6355	.6118	.5887
.90	.8254	.8016	.7777	.7537	.7298	.7062	.6827	.6596	.6369	.6145
.95	.8395	.8173	.7948	.7723	.7497	.7273	.7050	.6829	.6611	.6397
.00	0.8528	0.8321	0.8111	0.7899	0.7686	0.7474	0.7263	0.7053	0.6845	0.6640
.05	.8653	.8460	.8264	.8066	.7866	.7666	.7466	.7268	.7070	.6874
.10	.8769	.8590	.8408	.8223	.8036	.7848	.7660	.7472	.7285	.7099
.15	.8878	.8712	.8543	.8371	.8196	.8021	.7844	.7667	.7491	.7315
.20	.8978	.8826	.8669	.8509	.8347	.8183	.8018	.7852	.7686	.7520
.25	0.9072	0.8931	0.8787	0.8639	0.8488	0.8336	0.8182	0.8027	0.7871	0.7715
.30	.9158	.9029	.8896	.8760	.8620	.8479	.8336	.8191	.8046	.7899
.35	.9238	.9120	.8998	.8872	.8744	.8613	.8480	.8346	.8210	.8074
.40	.9312	.9204	.9092	.8976	.8858	.8737	.8615	.8490	.8364	.8237
.45	.9379	.9281	.9178	.9073	.8964	.8853	.8740	.8625	.8508	.8391
.50	0.9441	0.9352	0.9258	0.9162	0.9062	0.8960	0.8856	0.8750	0.8643	0.8534

134   *Range and Studentized Range*

PROBABILITY INTEGRAL OF THE RANGE

W \ n	2	3	4	5	6	7	8	9	10
4.50	0.9985	0.9958	0.9920	0.9873	0.9817	0.9754	0.9684	0.9608	0.9527
4.55	.9987	.9963	.9929	.9887	.9837	.9780	.9717	.9649	.9576
4.60	.9989	.9967	.9937	.9899	.9855	.9804	.9747	.9686	.9620
4.65	.9990	.9971	.9944	.9911	.9871	.9825	.9775	.9719	.9660
4.70	.9991	.9974	.9951	.9921	.9885	.9845	.9799	.9750	.9696
4.75	0.9992	0.9977	0.9956	0.9930	0.9898	0.9862	0.9822	0.9777	0.9729
4.80	.9993	.9980	.9962	.9938	.9910	.9878	.9842	.9802	.9759
4.85	.9994	.9982	.9966	.9945	.9920	.9892	.9860	.9824	.9786
4.90	.9995	.9985	.9970	.9952	.9930	.9904	.9876	.9844	.9810
4.95	.9995	.9986	.9974	.9958	.9938	.9916	.9890	.9862	.9832
5.00	0.9996	0.9988	0.9977	0.9963	0.9945	0.9926	0.9903	0.9878	0.9851
5.05	.9996	.9990	.9980	.9967	.9952	.9935	.9915	.9893	.9869
5.10	.9997	.9991	.9982	.9971	.9958	.9942	.9925	.9906	.9884
5.15	.9997	.9992	.9985	.9975	.9963	.9950	.9934	.9917	.9898
5.20	.9998	.9993	.9987	.9978	.9968	.9956	.9942	.9927	.9911
5.25	0.9998	0.9994	0.9988	0.9981	0.9972	0.9961	0.9949	0.9936	0.9922
5.30	.9998	.9995	.9990	.9983	.9975	.9966	.9956	.9944	.9931
5.35	.9998	.9995	.9991	.9985	.9979	.9971	.9961	.9951	.9940
5.40	.9999	.9996	.9992	.9987	.9981	.9974	.9966	.9957	.9948
5.45	.9999	.9997	.9993	.9989	.9984	.9978	.9971	.9963	.9954
5.50	0.9999	0.9997	0.9994	0.9990	0.9986	0.9981	0.9974	0.9968	0.9960
5.55	.9999	.9997	.9995	.9992	.9988	.9983	.9978	.9972	.9965
5.60	.9999	.9998	.9996	.9993	.9989	.9985	.9981	.9976	.9970
5.65	.9999	.9998	.9996	.9994	.9991	.9987	.9983	.9979	.9974
5.70	0.9999	.9998	.9997	.9995	.9992	.9989	.9986	.9982	.9977
5.75	1.0000	0.9999	0.9997	0.9995	0.9993	0.9991	0.9988	0.9984	0.9980
5.80		.9999	.9998	.9996	.9994	.9992	.9989	.9986	.9983
5.85		.9999	.9998	.9997	.9995	.9993	.9991	.9988	.9985
5.90		.9999	.9998	.9997	.9996	.9994	.9992	.9990	.9988
5.95		.9999	.9998	.9997	.9996	.9995	.9993	.9991	.9989
6.00		0.9999	0.9999	0.9998	0.9997	0.9996	0.9994	0.9993	0.9991

PROBABILITY INTEGRAL OF THE RANGE

W \ n	11	12	13	14	15	16	17	18	19	20
4.50	0.9441	0.9352	0.9258	0.9162	0.9062	0.8960	0.8856	0.8750	0.8643	0.8534
4.55	.9498	.9417	.9332	.9244	.9153	.9060	.8964	.8867	.8768	.8667
4.60	.9550	.9476	.9399	.9319	.9236	.9151	.9064	.8975	.8884	.8791
4.65	.9597	.9530	.9460	.9388	.9313	.9235	.9155	.9074	.8991	.8906
4.70	.9639	.9579	.9516	.9451	.9382	.9312	.9240	.9165	.9089	.9012
4.75	0.9678	0.9624	0.9567	0.9508	0.9446	0.9383	0.9317	0.9249	0.9180	0.9110
4.80	.9713	.9665	.9614	.9560	.9505	.9447	.9387	.9326	.9263	.9199
4.85	.9745	.9702	.9656	.9608	.9557	.9505	.9452	.9396	.9339	.9281
4.90	.9774	.9735	.9694	.9650	.9605	.9559	.9510	.9460	.9409	.9356
4.95	.9799	.9765	.9728	.9689	.9649	.9607	.9563	.9518	.9472	.9424
5.00	0.9822	0.9791	0.9759	0.9724	0.9688	0.9650	0.9611	0.9571	0.9529	0.9486
5.05	.9843	.9816	.9786	.9756	.9723	.9690	.9655	.9618	.9581	.9543
5.10	.9862	.9837	.9811	.9784	.9755	.9725	.9694	.9661	.9628	.9593
5.15	.9878	.9856	.9833	.9809	.9783	.9757	.9729	.9700	.9670	.9639
5.20	.9893	.9874	.9853	.9832	.9809	.9785	.9760	.9735	.9708	.9681
5.25	0.9906	0.9889	0.9871	0.9852	0.9832	0.9811	0.9789	0.9766	0.9742	0.9718
5.30	.9917	.9903	.9887	.9870	.9852	.9833	.9814	.9794	.9773	.9751
5.35	.9928	.9915	.9901	.9886	.9870	.9854	.9836	.9819	.9800	.9781
5.40	.9937	.9925	.9913	.9900	.9886	.9872	.9856	.9841	.9824	.9807
5.45	.9945	.9935	.9924	.9913	.9900	.9888	.9874	.9860	.9846	.9831
5.50	0.9952	0.9943	0.9934	0.9924	0.9913	0.9902	0.9890	0.9878	0.9865	0.9852
5.55	.9958	.9951	.9942	.9933	.9924	.9914	.9904	.9893	.9882	.9870
5.60	.9964	.9957	.9950	.9942	.9934	.9925	.9916	.9907	.9897	.9887
5.65	.9969	.9963	.9956	.9950	.9943	.9935	.9927	.9919	.9910	.9901
5.70	.9973	.9968	.9962	.9956	.9950	.9944	.9937	.9930	.9922	.9914
5.75	0.9976	0.9972	0.9967	0.9962	0.9957	0.9951	0.9945	0.9939	0.9932	0.9925
5.80	.9980	.9976	.9972	.9967	.9963	.9958	.9952	.9947	.9941	.9935
5.85	.9982	.9979	.9976	.9972	.9968	.9963	.9959	.9954	.9949	.9944
5.90	.9985	.9982	.9979	.9976	.9972	.9968	.9964	.9960	.9956	.9952
5.95	.9987	.9985	.9982	.9979	.9976	.9973	.9969	.9966	.9962	.9958
6.00	0.9989	0.9987	0.9984	0.9982	0.9979	0.9977	0.9974	0.9971	0.9967	0.9964

Range and Studentized Range

PROBABILITY INTEGRAL OF THE RANGE

W \ n	2	3	4	5	6	7	8	9	10
6.00		0.9999	0.9999	0.9998	0.9997	0.9996	0.9994	0.9993	0.9991
6.05		0.9999	.9999	.9998	.9997	.9996	.9995	.9994	.9992
6.10		1.0000	.9999	.9998	.9998	.9997	.9996	.9995	.9993
6.15			.9999	.9999	.9998	.9997	.9996	.9995	.9994
6.20			.9999	.9999	.9998	.9998	.9997	.9996	.9995
6.25			0.9999	0.9999	0.9999	0.9998	0.9997	0.9997	0.9996
6.30			1.0000	.9999	.9999	.9998	.9998	.9997	.9996
6.35				.9999	.9999	.9999	.9998	.9998	.9997
6.40				0.9999	.9999	.9999	.9998	.9998	.9997
6.45				0.9999	.9999	.9999	.9999	.9998	.9998
6.50				1.0000	0.9999	0.9999	0.9999	0.9999	0.9998
6.55					0.9999	.9999	.9999	.9999	.9998
6.60					1.0000	.9999	.9999	.9999	.9999
6.65						0.9999	.9999	.9999	.9999
6.70						1.0000	0.9999	.9999	.9999
6.75							1.0000	0.9999	0.9999
6.80								0.9999	.9999
6.85								1.0000	0.9999
6.90									1.0000
6.95									
7.00									
7.05									
7.10									
7.15									
7.20									
7.25									

PROBABILITY INTEGRAL OF THE RANGE

n / W	11	12	13	14	15	16	17	18	19	20
6.00	0.9989	0.9987	0.9984	0.9982	0.9979	0.9977	0.9974	0.9971	0.9967	0.9964
6.05	.9990	.9989	.9987	.9985	.9982	.9980	.9977	.9975	.9972	.9969
6.10	.9992	.9990	.9989	.9987	.9985	.9983	.9981	.9978	.9976	.9973
6.15	.9993	.9992	.9990	.9989	.9987	.9985	.9983	.9981	.9979	.9977
6.20	.9994	.9993	.9992	.9990	.9989	.9987	.9986	.9984	.9982	.9980
6.25	0.9995	0.9994	0.9993	0.9992	0.9990	0.9989	0.9988	0.9986	0.9985	0.9983
6.30	.9996	.9995	.9994	.9993	.9992	.9991	.9990	.9988	.9987	.9986
6.35	.9996	.9996	.9995	.9994	.9993	.9992	.9991	.9990	.9989	.9988
6.40	.9997	.9996	.9996	.9995	.9994	.9993	.9992	.9992	.9991	.9990
6.45	.9997	.9997	.9996	.9996	.9995	.9994	.9994	.9993	.9992	.9991
6.50	0.9998	0.9997	0.9997	0.9996	0.9996	0.9995	0.9995	0.9994	0.9993	0.9993
6.55	.9998	.9998	.9997	.9997	.9996	.9996	.9995	.9995	.9994	.9994
6.60	.9998	.9998	.9998	.9997	.9997	.9997	.9996	.9996	.9995	.9995
6.65	.9999	.9998	.9998	.9998	.9997	.9997	.9997	.9996	.9996	.9995
6.70	.9999	.9999	.9998	.9998	.9998	.9998	.9997	.9997	.9997	.9996
6.75	0.9999	0.9999	0.9999	0.9998	0.9998	0.9998	0.9998	0.9997	0.9997	0.9997
6.80	.9999	.9999	.9999	.9999	.9998	.9998	.9998	.9998	.9998	.9997
6.85	.9999	.9999	.9999	.9999	.9999	.9999	.9998	.9998	.9998	.9998
6.90	0.9999	.9999	.9999	.9999	.9999	.9999	.9999	.9998	.9998	.9998
6.95	1.0000	0.9999	.9999	.9999	.9999	.9999	.9999	.9999	.9999	.9998
7.00		1.0000	0.9999	0.9999	0.9999	0.9999	0.9999	0.9999	0.9999	0.9999
7.05			1.0000	0.9999	.9999	.9999	.9999	.9999	.9999	.9999
7.10				1.0000	0.9999	0.9999	.9999	.9999	.9999	.9999
7.15					1.0000	1.0000	0.9999	0.9999	.9999	.9999
7.20							0.9999	0.9999	0.9999	0.9999
7.25							1.0000	1.0000	1.0000	0.9999
7.26										1.0000

VIII.2   PERCENTAGE POINTS, DISTRIBUTION OF THE RANGE

Percentage points of the range are found by the use of inverse interpolation in the table for the probability integral of the range.

Size of sample n	Factor $1/d_n$	Lower percentage points						Upper percentage points					
		0.1	0.5	1.0	2.5	5.0	10.0	10.0	5.0	2.5	1.0	0.5	0.1
2	0.8862	0.00	0.01	0.02	0.04	0.09	0.18	2.33	2.77	3.17	3.64	3.97	4.65
3	.5908	0.06	0.13	0.19	0.30	0.43	0.62	2.90	3.31	3.68	4.12	4.42	5.06
4	.4857	0.20	0.34	0.43	0.59	0.76	0.98	3.24	3.63	3.98	4.40	4.69	5.31
5	.4299	0.37	0.55	0.67	0.85	1.03	1.26	3.48	3.86	4.20	4.60	4.89	5.48
6	0.3946	0.53	0.75	0.87	1.07	1.25	1.49	3.66	4.03	4.36	4.76	5.03	5.62
7	.3698	0.69	0.92	1.05	1.25	1.44	1.68	3.81	4.17	4.49	4.88	5.15	5.73
8	.3512	0.83	1.08	1.20	1.41	1.60	1.84	3.93	4.29	4.60	4.99	5.25	5.82
9	.3367	0.97	1.21	1.34	1.55	1.74	1.97	4.04	4.39	4.70	5.08	5.34	5.90
10	0.3249	1.08	1.33	1.47	1.67	1.86	2.09	4.13	4.47	4.78	5.16	5.42	5.97
11	.3152	1.19	1.45	1.58	1.78	1.97	2.20	4.21	4.55	4.86	5.23	5.49	6.04
12	.3069	1.29	1.55	1.68	1.88	2.07	2.30	4.28	4.62	4.92	5.29	5.55	6.09
13	.2998	1.39	1.64	1.77	1.98	2.16	2.39	4.35	4.68	4.99	5.35	5.60	6.14
14	.2935	1.47	1.72	1.86	2.06	2.24	2.47	4.41	4.74	5.04	5.40	5.65	6.19
15	0.2880	1.55	1.80	1.93	2.14	2.32	2.54	4.47	4.80	5.09	5.45	5.70	6.23
16	.2831	1.63	1.88	2.01	2.21	2.39	2.61	4.52	4.85	5.14	5.49	5.74	6.27
17	.2787	1.69	1.94	2.07	2.27	2.45	2.67	4.57	4.89	5.18	5.54	5.78	6.31
18	.2747	1.76	2.01	2.14	2.34	2.52	2.73	4.61	4.93	5.22	5.57	5.82	6.35
19	.2711	1.82	2.07	2.20	2.39	2.57	2.79	4.65	4.97	5.26	5.61	5.86	6.38
20	0.2677	1.88	2.12	2.25	2.45	2.63	2.84	4.69	5.01	5.30	5.65	5.89	6.41

The unit is the population standard deviation.
Estimate of σ = range (or mean range) in a sample of n observations $\times 1/d_n$.

VIII.3   PERCENTAGE POINTS, STUDENTIZED RANGE

If in the standardized range $W = \dfrac{w}{\sigma}$, the unknown population standard deviation σ is replaced by s, the sample standard deviation computed from another sample from the same population, then the studentized range q is given by

$$q = \frac{w}{s},$$

where w is the range from a sample of size n and s is independent of w and is based on ν degrees of freedom. The probability integral of the studentized range is given by

$$\Pr\left\{\frac{w}{s} \le q\right\} = \int_0^\infty \left[\Gamma\left(\frac{\nu}{2}\right)\right]^{-1} 2^{-\frac{1}{2}\nu+1}\nu^{\nu/2}s^{\nu-1}e^{-\frac{1}{2}\nu s^2}f(qs)\, ds$$

where $f(qs)$ is the probability integral of the range for samples of size n.

140   **Range and Studentized Range**

UPPER 1 PER CENT POINTS OF THE STUDENTIZED RANGE

The entries are $q_{.01}$, where $P(q < q_{.01}) = .99$

ν \\ n	2	3	4	5	6	7	8	9	10
1	90.03	135.0	164.3	185.6	202.2	215.8	227.2	237.0	245.6
2	14.04	19.02	22.29	24.72	26.63	28.20	29.53	30.68	31.69
3	8.26	10.62	12.17	13.33	14.24	15.00	15.64	16.20	16.69
4	6.51	8.12	9.17	9.96	10.58	11.10	11.55	11.93	12.27
5	5.70	6.98	7.80	8.42	8.91	9.32	9.67	9.97	10.24
6	5.24	6.33	7.03	7.56	7.97	8.32	8.61	8.87	9.10
7	4.95	5.92	6.54	7.01	7.37	7.68	7.94	8.17	8.37
8	4.75	5.64	6.20	6.62	6.96	7.24	7.47	7.68	7.86
9	4.60	5.43	5.96	6.35	6.66	6.91	7.13	7.33	7.49
10	4.48	5.27	5.77	6.14	6.43	6.67	6.87	7.05	7.21
11	4.39	5.15	5.62	5.97	6.25	6.48	6.67	6.84	6.99
12	4.32	5.05	5.50	5.84	6.10	6.32	6.51	6.67	6.81
13	4.26	4.96	5.40	5.73	5.98	6.19	6.37	6.53	6.67
14	4.21	4.89	5.32	5.63	5.88	6.08	6.26	6.41	6.54
15	4.17	4.84	5.25	5.56	5.80	5.99	6.16	6.31	6.44
16	4.13	4.79	5.19	5.49	5.72	5.92	6.08	6.22	6.35
17	4.10	4.74	5.14	5.43	5.66	5.85	6.01	6.15	6.27
18	4.07	4.70	5.09	5.38	5.60	5.79	5.94	6.08	6.20
19	4.05	4.67	5.05	5.33	5.55	5.73	5.89	6.02	6.14
20	4.02	4.64	5.02	5.29	5.51	5.69	5.84	5.97	6.09
24	3.96	4.55	4.91	5.17	5.37	5.54	5.69	5.81	5.92
30	3.89	4.45	4.80	5.05	5.24	5.40	5.54	5.65	5.76
40	3.82	4.37	4.70	4.93	5.11	5.26	5.39	5.50	5.60
60	3.76	4.28	4.59	4.82	4.99	5.13	5.25	5.36	5.45
120	3.70	4.20	4.50	4.71	4.87	5.01	5.12	5.21	5.30
∞	3.64	4.12	4.40	4.60	4.76	4.88	4.99	5.08	5.16

UPPER 1 PER CENT POINTS OF THE STUDENTIZED RANGE

n	11	12	13	14	15	16	17	18	19	20
1	253.2	260.0	266.2	271.8	277.0	281.8	286.3	290.4	294.3	298.0
2	32.59	33.40	34.13	34.81	35.43	36.00	36.53	37.03	37.50	37.95
3	17.13	17.53	17.89	18.22	18.52	18.81	19.07	19.32	19.55	19.77
4	12.57	12.84	13.09	13.32	13.53	13.73	13.91	14.08	14.24	14.40
5	10.48	10.70	10.89	11.08	11.24	11.40	11.55	11.68	11.81	11.93
6	9.30	9.48	9.65	9.81	9.95	10.08	10.21	10.32	10.43	10.54
7	8.55	8.71	8.86	9.00	9.12	9.24	9.35	9.46	9.55	9.65
8	8.03	8.18	8.31	8.44	8.55	8.66	8.76	8.85	8.94	9.03
9	7.65	7.78	7.91	8.03	8.13	8.23	8.33	8.41	8.49	8.57
10	7.36	7.49	7.60	7.71	7.81	7.91	7.99	8.08	8.15	8.23
11	7.13	7.25	7.36	7.46	7.56	7.65	7.73	7.81	7.88	7.95
12	6.94	7.06	7.17	7.26	7.36	7.44	7.52	7.59	7.66	7.73
13	6.79	6.90	7.01	7.10	7.19	7.27	7.35	7.42	7.48	7.55
14	6.66	6.77	6.87	6.96	7.05	7.13	7.20	7.27	7.33	7.39
15	6.55	6.66	6.76	6.84	6.93	7.00	7.07	7.14	7.20	7.26
16	6.46	6.56	6.66	6.74	6.82	6.90	6.97	7.03	7.09	7.15
17	6.38	6.48	6.57	6.66	6.73	6.81	6.87	6.94	7.00	7.05
18	6.31	6.41	6.50	6.58	6.65	6.73	6.79	6.85	6.91	6.97
19	6.25	6.34	6.43	6.51	6.58	6.65	6.72	6.78	6.84	6.89
20	6.19	6.28	6.37	6.45	6.52	6.59	6.65	6.71	6.77	6.82
24	6.02	6.11	6.19	6.26	6.33	6.39	6.45	6.51	6.56	6.61
30	5.85	5.93	6.01	6.08	6.14	6.20	6.26	6.31	6.36	6.41
40	5.69	5.76	5.83	5.90	5.96	6.02	6.07	6.12	6.16	6.21
60	5.53	5.60	5.67	5.73	5.78	5.84	5.89	5.93	5.97	6.01
120	5.37	5.44	5.50	5.56	5.61	5.66	5.71	5.75	5.79	5.83
∞	5.23	5.29	5.35	5.40	5.45	5.49	5.54	5.57	5.61	5.65

Range and Studentized Range

UPPER 5 PER CENT POINTS OF THE STUDENTIZED RANGE

The entries are $q_{.05}$, where $P(q < q_{.05}) = .95$

ν \ n	2	3	4	5	6	7	8	9	10
1	17.97	26.98	32.82	37.08	40.41	43.12	45.40	47.36	49.07
2	6.08	8.33	9.80	10.88	11.74	12.44	13.03	13.54	13.99
3	4.50	5.91	6.82	7.50	8.04	8.48	8.85	9.18	9.46
4	3.93	5.04	5.76	6.29	6.71	7.05	7.35	7.60	7.83
5	3.64	4.60	5.22	5.67	6.03	6.33	6.58	6.80	6.99
6	3.46	4.34	4.90	5.30	5.63	5.90	6.12	6.32	6.49
7	3.34	4.16	4.68	5.06	5.36	5.61	5.82	6.00	6.16
8	3.26	4.04	4.53	4.89	5.17	5.40	5.60	5.77	5.92
9	3.20	3.95	4.41	4.76	5.02	5.24	5.43	5.59	5.74
10	3.15	3.88	4.33	4.65	4.91	5.12	5.30	5.46	5.60
11	3.11	3.82	4.26	4.57	4.82	5.03	5.20	5.35	5.49
12	3.08	3.77	4.20	4.51	4.75	4.95	5.12	5.27	5.39
13	3.06	3.73	4.15	4.45	4.69	4.88	5.05	5.19	5.32
14	3.03	3.70	4.11	4.41	4.64	4.83	4.99	5.13	5.25
15	3.01	3.67	4.08	4.37	4.59	4.78	4.94	5.08	5.20
16	3.00	3.65	4.05	4.33	4.56	4.74	4.90	5.03	5.15
17	2.98	3.63	4.02	4.30	4.52	4.70	4.86	4.99	5.11
18	2.97	3.61	4.00	4.28	4.49	4.67	4.82	4.96	5.07
19	2.96	3.59	3.98	4.25	4.47	4.65	4.79	4.92	5.04
20	2.95	3.58	3.96	4.23	4.45	4.62	4.77	4.90	5.01
24	2.92	3.53	3.90	4.17	4.37	4.54	4.68	4.81	4.92
30	2.89	3.49	3.85	4.10	4.30	4.46	4.60	4.72	4.82
40	2.86	3.44	3.79	4.04	4.23	4.39	4.52	4.63	4.73
60	2.83	3.40	3.74	3.98	4.16	4.31	4.44	4.55	4.65
120	2.80	3.36	3.68	3.92	4.10	4.24	4.36	4.47	4.56
∞	2.77	3.31	3.63	3.86	4.03	4.17	4.29	4.39	4.47

UPPER 5 PER CENT POINTS OF THE STUDENTIZED RANGE

n	11	12	13	14	15	16	17	18	19	20
1	50.59	51.96	53.20	54.33	55.36	56.32	57.22	58.04	58.83	59.56
2	14.39	14.75	15.08	15.38	15.65	15.91	16.14	16.37	16.57	16.77
3	9.72	9.95	10.15	10.35	10.53	10.69	10.84	10.98	11.11	11.24
4	8.03	8.21	8.37	8.52	8.66	8.79	8.91	9.03	9.13	9.23
5	7.17	7.32	7.47	7.60	7.72	7.83	7.93	8.03	8.12	8.21
6	6.65	6.79	6.92	7.03	7.14	7.24	7.34	7.43	7.51	7.59
7	6.30	6.43	6.55	6.66	6.76	6.85	6.94	7.02	7.10	7.17
8	6.05	6.18	6.29	6.39	6.48	6.57	6.65	6.73	6.80	6.87
9	5.87	5.98	6.09	6.19	6.28	6.36	6.44	6.51	6.58	6.64
10	5.72	5.83	5.93	6.03	6.11	6.19	6.27	6.34	6.40	6.47
11	5.61	5.71	5.81	5.90	5.98	6.06	6.13	6.20	6.27	6.33
12	5.51	5.61	5.71	5.80	5.88	5.95	6.02	6.09	6.15	6.21
13	5.43	5.53	5.63	5.71	5.79	5.86	5.93	5.99	6.05	6.11
14	5.36	5.46	5.55	5.64	5.71	5.79	5.85	5.91	5.97	6.03
15	5.31	5.40	5.49	5.57	5.65	5.72	5.78	5.85	5.90	5.96
16	5.26	5.35	5.44	5.52	5.59	5.66	5.73	5.79	5.84	5.90
17	5.21	5.31	5.39	5.47	5.54	5.61	5.67	5.73	5.79	5.84
18	5.17	5.27	5.35	5.43	5.50	5.57	5.63	5.69	5.74	5.79
19	5.14	5.23	5.31	5.39	5.46	5.53	5.59	5.65	5.70	5.75
20	5.11	5.20	5.28	5.36	5.43	5.49	5.55	5.61	5.66	5.71
24	5.01	5.10	5.18	5.25	5.32	5.38	5.44	5.49	5.55	5.59
30	4.92	5.00	5.08	5.15	5.21	5.27	5.33	5.38	5.43	5.47
40	4.82	4.90	4.98	5.04	5.11	5.16	5.22	5.27	5.31	5.36
60	4.73	4.81	4.88	4.94	5.00	5.06	5.11	5.15	5.20	5.24
120	4.64	4.71	4.78	4.84	4.90	4.95	5.00	5.04	5.09	5.13
∞	4.55	4.62	4.68	4.74	4.80	4.85	4.89	4.93	4.97	5.01

UPPER 10 PER CENT POINTS OF THE STUDENTIZED RANGE

The entries are $q_{.10}$, where $P(q < q_{.10}) = .90$

ν \ n	2	3	4	5	6	7	8	9	10
1	8.93	13.44	16.36	18.49	20.15	21.51	22.64	23.62	24.48
2	4.13	5.73	6.77	7.54	8.14	8.63	9.05	9.41	9.72
3	3.33	4.47	5.20	5.74	6.16	6.51	6.81	7.06	7.29
4	3.01	3.98	4.59	5.03	5.39	5.68	5.93	6.14	6.33
5	2.85	3.72	4.26	4.66	4.98	5.24	5.46	5.65	5.82
6	2.75	3.56	4.07	4.44	4.73	4.97	5.17	5.34	5.50
7	2.68	3.45	3.93	4.28	4.55	4.78	4.97	5.14	5.28
8	2.63	3.37	3.83	4.17	4.43	4.65	4.83	4.99	5.13
9	2.59	3.32	3.76	4.08	4.34	4.54	4.72	4.87	5.01
10	2.56	3.27	3.70	4.02	4.26	4.47	4.64	4.78	4.91
11	2.54	3.23	3.66	3.96	4.20	4.40	4.57	4.71	4.84
12	2.52	3.20	3.62	3.92	4.16	4.35	4.51	4.65	4.78
13	2.50	3.18	3.59	3.88	4.12	4.30	4.46	4.60	4.72
14	2.49	3.16	3.56	3.85	4.08	4.27	4.42	4.56	4.68
15	2.48	3.14	3.54	3.83	4.05	4.23	4.39	4.52	4.64
16	2.47	3.12	3.52	3.80	4.03	4.21	4.36	4.49	4.61
17	2.46	3.11	3.50	3.78	4.00	4.18	4.33	4.46	4.58
18	2.45	3.10	3.49	3.77	3.98	4.16	4.31	4.44	4.55
19	2.45	3.09	3.47	3.75	3.97	4.14	4.29	4.42	4.53
20	2.44	3.08	3.46	3.74	3.95	4.12	4.27	4.40	4.51
24	2.42	3.05	3.42	3.69	3.90	4.07	4.21	4.34	4.44
30	2.40	3.02	3.39	3.65	3.85	4.02	4.16	4.28	4.38
40	2.38	2.99	3.35	3.60	3.80	3.96	4.10	4.21	4.32
60	2.36	2.96	3.31	3.56	3.75	3.91	4.04	4.16	4.25
120	2.34	2.93	3.28	3.52	3.71	3.86	3.99	4.10	4.19
∞	2.33	2.90	3.24	3.48	3.66	3.81	3.93	4.04	4.13

UPPER 10 PER CENT POINTS OF THE STUDENTIZED RANGE

n	11	12	13	14	15	16	17	18	19	20
1	25.24	25.92	26.54	27.10	27.62	28.10	28.54	28.96	29.35	29.71
2	10.01	10.26	10.49	10.70	10.89	11.07	11.24	11.39	11.54	11.68
3	7.49	7.67	7.83	7.98	8.12	8.25	8.37	8.48	8.58	8.68
4	6.49	6.65	6.78	6.91	7.02	7.13	7.23	7.33	7.41	7.50
5	5.97	6.10	6.22	6.34	6.44	6.54	6.63	6.71	6.79	6.86
6	5.64	5.76	5.87	5.98	6.07	6.16	6.25	6.32	6.40	6.47
7	5.41	5.53	5.64	5.74	5.83	5.91	5.99	6.06	6.13	6.19
8	5.25	5.36	5.46	5.56	5.64	5.72	5.80	5.87	5.93	6.00
9	5.13	5.23	5.33	5.42	5.51	5.58	5.66	5.72	5.79	5.85
10	5.03	5.13	5.23	5.32	5.40	5.47	5.54	5.61	5.67	5.73
11	4.95	5.05	5.15	5.23	5.31	5.38	5.45	5.51	5.57	5.63
12	4.89	4.99	5.08	5.16	5.24	5.31	5.37	5.44	5.49	5.55
13	4.83	4.93	5.02	5.10	5.18	5.25	5.31	5.37	5.43	5.48
14	4.79	4.88	4.97	5.05	5.12	5.19	5.26	5.32	5.37	5.43
15	4.75	4.84	4.93	5.01	5.08	5.15	5.21	5.27	5.32	5.38
16	4.71	4.81	4.89	4.97	5.04	5.11	5.17	5.23	5.28	5.33
17	4.68	4.77	4.86	4.93	5.01	5.07	5.13	5.19	5.24	5.30
18	4.65	4.75	4.83	4.90	4.98	5.04	5.10	5.16	5.21	5.26
19	4.63	4.72	4.80	4.88	4.95	5.01	5.07	5.13	5.18	5.23
20	4.61	4.70	4.78	4.85	4.92	4.99	5.05	5.10	5.16	5.20
24	4.54	4.63	4.71	4.78	4.85	4.91	4.97	5.02	5.07	5.12
30	4.47	4.56	4.64	4.71	4.77	4.83	4.89	4.94	4.99	5.03
40	4.41	4.49	4.56	4.63	4.69	4.75	4.81	4.86	4.90	4.95
60	4.34	4.42	4.49	4.56	4.62	4.67	4.73	4.78	4.82	4.86
20	4.28	4.35	4.42	4.48	4.54	4.60	4.65	4.69	4.74	4.78
∞	4.21	4.28	4.35	4.41	4.47	4.52	4.57	4.61	4.65	4.69

VIII.4   SUBSTITUTE *t*-RATIOS

A. The statistic $\tau_1 = \dfrac{\bar{x} - \mu}{w}$, where w is the range of the observations can be used t

test hypotheses about μ. This table gives percentage points of the distribution of τ_1 fo
sample sizes up to 20. The percentage points at the top of the table are upper percentag
points. Lower percentage points are obtained by entering the bottom of the table an
prefixing the tabulated value with a minus sign.

B. The statistic $\tau_d = \dfrac{\bar{x}_1 - \bar{x}_2}{\frac{1}{2}(w_1 + w_2)}$ may be used to test hypotheses about the difference

between means. Percentage points of the distribution of τ_d are given for samples of equa
size up to 20.

C. The statistic $\tau_2 = \dfrac{\frac{1}{2}[x_{(1)} + x_{(n)}] - \mu}{w}$ can be used to test hypotheses about μ

This table gives percentage points of the distribution of τ_2 for samples of size 10 or less

SUBSTITUTE t-RATIOS

A. Percentiles* for $\tau_1 = \dfrac{\bar{x} - \mu}{w}$

Sample size	P_{95}	$P_{97.5}$	P_{99}	$P_{99.5}$	$P_{99.9}$	$P_{99.95}$
2	3.175	6.353	15.910	31.828	159.16	318.31
3	.885	1.304	2.111	3.008	6.77	9.58
4	.529	.717	1.023	1.316	2.29	2.85
5	.388	.507	.685	.843	1.32	1.58
6	.312	.399	.523	.628	.92	1.07
7	.263	.333	.429	.507	.71	.82
8	.230	.288	.366	.429	.59	.67
9	.205	.255	.322	.374	.50	.57
10	.186	.230	.288	.333	.44	.50
11	.170	.210	.262	.302	.40	.44
12	.158	.194	.241	.277	.36	.40
13	.147	.181	.224	.256	.33	.37
14	.138	.170	.209	.239	.31	.34
15	.131	.160	.197	.224	.29	.32
16	.124	.151	.186	.212	.27	.30
17	.118	.144	.177	.201	.26	.28
18	.113	.137	.168	.191	.24	.26
19	.108	.131	.161	.182	.23	.25
20	.104	.126	.154	.175	.22	.24
	$-P_{05}$	$-P_{02.5}$	$-P_{01}$	$-P_{0.5}$	$-P_{0.1}$	$-P_{0.05}$

* When the table is read from the foot, the tabled values are to be prefixed with a negative sign.

SUBSTITUTE *t*-RATIOS

B. Percentiles* for $\tau_d = \dfrac{\bar{x}_1 - \bar{x}_2}{\frac{1}{2}(w_1 + w_2)}$

Sample sizes $N_1 = N_2$	P_{95}	$P_{97.5}$	P_{99}	$P_{99.5}$	$P_{99.9}$	$P_{99.95}$
2	2.322	3.427	5.553	7.916	17.81	25.23
3	.974	1.272	1.715	2.093	3.27	4.18
4	.644	.813	1.047	1.237	1.74	1.99
5	.493	.613	.772	.896	1.21	1.35
6	.405	.499	.621	.714	.94	1.03
7	.347	.426	.525	.600	.77	.85
8	.306	.373	.459	.521	.67	.73
9	.275	.334	.409	.464	.59	.64
10	.250	.304	.371	.419	.53	.58
11	.233	.280	.340	.384	.48	.52
12	.214	.260	.315	.355	.44	.48
13	.201	.243	.294	.331	.41	.45
14	.189	.228	.276	.311	.39	.42
15	.179	.216	.261	.293	.36	.39
16	.170	.205	.247	.278	.34	.37
17	.162	.195	.236	.264	.33	.35
18	.155	.187	.225	.252	.31	.34
19	.149	.179	.216	.242	.30	.32
20	.143	.172	.207	.232	.29	.31
	$-P_{05}$	$-P_{02.5}$	$-P_{01}$	$-P_{0.5}$	$-P_{0.1}$	$-P_{0.05}$

* When the table is read from the foot, the tabled values are to be prefixed with a negative sign.

C. Percentiles for $\tau_2 = \dfrac{\frac{1}{2}[x_{(1)} + x_{(n)}] - \mu}{w}$

Sample size	P_{95}	$P_{97.5}$	P_{99}	$P_{99.5}$
2	3.16	6.35	15.91	31.83
3	.90	1.30	2.11	3.02
4	.55	.74	1.04	1.37
5	.42	.52	.71	.85
6	.35	.43	.56	.66
7	.30	.37	.47	.55
8	.26	.33	.42	.47
9	.24	.30	.38	.42
10	.22	.27	.35	.39
	$-P_{05}$	$-P_{02.5}$	$-P_{01}$	$-P_{0.5}$

VIII.5 SUBSTITUTE F-RATIO

The ratio $F' = w_1/w_2$ of two ranges can be used as a substitute for the ratio of two variances. This table gives percentage points of the ratio of two ranges for respective sample sizes n_1 and n_2 less than or equal to 10. The hypothesis $\sigma_1 = \sigma_2$ is rejected if F' is significantly large or small. To test the hypothesis $\sigma_1 = \sigma_2$ at the α-level of significance, use the critical region $F' < F'_{\frac{1}{2}\alpha}$ and $F' > F'_{1-\frac{1}{2}\alpha}$. For a one-sided test $\sigma_1 \leq \sigma_2$ the α-critical region is $F' > F'_{1-\alpha}$.

Range and Studentized Range

SUBSTITUTE F-RATIO

Sample size for denominator	Cum. prop.	Sample size for numerator								
		2	3	4	5	6	7	8	9	10
2	.005	.0078	.096	.21	.30	.38	.44	.49	.54	.5?
	.01	.0157	.136	.26	.38	.46	.53	.59	.64	.6?
	.025	.039	.217	.37	.50	.60	.68	.74	.79	.8?
	.05	.079	.31	.50	.62	.74	.80	.86	.91	.9?
	.95	12.7	19.1	23	26	29	30	32	34	35
	.975	25.5	38.2	52	57	60	62	64	67	68
	.99	63.7	95	116	132	142	153	160	168	174
	.995	127	191	230	250	260	270	280	290	290
3	.005	.0052	.071	.16	.24	.32	.38	.43	.47	.5?
	.01	.0105	.100	.20	.30	.37	.43	.49	.53	.5?
	.025	.026	.160	.28	.39	.47	.54	.59	.64	.6?
	.05	.052	.23	.37	.49	.57	.64	.70	.75	.8?
	.95	3.19	4.4	5.0	5.7	6.2	6.6	6.9	7.2	7.4
	.975	4.61	6.3	7.3	8.0	8.7	9.3	9.8	10.2	10.5
	.99	7.37	10	12	13	14	15	15	16	17
	.995	10.4	14	17	18	20	21	22	23	25
4	.005	.0043	.059	.14	.22	.28	.34	.39	.43	.4?
	.01	.0086	.084	.18	.26	.33	.39	.44	.48	.5?
	.025	.019	.137	.25	.34	.42	.48	.53	.57	.6?
	.05	.043	.20	.32	.42	.50	.57	.62	.67	.7?
	.95	2.02	2.7	3.1	3.4	3.6	3.8	4.0	4.2	4.4
	.975	2.72	3.5	4.0	4.4	4.7	5.0	5.2	5.4	5.6
	.99	3.83	5.0	5.5	6.0	6.4	6.7	7.0	7.2	7.5
	.995	4.85	6.1	7.0	7.6	8.1	8.5	8.8	9.3	9.6
5	.005	.0039	.054	.13	.20	.26	.32	.36	.40	.4?
	.01	.0076	.079	.17	.24	.31	.36	.41	.45	.4?
	.025	.018	.124	.23	.32	.38	.44	.49	.53	.5?
	.05	.038	.18	.29	.40	.46	.52	.57	.61	.6?
	.95	1.61	2.1	2.4	2.6	2.8	2.9	3.0	3.1	3.2
	.975	2.01	2.6	2.9	3.2	3.4	3.6	3.7	3.8	3.9?
	.99	2.64	3.4	3.8	4.1	4.3	4.6	4.7	4.9	5.0?
	.995	3.36	4.1	4.6	4.9	5.2	5.5	5.7	5.9	6.1
6	.005	.0038	.051	.12	.19	.25	.30	.35	.38	.4?
	.01	.0070	.073	.16	.23	.29	.34	.39	.43	.4?
	.025	.017	.115	.21	.30	.36	.42	.46	.50	.5?
	.05	.035	.16	.27	.36	.43	.49	.54	.58	.6?
	.95	1.36	1.8	2.0	2.2	2.3	2.4	2.5	2.6	2.7
	.975	1.67	2.1	2.4	2.6	2.8	2.9	3.0	3.1	3.2
	.99	2.16	2.7	3.0	3.2	3.4	3.6	3.7	3.8	3.9
	.995	2.67	3.1	3.5	3.8	4.0	4.1	4.3	4.5	4.6

SUBSTITUTE *F*-RATIO

nple e for omi-tor	Cum. prop.	Sample size for numerator								
		2	3	4	5	6	7	8	9	10
7	.005	.0037	.048	.12	.18	.24	.29	.33	.37	.40
	.01	.0066	.069	.15	.22	.28	.33	.37	.41	.45
	.025	.016	.107	.20	.28	.34	.40	.44	.48	.52
	.05	.032	.15	.26	.35	.41	.47	.51	.55	.59
	.95	1.26	1.6	1.8	1.9	2.0	2.1	2.2	2.3	2.4
	.975	1.48	1.9	2.1	2.3	2.4	2.5	2.6	2.7	2.8
	.99	1.87	2.3	2.6	2.8	2.9	3.0	3.1	3.2	3.3
	.995	2.28	2.7	2.9	2.9	3.3	3.5	3.6	3.7	3.8
8	.005	.0036	.045	.11	.18	.23	.28	.32	.36	.39
	.01	.0063	.065	.14	.21	.27	.32	.36	.40	.43
	.025	.016	.102	.19	.27	.33	.38	.43	.47	.50
	.05	.031	.14	.25	.33	.40	.45	.50	.53	.57
	.95	1.17	1.4	1.6	1.8	1.9	1.9	2.0	2.1	2.1
	.975	1.36	1.7	1.9	2.0	2.2	2.3	2.3	2.4	2.5
	.99	1.69	2.1	2.3	2.4	2.6	2.7	2.8	2.8	2.9
	.995	2.03	2.3	2.6	2.7	2.9	3.0	3.1	3.2	3.3
9	.005	.0035	.042	.11	.17	.22	.27	.31	.35	.38
	.01	.0060	.062	.14	.21	.26	.31	.35	.39	.42
	.025	.015	.098	.18	.26	.32	.37	.42	.46	.49
	.05	.030	.14	.24	.32	.38	.44	.48	.52	.55
	.95	1.10	1.3	1.5	1.6	1.7	1.8	1.9	1.9	2.0
	.975	1.27	1.6	1.8	1.9	2.0	2.1	2.1	2.2	2.3
	.99	1.56	1.9	2.1	2.2	2.3	2.4	2.5	2.6	2.6
	.995	1.87	2.1	2.3	2.5	2.6	2.7	2.8	2.9	3.0
10	.005	.0034	.041	.10	.16	.22	.26	.30	.34	.37
	.01	.0058	.060	.13	.20	.26	.30	.34	.38	.41
	.025	.015	.095	.18	.25	.31	.36	.41	.44	.48
	.05	.029	.13	.23	.31	.37	.43	.47	.51	.54
	.95	1.05	1.3	1.4	1.5	1.6	1.7	1.8	1.8	1.9
	.975	1.21	1.5	1.6	1.8	1.9	1.9	2.0	2.0	2.1
	.99	1.47	1.8	1.9	2.1	2.2	2.2	2.3	2.4	2.4
	.995	1.75	2.0	2.2	2.3	2.4	2.5	2.6	2.6	2.7

VIII.6   ANALYSIS OF VARIANCE BASED ON RANGE

Standard tests of significance in the analysis of variance are often F-tests based on t ratio of a "treatment" mean square to an error mean square s^2. If the treatment mea $\bar{x}_t$, $(t = 1, 2, \ldots, k)$, are all calculated from the same number of observations n possible alternative criterion is

$$(1) \qquad\qquad \sqrt{n} \text{ range } \bar{x}_t/s \; .$$

The independence of numerator and denominator may be proved, so that (1) is a stude ized range q. In an overall analysis of variance its use would save very little work; howe in procedures for ranking treatment means it plays a fundamental role.

For the case of a one-way classification a computationally simple criterion is obtai if s in (1) is replaced by $\dfrac{\bar{w}}{c}$. The ratio

$$(2) \qquad\qquad c \sqrt{n} \text{ range } \bar{x}_t/\bar{w}$$

is then distributed approximately as q with degrees of freedom and scale factor c obtai from table (ii). Simpler still, an immediate test can be made with the help of table which gives upper 5% and 1% points of

$$Q = \text{range } X_t/W,$$

where

$$X_t = \sum_{i=1}^{n} x_{ti} \quad \text{and} \quad W = \sum_{t=1}^{k} w_t \; .$$

(i) UPPER PERCENTAGE POINTS OF Q = RANGE OF GROUP TOTALS/SUM OF GROUP RANGES IN A ONE-WAY CLASSIFICATION INTO k GROUPS OF n OBSERVATIONS

k \ n	2 5%	2 1%	3 5%	3 1%	4 5%	4 1%	5 5%	5 1%	6 5%	6 1%	7 5%	7 1%	8 5%	8 1%	9 5%	9 1%	10 5%	10 1%
2	3.5	8.3	1.91	3.2	1.63	2.5	1.54	2.3	1.50	2.2	1.49	2.1	1.49	2.1	1.51	2.1	1.52	2.1
3	2.4	4.4	1.44	2.1	1.26	1.74	1.19	1.61	1.17	1.55	1.17	1.53	1.17	1.53	1.19	1.54	1.20	1.55
4	1.75	2.9	1.14	1.57	1.01	1.33	0.97	1.24	0.95	1.21	0.95	1.20	0.96	1.20	0.97	1.21	0.98	1.22
5	1.40	2.1	0.94	1.25	0.84	1.07	.81	1.01	.80	1.00	.80	0.99	.81	0.99	.82	1.00	.83	1.01
6	1.16	1.68	.80	1.04	.72	0.90	.70	0.86	.69	0.84	.69	0.84	.70	0.84	.71	0.85	.72	0.86
7	1.00	1.39	.70	0.89	.64	.78	.61	.75	.61	.74	.61	.74	.62	.74	.62	.74	.63	.75
8	0.87	1.18	.62	.78	.57	.69	.55	.66	.55	.65	.55	.66	.56	.66	.56	.67	.57	.67
9	.78	1.03	.56	.69	.51	.62	.50	.59	.50	.59	.50	.59	.50	.59	.51	.60	.52	.61
10	.70	0.91	.51	.62	.47	.56	.46	.54	.45	.53	.45	.53	.46	.54	.47	.55	.47	.55

(ii) SCALE FACTOR, c, AND EQUIVALENT DEGREES OF FREEDOM, ν, APPROPRIATE TO A ONE-WAY CLASSIFICATION INTO k GROUPS OF n OBSERVATIONS

k \ n	2 ν	2 c	3 ν	3 c	4 ν	4 c	5 ν	5 c	6 ν	6 c	7 ν	7 c	8 ν	8 c	9 ν	9 c	10 ν	10 c
1	1.00	1.41	1.98	1.91	2.93	2.24	3.83	2.48	4.68	2.67	5.48	2.83	6.25	2.96	6.98	3.08	7.68	3.18
2	1.92	1.28	3.83	1.81	5.69	2.15	7.47	2.40	9.16	2.60	10.8	2.77	12.3	2.91	13.8	3.02	15.1	3.13
3	2.82	1.23	5.66	1.77	8.44	2.12	11.1	2.38	13.6	2.58	16.0	2.75	18.3	2.89	20.5	3.01	22.6	3.11
4	3.71	1.21	7.49	1.75	11.2	2.11	14.7	2.37	18.1	2.57	21.3	2.74	24.4	2.88	27.3	3.00	30.1	3.10
5	4.59	1.19	9.30	1.74	13.9	2.10	18.4	2.36	22.6	2.56	26.6	2.73	30.4	2.87	34.0	2.99	37.5	3.10
6	5.47	1.18	11.1	1.73	16.7	2.09	22.0	2.35	27.0	2.56	31.8	2.73	36.4	2.87	40.8	2.99	45.0	3.09
7	6.35	1.17	12.9	1.73	19.4	2.09	25.6	2.35	31.5	2.55	37.1	2.72	42.5	2.86	47.6	2.99	52.4	3.09
8	7.23	1.17	14.8	1.72	22.1	2.08	29.2	2.35	36.0	2.55	42.4	2.72	48.5	2.86	54.3	2.98	59.9	3.09
9	8.11	1.16	16.6	1.72	24.9	2.08	32.9	2.34	40.4	2.55	47.6	2.72	54.5	2.86	61.1	2.98	67.3	3.09
10	8.99	1.16	18.4	1.72	27.6	2.08	36.5	2.34	44.9	2.55	52.9	2.72	60.6	2.86	67.8	2.98	74.8	3.09
d_n		1.13		1.69		2.06		2.33		2.53		2.70		2.85		2.97		3.08
C.D.	0.88		1.82		2.74		3.62		4.47		5.27		6.03		6.76		7.45	

N.B.: C.D. = constant difference

IX. Correlation Coefficient

IX.1 PERCENTAGE POINTS, DISTRIBUTION OF THE CORRELATION COEFFICIENT, WHEN $\rho = 0$

The bivariate normal probability function is given by

$$f(x,y) = \frac{1}{2\pi\sigma_x\sigma_y \sqrt{1 - \rho^2}} \exp\left\{- \frac{1}{2(1 - \rho^2)}\right.$$

$$\left. \left[\left(\frac{x - \mu_x}{\sigma_x}\right)^2 - 2\rho\left(\frac{x - \mu_x}{\sigma_x}\right)\left(\frac{y - \mu_y}{\sigma_y}\right) + \left(\frac{y - \mu_y}{\sigma_y}\right)^2\right]\right\}$$

where μ_x = mean of x
μ_y = mean of y
σ_x = standard deviation of x
σ_y = standard deviation of y
ρ = correlation coefficient between x and y.

If (x_i, y_i) $(i = 1, 2, \ldots, n)$ denote a random sample of n ordered observations drawn from a bivariate normal distribution, then an estimate of ρ is given by the sample product moment correlation coefficient r, given by

$$r = \frac{\sum_i (x_i - \bar{x})(y_i - \bar{y})}{\sqrt{\sum_i (x_i - \bar{x})^2 \cdot \sum_i (y_i - \bar{y})^2}} .$$

The frequency function of r is given by

$$f(r) = \frac{(1 - \rho^2)^{\frac{n-1}{2}}}{\pi(n - 3)!} (1 - r^2)^{\frac{n-4}{2}} \frac{d^{n-2}}{d(r\rho)^{n-2}} \left[\frac{\text{Arccos} (-r\rho)}{\sqrt{1 - r^2\rho^2}}\right]$$

which can also be written as

$$f(r) = \frac{(1 - \rho^2)^{\frac{n-1}{2}} (1 - r^2)^{\frac{n-4}{2}}}{\sqrt{\pi} \, \Gamma\left(\frac{n - 1}{2}\right) \Gamma\left(\frac{n - 2}{2}\right)} \sum_{i=0}^{\infty} \frac{(2r\rho)^i}{i!} \Gamma^2\left(\frac{n - 1 + i}{2}\right) .$$

In the special case where $\rho = 0$, the frequency function of r becomes

$$f(r) = \frac{\Gamma\left(\frac{n - 1}{2}\right)}{\sqrt{\pi} \, \Gamma\left(\frac{n}{2} - 1\right)} (1 - r^2)^{\frac{n-4}{2}} .$$

Under the transformation

$$r^2 = \frac{t^2}{t^2 + \nu} ,$$

$f(r)$ is transformed into the t-distribution with $\nu = n - 2$ degrees of freedom. This table gives percentage points of the distribution of the correlation coefficient when $\rho = 0$.

PERCENTAGE POINTS, DISTRIBUTION OF THE CORRELATION COEFFICIENT,
WHEN $\rho = 0$

$\Pr\{r \leq \text{tabular value}|\rho = 0\} = \alpha$

α = 0.05 2α = 0.1	0.025 0.05	0.01 0.02	0.005 0.01	0.0025 0.005	0.0005 0.001	ν	α = 0.05 2α = 0.1	0.025 0.05	0.01 0.02	0.005 0.01	0.0025 0.005	0.0005 0.001
0.9877	$0.9^{2}692$	$0.9^{3}507$	$0.9^{3}877$	$0.9^{4}692$	$0.9^{5}877$	16	0.400	0.468	0.543	0.590	0.631	0.708
.9000	.9500	.9800	$.9^{2}000$	$.9^{2}500$	$.9^{3}000$	17	.389	.456	.529	.575	.616	.693
.805	.878	.9343	.9587	.9740	$.9^{2}114$	18	.378	.444	.516	.561	.602	.679
.729	.811	.882	.9172	.9417	.9741	19	.369	.433	.503	.549	.589	.665
.669	.754	.833	.875	.9056	.9509	20	.360	.423	.492	.537	.576	.652
0.621	0.707	0.789	0.834	0.870	0.9249	25	0.323	0.381	0.445	0.487	0.524	0.597
.582	.666	.750	.798	.836	.898	30	.296	.349	.409	.449	.484	.554
.549	.632	.715	.765	.805	.872	35	.275	.325	.381	.418	.452	.519
.521	.602	.685	.735	.776	.847	40	.257	.304	.358	.393	.425	.490
.497	.576	.658	.708	.750	.823	45	.243	.288	.338	.372	.403	.465
0.476	0.553	0.634	0.684	0.726	0.801	50	0.231	0.273	0.322	0.354	0.384	0.443
.457	.532	.612	.661	.703	.780	60	.211	.250	.295	.325	.352	.408
.441	.514	.592	.641	.683	.760	70	.195	.232	.274	.302	.327	.380
.426	.497	.574	.623	.664	.742	80	.183	.217	.257	.283	.307	.357
.412	.482	.558	.606	.647	.725	90	.173	.205	.242	.267	.290	.338
						100	.164	.195	.230	.254	.276	.321

$\alpha = 1 - F(r|\nu, \rho = 0)$ is the upper-tail area of the distribution of r appropriate for use in a single-tail For a two-tail test, 2α must be used. If r is calculated from n paired observations, enter the table $\nu = n - 2$. For partial correlations enter with $\nu = n - k - 2$, where k is the number of variables constant.

Correlation Coefficient

IX.2 CONFIDENCE LIMITS FOR THE POPULATION CORRELATION COEFFICIENT

The cumulative distribution of the correlation coefficient is given by

$$F(r;n,\rho) = \int_{-1}^{r} \frac{(1 - \rho^2)^{\frac{n-1}{2}}}{\pi(n - 3)!} (1 - u^2)^{\frac{n-4}{2}} \frac{d^{n-2}}{d(u\rho)^{n-2}} \left[\frac{\text{Arccos } (-u\rho)}{\sqrt{1 - u^2\rho^2}} \right] du .$$

This table shows graphically the roots ρ_1 and $\rho_2(\rho_2 > \rho_1)$ of

$$\alpha = F(r;n,\rho_2) \quad \text{and} \quad 1 - \alpha = F(r;n,\rho_1)$$

plotted against r for selected sample sizes n for the values $\alpha = 0.025$ and 0.005.

GRAPHS SHOWING CONFIDENCE LIMITS FOR THE POPULATION CORRELATION COEFFICIENT, ρ, GIVEN THE SAMPLE COEFFICIENT, r. CONFIDENCE COEFFICIENT, $1 - 2\alpha = 0.95$

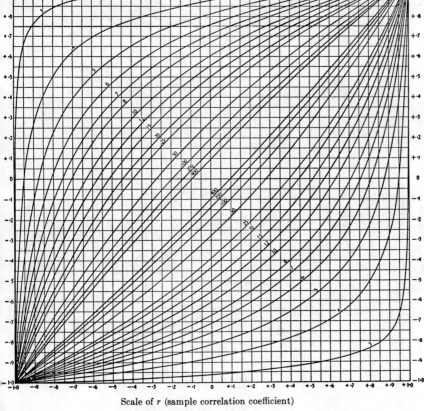

Scale of r (sample correlation coefficient)

numbers on the curves indicate sample size. The chart can also be used to determine upper and lower 2.5% significance points for r, given ρ.

158 *Correlation Coefficient*

GRAPHS SHOWING CONFIDENCE LIMITS FOR THE POPULATION CORRELATION COEFFICIENT, ρ, GIVEN THE SAMPLE COEFFICIENT, r. CONFIDENCE COEFFICIENT, $1 - 2\alpha = 0.99$

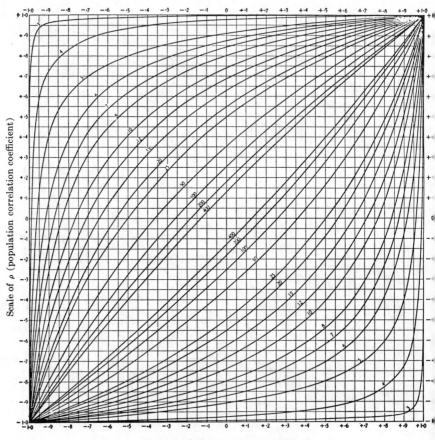

Scale of r (sample correlation coefficient)

The numbers on the curves indicate sample size. The chart can also be used to determine upper and low 0.5% significance points for r, given ρ.

IX.3 THE TRANSFORMATION $Z = \text{TANH}^{-1}\ r$ FOR THE CORRELATION COEFFICIENT

If one introduces the transformation

$$Z = \frac{1}{2}\log_e \frac{1+r}{1-r} = \tanh^{-1} r,$$

then Z is approximately normally distributed with mean

$$\frac{1}{2}\log_e \frac{1+\rho}{1-\rho} = \tanh^{-1} \rho$$

and variance

$$\frac{1}{n-3}.$$

This table gives the function $Z = \tanh^{-1} r$. Methods for interpolation in the table are given following the table.

Correlation Coefficient

THE TRANSFORMATION $Z = \tanh^{-1} r$ FOR THE CORRELATION COEFFICIENT

r	.000	.002	.004	.006	.008	1 2 3 4 5 6 7 8 9 10	.000	.002	.004	.006	.008	r
.00	.0000	.0020	.0040	.0060	.0080	1 3 4 5 7 8 9 11 12 13	.5493	.5520	.5547	.5573	.5600	.50
1	.0100	.0120	.0140	.0160	.0180	1 3 4 5 7 8 10 11 12 14	.5627	.5654	.5682	.5709	.5736	1
2	.0200	.0220	.0240	.0260	.0280	1 3 4 6 7 8 10 11 13 14	.5763	.5791	.5818	.5846	.5874	2
3	.0300	.0320	.0340	.0360	.0380	1 3 4 6 7 8 10 11 13 14	.5901	.5929	.5957	.5985	.6013	3
4	.0400	.0420	.0440	.0460	.0480	1 3 4 6 7 9 10 11 13 14	.6042	.6070	.6098	.6127	.6155	4
.05	.0500	.0520	.0541	.0561	.0581	1 3 4 6 7 9 10 12 13 14	.6184	.6213	.6241	.6270	.6299	.55
6	.0601	.0621	.0641	.0661	.0681	1 3 4 6 7 9 10 12 13 15	.6328	.6358	.6387	.6416	.6446	6
7	.0701	.0721	.0741	.0761	.0782	1 3 4 6 7 9 10 12 14 15	.6475	.6505	.6535	.6565	.6595	7
8	.0802	.0822	.0842	.0862	.0882	2 3 5 6 8 9 11 12 14 15	.6625	.6655	.6685	.6716	.6746	8
9	.0902	.0923	.0943	.0963	.0983	2 3 5 6 8 9 11 12 14 15	.6777	.6807	.6838	.6869	.6900	9
.10	.1003	.1024	.1044	.1064	.1084	2 3 5 6 8 9 11 13 14 16	.6931	.6963	.6994	.7026	.7057	.60
1	.1104	.1125	.1145	.1165	.1186	2 3 5 6 8 10 11 13 14 16	.7089	.7121	.7153	.7185	.7218	1
2	.1206	.1226	.1246	.1267	.1287	2 3 5 7 8 10 11 13 15 16	.7250	.7283	.7315	.7348	.7381	2
3	.1307	.1328	.1348	.1368	.1389	2 3 5 7 8 10 12 13 15 17	.7414	.7447	.7481	.7514	.7548	3
4	.1409	.1430	.1450	.1471	.1491	2 3 5 7 9 10 12 14 15 17	.7582	.7616	.7650	.7684	.7718	4
.15	.1511	.1532	.1552	.1573	.1593	2 4 5 7 9 11 12 14 16 18	.7753	.7788	.7823	.7858	.7893	.65
6	.1614	.1634	.1655	.1676	.1696	2 4 5 7 9 11 13 14 16 18	.7928	.7964	.7999	.8035	.8071	6
7	.1717	.1737	.1758	.1779	.1799	2 4 6 7 9 11 13 15 17 18	.8107	.8144	.8180	.8217	.8254	7
8	.1820	.1841	.1861	.1882	.1903	2 4 6 8 9 11 13 15 17 19	.8291	.8328	.8366	.8404	.8441	8
9	.1923	.1944	.1965	.1986	.2007	2 4 6 8 10 12 14 15 17 19	.8480	.8518	.8556	.8595	.8634	9
.20	.2027	.2048	.2069	.2090	.2111	2 4 6 8 10 12 14 16 18 20	.8673	.8712	.8752	.8792	.8832	.70
1	.2132	.2153	.2174	.2195	.2216	2 4 6 8 10 12 15 16 18 20	.8872	.8912	.8953	.8994	.9035	1
2	.2237	.2258	.2279	.2300	.2321	2 4 6 8 11 13 15 17 19 21	.9076	.9118	.9160	.9202	.9245	2
3	.2342	.2363	.2384	.2405	.2427	2 4 7 9 11 13 15 17 20 22	.9287	.9330	.9373	.9417	.9461	3
4	.2448	.2469	.2490	.2512	.2533	2 4 7 9 11 13 16 18 20 22	.9505	.9549	.9594	.9639	.9684	4
r	.000	.002	.004	.006	.008	1 2 3 4 5 6 7 8 9 10	.000	.002	.004	.006	.008	r

r (3rd decimal)   ←Proportional parts, for left side   r (3rd decimal)

Interpolation

(1) $0 \leq r \leq 0.25$: find argument r_0 nearest to r and form $z = z(r_0) + \Delta r$ (where $\Delta r = r - r_0$), e.g. for $r = 0.2042$, $z = 0.2069 + 0.0002 = 0.2071$.

(2) $0.25 \leq r \leq 0.75$: find argument r_0 nearest to r and form $z = z(r_0) \pm P$, where P is the proportional part for $\Delta r = r - r_0$, e.g. for $r = 0.5146$, $z = 0.5682 + 0.0008 = 0.5690$; for $r = 0.5372$ $z = 0.6013 - 0.0011 = 0.6002$.

(3) $0.75 \leq r \leq 0.98$: use linear interpolation to get 3-decimal place accuracy.

(4) $0.98 \leq r < 1$: form $z = -\frac{1}{2}\log_e (1-r) + 0.097 + \frac{1}{4}r$, with the help of table of natural logarithms.

THE TRANSFORMATION $Z = \mathrm{TANH}^{-1}\ r$ FOR THE CORRELATION COEFFICIENT

r	.000	.002	.004	.006	.008	Proportional parts, for right side→ (1 2 3 4 5 6 7 8 9 10)	.000	.002	.004	.006	.008	r
.25	.2554	.2575	.2597	.2618	.2640	1 2 3 4 5 6 7 9 10 11	0.973	0.978	0.982	0.987	0.991	.75
6	.2661	.2683	.2704	.2726	.2747	1 2 3 4 5 6 8 9 10 11	0.996	1.001	1.006	1.011	1.015	6
7	.2769	.2790	.2812	.2833	.2855	1 2 3 4 5 6 8 9 10 11	1.020	1.025	1.030	1.035	1.040	7
8	.2877	.2899	.2920	.2942	.2964	1 2 3 4 5 7 8 9 10 11	1.045	1.050	1.056	1.061	1.066	8
9	.2986	.3008	.3029	.3051	.3073	1 2 3 4 5 7 8 9 10 11	1.071	1.077	1.082	1.088	1.093	9
.30	.3095	.3117	.3139	.3161	.3183	1 2 3 4 6 7 8 9 10 11	1.099	1.104	1.110	1.116	1.121	.80
1	.3205	.3228	.3250	.3272	.3294	1 2 3 4 6 7 8 9 10 11	1.127	1.133	1.139	1.145	1.151	1
2	.3316	.3339	.3361	.3383	.3406	1 2 3 4 6 7 8 9 10 11	1.157	1.163	1.169	1.175	1.182	2
3	.3428	.3451	.3473	.3496	.3518	1 2 3 5 6 7 8 9 10 11	1.188	1.195	1.201	1.208	1.214	3
4	.3541	.3564	.3586	.3609	.3632	1 2 3 5 6 7 8 9 10 11	1.221	1.228	1.235	1.242	1.249	4
.35	.3654	.3677	.3700	.3723	.3746	1 2 3 5 6 7 8 9 10 11	1.256	1.263	1.271	1.278	1.286	.85
6	.3769	.3792	.3815	.3838	.3861	1 2 3 5 6 7 8 9 10 12	1.293	1.301	1.309	1.317	1.325	6
7	.3884	.3907	.3931	.3954	.3977	1 2 3 5 6 7 8 9 10 12	1.333	1.341	1.350	1.358	1.367	7
8	.4001	.4024	.4047	.4071	.4094	1 2 4 5 6 7 8 9 11 12	1.376	1.385	1.394	1.403	1.412	8
9	.4118	.4142	.4165	.4189	.4213	1 2 4 5 6 7 8 9 11 12	1.422	1.432	1.442	1.452	1.462	9
.40	.4236	.4260	.4284	.4308	.4332	1 2 4 5 6 7 8 10 11 12	1.472	1.483	1.494	1.505	1.516	.90
1	.4356	.4380	.4404	.4428	.4453	1 2 4 5 6 7 8 10 11 12	1.528	1.539	1.551	1.564	1.576	1
2	.4477	.4501	.4526	.4550	.4574	1 2 4 5 6 7 9 10 11 12	1.589	1.602	1.616	1.630	1.644	2
3	.4599	.4624	.4648	.4673	.4698	1 2 4 5 6 7 9 10 11 12	1.658	1.673	1.689	1.705	1.721	3
4	.4722	.4747	.4772	.4797	.4822	1 2 4 5 6 7 9 10 11 12	1.738	1.756	1.774	1.792	1.812	4
.45	.4847	.4872	.4897	.4922	.4948	1 3 4 5 6 8 9 10 11 13	1.832	1.853	1.874	1.897	1.921	.95
6	.4973	.4999	.5024	.5049	.5075	1 3 4 5 6 8 9 10 11 13	1.946	1.972	2.000	2.029	2.060	6
7	.5101	.5126	.5152	.5178	.5204	1 3 4 5 6 8 9 10 12 13	2.092	2.127	2.165	2.205	2.249	7
8	.5230	.5256	.5282	.5308	.5334	1 3 4 5 7 8 9 10 12 13	2.298	2.351	2.410	2.477	2.555	8
9	.5361	.5387	.5413	.5440	.5466	1 3 4 5 7 8 9 11 12 13	2.647	2.759	2.903	3.106	3.453	9
r	.000	.002	.004	.006	.008	←Proportional parts, for left side (1 2 3 4 5 6 7 8 9 10)	.000	.002	.004	.006	.008	r

r (3rd decimal) *r* (3rd decimal)

Interpolation

(1) $0 \leq r \leq 0.25$: find argument r_0 nearest to r and form $z = z(r_0) + \Delta r$ (where $\Delta r = r - r_0$), e.g. or $r = 0.2042$, $z = 0.2069 + 0.0002 = 0.2071$.

(2) $0.25 \leq r \leq 0.75$: find argument r_0 nearest to r and form $z = z(r_0) \pm P$, where P is the proportional part for $\Delta r = r - r_0$, e.g. for $r = 0.5146$, $z = 0.5682 + 0.0008 = 0.5690$; for $r = 0.5372$, $= 0.6013 - 0.0011 = 0.6002$.

(3) $0.75 \leq r \leq 0.98$: use linear interpolation to get 3-decimal place accuracy.

(4) $0.98 \leq r < 1$: form $z = -\frac{1}{2} \log_e (1 - r) + 0.097 + \frac{1}{4} r$, with the help of table of natural logarithms.

X. Non-Parametric Statistics

X.1 CRITICAL VALUES FOR THE SIGN TEST

The observations in a random sample of size n from X and those of the same size from Y are paired according to the order of observation: (X_i, Y_i), $i = 1, 2, \ldots, n$. The differences $d_i = X_i - Y_i$ are calculated for each of the n pairs. The null hypothesis is that the difference d_i has a distribution with median zero, i.e., the true proportion of positive (negative) signs is equal to $p = \frac{1}{2}$. Thus the test is whether X and Y have the same median. The probability of x positive (negative) signs is given by the binomial probability function

$$f(x) = f(x;n, \ p = \tfrac{1}{2}) = \binom{n}{x}\left(\frac{1}{2}\right)^n .$$

This table gives the critical value k such that

$$P(x \leq k) = \sum_{x=0}^{k} \binom{n}{x}\left(\frac{1}{2}\right)^n < \frac{\alpha}{2} .$$

For a one-tailed test with significance level α enter the table in the column headed by 2α.

CRITICAL VALUES FOR THE SIGN TEST

(Two-tail percentage points for the binomial for $p = .5$)

n	1%	5%	10%	25%	n	1%	5%	10%	25%
1					46	13	15	16	18
2					47	14	16	17	19
3				0	48	14	16	17	19
4				0	49	15	17	18	19
5			0	0	50	15	17	18	20
6		0	0	1	51	15	18	19	20
7		0	0	1	52	16	18	19	21
8	0	0	1	1	53	16	18	20	21
9	0	1	1	2	54	17	19	20	22
10	0	1	1	2	55	17	19	20	22
11	0	1	2	3	56	17	20	21	23
12	1	2	2	3	57	18	20	21	23
13	1	2	3	3	58	18	21	22	24
14	1	2	3	4	59	19	21	22	24
15	2	3	3	4	60	19	21	23	25
16	2	3	4	5	61	20	22	23	25
17	2	4	4	5	62	20	22	24	25
18	3	4	5	6	63	20	23	24	26
19	3	4	5	6	64	21	23	24	26
20	3	5	5	6	65	21	24	25	27
21	4	5	6	7	66	22	24	25	27
22	4	5	6	7	67	22	25	26	28
23	4	6	7	8	68	22	25	26	28
24	5	6	7	8	69	23	25	27	29
25	5	7	7	9	70	23	26	27	29
26	6	7	8	9	71	24	26	28	30
27	6	7	8	10	72	24	27	28	30
28	6	8	9	10	73	25	27	28	31
29	7	8	9	10	74	25	28	29	31
30	7	9	10	11	75	25	28	29	32
31	7	9	10	11	76	26	28	30	32
32	8	9	10	12	77	26	29	30	32
33	8	10	11	12	78	27	29	31	33
34	9	10	11	13	79	27	30	31	33
35	9	11	12	13	80	28	30	32	34
36	9	11	12	14	81	28	31	32	34
37	10	12	13	14	82	28	31	33	35
38	10	12	13	14	83	29	32	33	35
39	11	12	13	15	84	29	32	33	36
40	11	13	14	15	85	30	32	34	36
41	11	13	14	16	86	30	33	34	37
42	12	14	15	16	87	31	33	35	37
43	12	14	15	17	88	31	34	35	38
44	13	15	16	17	89	31	34	36	38
45	13	15	16	18	90	32	35	36	39

For values of n larger than 90, approximate values of r may be found by taking the nearest integer ess than $(n - 1)/2 - k \sqrt{n + 1}$, where k is 1.2879, 0.9800, 0.8224, 0.5752 for the 1, 5, 10, 25% values, espectively.

Non-Parametric Statistics

X.2 CRITICAL VALUES OF T IN THE WILCOXON MATCHED-PAIRS SIGNED-RANKS TEST

Let d_i denote the difference score for any matched pair of a set of n pairs of observations: $d_i = x_i - y_i$. Rank all the d_i's without regard to sign: give the rank of 1 to the smallest d_i, the rank of 2 to the next smallest, etc. After the ranking is completed, affix the sign of the difference to each rank. Let T equal the smaller sum of the like-signed ranks. This table gives approximate 1%, 2%, and 5% points of T for various values of n. The hypothesis tested is that there is no difference between the distributions of x and y. The table is adapted for use with both one-tailed and two-tailed tests.

CRITICAL VALUES OF T IN THE WILCOXON MATCHED-PAIRS
SIGNED-RANKS TEST
$$n = 5(1)50$$

One-sided	Two-sided	$n = 5$	$n = 6$	$n = 7$	$n = 8$	$n = 9$	$n = 10$
$P = .05$	$P = .10$	1	2	4	6	8	11
$P = .025$	$P = .05$		1	2	4	6	8
$P = .01$	$P = .02$			0	2	3	5
$P = .005$	$P = .01$				0	2	3

One-sided	Two-sided	$n = 11$	$n = 12$	$n = 13$	$n = 14$	$n = 15$	$n = 16$
$P = .05$	$P = .10$	14	17	21	26	30	36
$P = .025$	$P = .05$	11	14	17	21	25	30
$P = .01$	$P = .02$	7	10	13	16	20	24
$P = .005$	$P = .01$	5	7	10	13	16	19

One-sided	Two-sided	$n = 17$	$n = 18$	$n = 19$	$n = 20$	$n = 21$	$n = 22$
$P = .05$	$P = .10$	41	47	54	60	68	75
$P = .025$	$P = .05$	35	40	46	52	59	66
$P = .01$	$P = .02$	28	33	38	43	49	56
$P = .005$	$P = .01$	23	28	32	37	43	49

One-sided	Two-sided	$n = 23$	$n = 24$	$n = 25$	$n = 26$	$n = 27$	$n = 28$
$P = .05$	$P = .10$	83	92	101	110	120	130
$P = .025$	$P = .05$	73	81	90	98	107	117
$P = .01$	$P = .02$	62	69	77	85	93	102
$P = .005$	$P = .01$	55	61	68	76	84	92

One-sided	Two-sided	$n = 29$	$n = 30$	$n = 31$	$n = 32$	$n = 33$	$n = 34$
$P = .05$	$P = .10$	141	152	163	175	188	201
$P = .025$	$P = .05$	127	137	148	159	171	183
$P = .01$	$P = .02$	111	120	130	141	151	162
$P = .005$	$P = .01$	100	109	118	128	138	149

One-sided	Two-sided	$n = 35$	$n = 36$	$n = 37$	$n = 38$	$n = 39$	
$P = .05$	$P = .10$	214	228	242	256	271	
$P = .025$	$P = .05$	195	208	222	235	250	
$P = .01$	$P = .02$	174	186	198	211	224	
$P = .005$	$P = .01$	160	171	183	195	208	

One-sided	Two-sided	$n = 40$	$n = 41$	$n = 42$	$n = 43$	$n = 44$	$n = 45$
$P = .05$	$P = .10$	287	303	319	336	353	371
$P = .025$	$P = .05$	264	279	295	311	327	344
$P = .01$	$P = .02$	238	252	267	281	297	313
$P = .005$	$P = .01$	221	234	248	262	277	292

One-sided	Two-sided	$n = 46$	$n = 47$	$n = 48$	$n = 49$	$n = 50$	
$P = .05$	$P = .10$	389	408	427	446	466	
$P = .025$	$P = .05$	361	379	397	415	434	
$P = .01$	$P = .02$	329	345	362	380	398	
$P = .005$	$P = .01$	307	323	339	356	373	

Non-Parametric Statistics

X.3 PROBABILITIES FOR THE WILCOXON (MANN-WHITNEY) TWO-SAMPLE STATISTIC

Given two samples of size m and n, $m \leq n$, the Mann-Whitney U-Statistic is used to test the hypothesis that the two samples are from populations with the same median. Rank all the observations in ascending order of magnitude. Let T be the sum of the ranks assigned to the sample of size m. Then U is defined as

$$U = mn + \frac{m(m + 1)}{2} - T .$$

This table is used to determine the exact probability associated with the occurrence under the null hypothesis of any U as extreme as an observed value of U.

The probabilities given in this table are one-tailed. For a two-tailed test, the value of p given in the table should be doubled. The table is made up of six separate subtables, one for each value of n.

PROBABILITIES ASSOCIATED WITH VALUES AS SMALL AS OBSERVED VALUES
OF U IN THE MANN-WHITNEY TEST

$n_2 = 3$

U \ n_1	1	2	3
0	.250	.100	.050
1	.500	.200	.100
2	.750	.400	.200
3		.600	.350
4			.500
5			.650

$n_2 = 4$

U \ n_1	1	2	3	4
0	.200	.067	.028	.014
1	.400	.133	.057	.029
2	.600	.267	.114	.057
3		.400	.200	.100
4		.600	.314	.171
5			.429	.243
6			.571	.343
7				.443
8				.557

$n_2 = 5$

U \ n_1	1	2	3	4	5
0	.167	.047	.018	.008	.004
1	.333	.095	.036	.016	.008
2	.500	.190	.071	.032	.016
3	.667	.286	.125	.056	.028
4		.429	.196	.095	.048
5		.571	.286	.143	.075
6			.393	.206	.111
7			.500	.278	.155
8			.607	.365	.210
9				.452	.274
10				.548	.345
11					.421
12					.500
13					.579

$n_2 = 6$

U \ n_1	1	2	3	4	5	6
0	.143	.036	.012	.005	.002	.001
1	.286	.071	.024	.010	.004	.002
2	.428	.143	.048	.019	.009	.004
3	.571	.214	.083	.033	.015	.008
4		.321	.131	.057	.026	.013
5		.429	.190	.086	.041	.021
6		.571	.274	.129	.063	.032
7			.357	.176	.089	.047
8			.452	.238	.123	.066
9			.548	.305	.165	.090
10				.381	.214	.120
11				.457	.268	.155
12				.545	.331	.197
13					.396	.242
14					.465	.294
15					.535	.350
16						.409
17						.469
18						.531

PROBABILITIES ASSOCIATED WITH VALUES AS SMALL AS OBSERVED VALUES
OF U IN THE MANN-WHITNEY TEST

$n_2 = 7$

n_1 / U	1	2	3	4	5	6	7
0	.125	.028	.008	.003	.001	.001	.000
1	.250	.056	.017	.006	.003	.001	.001
2	.375	.111	.033	.012	.005	.002	.001
3	.500	.167	.058	.021	.009	.004	.002
4	.625	.250	.092	.036	.015	.007	.003
5		.333	.133	.055	.024	.011	.006
6		.444	.192	.082	.037	.017	.009
7		.556	.258	.115	.053	.026	.013
8			.333	.158	.074	.037	.019
9			.417	.206	.101	.051	.027
10			.500	.264	.134	.069	.036
11			.583	.324	.172	.090	.049
12				.394	.216	.117	.064
13				.464	.265	.147	.082
14				.538	.319	.183	.104
15					.378	.223	.130
16					.438	.267	.159
17					.500	.314	.191
18					.562	.365	.228
19						.418	.267
20						.473	.310
21						.527	.355
22							.402
23							.451
24							.500
25							.549

PROBABILITIES ASSOCIATED WITH VALUES AS SMALL AS OBSERVED VALUES OF U IN THE MANN-WHITNEY TEST

$$n_2 = 8$$

n_1 U	1	2	3	4	5	6	7	8	t	Normal
0	.111	.022	.006	.002	.001	.000	.000	.000	3.308	.001
1	.222	.044	.012	.004	.002	.001	.000	.000	3.203	.001
2	.333	.089	.024	.008	.003	.001	.001	.000	3.098	.001
3	.444	.133	.042	.014	.005	.002	.001	.001	2.993	.001
4	.556	.200	.067	.024	.009	.004	.002	.001	2.888	.002
5		.267	.097	.036	.015	.006	.003	.001	2.783	.003
6		.356	.139	.055	.023	.010	.005	.002	2.678	.004
7		.444	.188	.077	.033	.015	.007	.003	2.573	.005
8		.556	.248	.107	.047	.021	.010	.005	2.468	.007
9			.315	.141	.064	.030	.014	.007	2.363	.009
10			.387	.184	.085	.041	.020	.010	2.258	.012
11			.461	.230	.111	.054	.027	.014	2.153	.016
12			.539	.285	.142	.071	.036	.019	2.048	.020
13				.341	.177	.091	.047	.025	1.943	.026
14				.404	.217	.114	.060	.032	1.838	.033
15				.467	.262	.141	.076	.041	1.733	.041
16				.533	.311	.172	.095	.052	1.628	.052
17					.362	.207	.116	.065	1.523	.064
18					.416	.245	.140	.080	1.418	.078
19					.472	.286	.168	.097	1.313	.094
20					.528	.331	.198	.117	1.208	.113
21						.377	.232	.139	1.102	.135
22						.426	.268	.164	.998	.159
23						.475	.306	.191	.893	.185
24						.525	.347	.221	.788	.215
25							.389	.253	.683	.247
26							.433	.287	.578	.282
27							.478	.323	.473	.318
28							.522	.360	.368	.356
29								.399	.263	.396
30								.439	.158	.437
31								.480	.052	.481
32								.520		

X.4 CRITICAL VALUES OF U IN THE WILCOXON (MANN-WHITNEY) TWO-SAMPLE STATISTIC

This table gives critical values of U for significance levels 0.001, 0.01, 0.025, and 0.05 for a one-tailed test. For a two-tailed test, the significance levels are 0.002, 0.02, 0.05, and 0.01. If an observed U is equal to or less than the tabular value, the null hypothesis may be rejected at the level of significance indicated at the head of that table.

CRITICAL VALUES OF U IN THE MANN-WHITNEY TEST

Critical Values of U for a One-tailed Test at $\alpha = .001$ or for a Two-tailed Test at $\alpha = .002$

n_1 \ n_2	9	10	11	12	13	14	15	16	17	18	19	20
1												
2												
3									0	0	0	0
4		0	0	0	1	1	1	2	2	3	3	3
5	1	1	2	2	3	3	4	5	5	6	7	7
6	2	3	4	4	5	6	7	8	9	10	11	12
7	3	5	6	7	8	9	10	11	13	14	15	16
8	5	6	8	9	11	12	14	15	17	18	20	21
9	7	8	10	12	14	15	17	19	21	23	25	26
10	8	10	12	14	17	19	21	23	25	27	29	32
11	10	12	15	17	20	22	24	27	29	32	34	37
12	12	14	17	20	23	25	28	31	34	37	40	42
13	14	17	20	23	26	29	32	35	38	42	45	48
14	15	19	22	25	29	32	36	39	43	46	50	54
15	17	21	24	28	32	36	40	43	47	51	55	59
16	19	23	27	31	35	39	43	48	52	56	60	65
17	21	25	29	34	38	43	47	52	57	61	66	70
18	23	27	32	37	42	46	51	56	61	66	71	76
19	25	29	34	40	45	50	55	60	66	71	77	82
20	26	32	37	42	48	54	59	65	70	76	82	88

Critical Values of U for a One-tailed Test at $\alpha = .01$ or for a Two-tailed Test at $\alpha = .02$

n_1 \ n_2	9	10	11	12	13	14	15	16	17	18	19	20
1												
2					0	0	0	0	0	0	1	1
3	1	1	1	2	2	2	3	3	4	4	4	5
4	3	3	4	5	5	6	7	7	8	9	9	10
5	5	6	7	8	9	10	11	12	13	14	15	16
6	7	8	9	11	12	13	15	16	18	19	20	22
7	9	11	12	14	16	17	19	21	23	24	26	28
8	11	13	15	17	20	22	24	26	28	30	32	34
9	14	16	18	21	23	26	28	31	33	36	38	40
10	16	19	22	24	27	30	33	36	38	41	44	47
11	18	22	25	28	31	34	37	41	44	47	50	53
12	21	24	28	31	35	38	42	46	49	53	56	60
13	23	27	31	35	39	43	47	51	55	59	63	67
14	26	30	34	38	43	47	51	56	60	65	69	73
15	28	33	37	42	47	51	56	61	66	70	75	80
16	31	36	41	46	51	56	61	66	71	76	82	87
17	33	38	44	49	55	60	66	71	77	82	88	93
18	36	41	47	53	59	65	70	76	82	88	94	100
19	38	44	50	56	63	69	75	82	88	94	101	107
20	40	47	53	60	67	73	80	87	93	100	107	114

CRITICAL VALUES OF U IN THE MANN-WHITNEY TEST

Critical Values of U for a One-tailed Test at $\alpha = .025$ or for a Two-tailed Test at $\alpha = .05$

n_1 \ n_2	9	10	11	12	13	14	15	16	17	18	19	20
1												
2	0	0	0	1	1	1	1	1	2	2	2	2
3	2	3	3	4	4	5	5	6	6	7	7	8
4	4	5	6	7	8	9	10	11	11	12	13	13
5	7	8	9	11	12	13	14	15	17	18	19	20
6	10	11	13	14	16	17	19	21	22	24	25	27
7	12	14	16	18	20	22	24	26	28	30	32	34
8	15	17	19	22	24	26	29	31	34	36	38	41
9	17	20	23	26	28	31	34	37	39	42	45	48
10	20	23	26	29	33	36	39	42	45	48	52	55
11	23	26	30	33	37	40	44	47	51	55	58	62
12	26	29	33	37	41	45	49	53	57	61	65	69
13	28	33	37	41	45	50	54	59	63	67	72	76
14	31	36	40	45	50	55	59	64	67	74	78	83
15	34	39	44	49	54	59	64	70	75	80	85	90
16	37	42	47	53	59	64	70	75	81	86	92	98
17	39	45	51	57	63	67	75	81	87	93	99	105
18	42	48	55	61	67	74	80	86	93	99	106	112
19	45	52	58	65	72	78	85	92	99	106	113	119
20	48	55	62	69	76	83	90	98	105	112	119	127

Critical Values of U for a One-tailed Test at $\alpha = .05$ or for a Two-tailed Test at $\alpha = .10$

n_2 \ n_1	9	10	11	12	13	14	15	16	17	18	19	20
1											0	0
2	1	1	1	2	2	2	3	3	3	4	4	4
3	3	4	5	5	6	7	7	8	9	9	10	11
4	6	7	8	9	10	11	12	14	15	16	17	18
5	9	11	12	13	15	16	18	19	20	22	23	25
6	12	14	16	17	19	21	23	25	26	28	30	32
7	15	17	19	21	24	26	28	30	33	35	37	39
8	18	20	23	26	28	31	33	36	39	41	44	47
9	21	24	27	30	33	36	39	42	45	48	51	54
10	24	27	31	34	37	41	44	48	51	55	58	62
11	27	31	34	38	42	46	50	54	57	61	65	69
12	30	34	38	42	47	51	55	60	64	68	72	77
13	33	37	42	47	51	56	61	65	70	75	80	84
14	36	41	46	51	56	61	66	71	77	82	87	92
15	39	44	50	55	61	66	72	77	83	88	94	100
16	42	48	54	60	65	71	77	83	89	95	101	107
17	45	51	57	64	70	77	83	89	96	102	109	115
18	48	55	61	68	75	82	88	95	102	109	116	123
19	51	58	65	72	80	87	94	101	109	116	123	130
20	54	62	69	77	84	92	100	107	115	123	130	138

X.5 DISTRIBUTION OF THE TOTAL NUMBER OF RUNS FOR UNEQUAL-SIZE SAMPLES

The theory of runs can be used to test data for randomness or to test the hypothesis that two samples come from the same population. A run is defined as a succession of identical elements which are followed and preceded by different elements or by no elements at all. Let N_1 be the number of elements of one kind and N_2 be the number of elements of the other kind. Let u equal the total number of runs among the $N_1 + N_2$ elements. Table (a) gives the sampling distribution for u for values of N_1 and N_2 less than or equal to 10 and Table (b) gives a number of percentage points of the distribution for larger sample sizes. The values listed in Table (a) give the probability that u or fewer runs will occur. In Table (b), the columns headed 0.5, 1, 2.5, 5 gives values of u such that u or fewer runs occur with probability less than that indicated; the columns headed 95, 97.5, 99, 99.5 gives values of u for which the probability of u or more runs is less than 0.05, 0.025, 0.01, 0.005. For large values of N_1 and N_2, particularly for $N_1 = N_2$ greater than 10, a normal approximation may be used, with

$$\text{mean} = \frac{2N_1N_2}{N_1 + N_2} + 1$$

and

$$\text{variance} = \frac{2N_1N_2(2N_1N_2 - N_1 - N_2)}{(N_1 + N_2)^2(N_1 + N_2 - 1)} \ .$$

Non-Parametric Statistics

DISTRIBUTION OF THE TOTAL NUMBER OF RUNS u IN SAMPLES OF SIZE (N_1,N_2)

(N_1,N_2) \ u	2	3	4	5	6	7	8	9	10
(2,3)	.200	.500	.900	1.000					
(2,4)	.133	.400	.800	1.000					
(2,5)	.095	.333	.714	1.000					
(2,6)	.071	.286	.643	1.000					
(2,7)	.056	.250	.583	1.000					
(2,8)	.044	.222	.533	1.000					
(2,9)	.036	.200	.491	1.000					
(2,10)	.030	.182	.455	1.000					
(3,3)	.100	.300	.700	.900	1.000				
(3,4)	.057	.200	.543	.800	.971	1.000			
(3,5)	.036	.143	.429	.714	.929	1.000			
(3,6)	.024	.107	.345	.643	.881	1.000			
(3,7)	.017	.083	.283	.583	.833	1.000			
(3,8)	.012	.067	.236	.533	.788	1.000			
(3,9)	.009	.055	.200	.491	.745	1.000			
(3,10)	.007	.045	.171	.455	.706	1.000			
(4,4)	.029	.114	.371	.629	.886	.971	1.000		
(4,5)	.016	.071	.262	.500	.786	.929	.992	1.000	
(4,6)	.010	.048	.190	.405	.690	.881	.976	1.000	
(4,7)	.006	.033	.142	.333	.606	.833	.954	1.000	
(4,8)	.004	.024	.109	.279	.533	.788	.929	1.000	
(4,9)	.003	.018	.085	.236	.471	.745	.902	1.000	
(4,10)	.002	.014	.068	.203	.419	.706	.874	1.000	
(5,5)	.008	.040	.167	.357	.643	.833	.960	.992	1.000
(5,6)	.004	.024	.110	.262	.522	.738	.911	.976	.998
(5,7)	.003	.015	.076	.197	.424	.652	.854	.955	.992
(5,8)	.002	.010	.054	.152	.347	.576	.793	.929	.984
(5,9)	.001	.007	.039	.119	.287	.510	.734	.902	.972
(5,10)	.001	.005	.029	.095	.239	.455	.678	.874	.958
(6,6)	.002	.013	.067	.175	.392	.608	.825	.933	.987
(6,7)	.001	.008	.043	.121	.296	.500	.733	.879	.966
(6,8)	.001	.005	.028	.086	.226	.413	.646	.821	.937
(6,9)	.000	.003	.019	.063	.175	.343	.566	.762	.902
(6,10)	.000	.002	.013	.047	.137	.288	.497	.706	.864
(7,7)	.001	.004	.025	.078	.209	.383	.617	.791	.922
(7,8)	.000	.002	.015	.051	.149	.296	.514	.704	.867
(7,9)	.000	.001	.010	.035	.108	.231	.427	.622	.806
(7,10)	.000	.001	.006	.024	.080	.182	.355	.549	.743
(8,8)	.000	.001	.009	.032	.100	.214	.405	.595	.786
(8,9)	.000	.001	.005	.020	.069	.157	.319	.500	.702
(8,10)	.000	.000	.003	.013	.048	.117	.251	.419	.621
(9,9)	.000	.000	.003	.012	.044	.109	.238	.399	.601
(9,10)	.000	.000	.002	.008	.029	.077	.179	.319	.510
(10,10)	.000	.000	.001	.004	.019	.051	.128	.242	.414

DISTRIBUTION OF THE TOTAL NUMBER OF RUNS u IN SAMPLES OF SIZE (N_1,N_2)

(N_1,N_2) \ u	11	12	13	14	15	16	17	18	19	20
(2,3)										
(2,4)										
(2,5)										
(2,6)										
(2,7)										
(2,8)										
(2,9)										
(2,10)										
(3,3)										
(3,4)										
(3,5)										
(3,6)										
(3,7)										
(3,8)										
(3,9)										
(3,10)										
(4,4)										
(4,5)										
(4,6)										
(4,7)										
(4,8)										
(4,9)										
(4,10)										
(5,5)										
(5,6)	1.000									
(5,7)	1.000									
(5,8)	1.000									
(5,9)	1.000									
(5,10)	1.000									
(6,6)	.998	1.000								
(6,7)	.992	.999	1.000							
(6,8)	.984	.998	1.000							
(6,9)	.972	.994	1.000							
(6,10)	.958	.990	1.000							
(7,7)	.975	.996	.999	1.000						
(7,8)	.949	.988	.998	1.000	1.000					
(7,9)	.916	.975	.994	.999	1.000					
(7,10)	.879	.957	.990	.998	1.000					
(8,8)	.900	.968	.991	.999	1.000	1.000				
(8,9)	.843	.939	.980	.996	.999	1.000	1.000			
(8,10)	.782	.903	.964	.990	.998	1.000	1.000			
(9,9)	.762	.891	.956	.988	.997	1.000	1.000	1.000		
(9,10)	.681	.834	.923	.974	.992	.999	1.000	1.000	1.000	
(10,10)	.586	.758	.872	.949	.981	.996	.999	1.000	1.000	1.000

Non-Parametric Statistics

DISTRIBUTION OF THE TOTAL NUMBER OF RUNS u IN SAMPLES OF SIZE (N_1,N_2)

The values listed on the previous pages give the chance that u or fewer runs will occur. For example, for two samples of size 4, the chance of three or fewer runs is .114. For sample sizes $N_1 = N_2$ larger than 10 the following table can be used. The columns headed 0.5, 1, 2.5, 5 give values of u such that u or fewer runs occur with chance less than the indicated percentage. For example, for $N_1 = N_2 = 12$ the chance of 8 or fewer runs is about .05. The columns headed 95, 97.5, 99, 99.5 give values of u for which the chance of u or more runs is less than 5, 2.5, 1, 0.5 per cent.

$N_1 = N_2$	0.5	1	2.5	5	95	97.5	99	99.5	Mean	Var.	s.d.
11	5	6	7	7	16	16	17	18	12	5.24	2.29
12	6	7	7	8	17	18	18	19	13	5.74	2.40
13	7	7	8	9	18	19	20	20	14	6.24	2.50
14	7	8	9	10	19	20	21	22	15	6.74	2.60
15	8	9	10	11	20	21	22	23	16	7.24	2.69
16	9	10	11	11	22	22	23	24	17	7.74	2.78
17	10	10	11	12	23	24	25	25	18	8.24	2.87
18	10	11	12	13	24	25	26	27	19	8.74	2.96
19	11	12	13	14	25	26	27	28	20	9.24	3.04
20	12	13	14	15	26	27	28	29	21	9.74	3.12
25	16	17	18	19	32	33	34	35	26	12.24	3.50
30	20	21	22	24	37	39	40	41	31	14.75	3.84
35	24	25	27	28	43	44	46	47	36	17.25	4.15
40	29	30	31	33	48	50	51	52	41	19.75	4.44
45	33	34	36	37	54	55	57	58	46	22.25	4.72
50	37	38	40	42	59	61	63	64	51	24.75	4.97
55	42	43	45	46	65	66	68	69	56	27.25	5.22
60	46	47	49	51	70	72	74	75	61	29.75	5.45
65	50	52	54	56	75	77	79	81	66	32.25	5.68
70	55	56	58	60	81	83	85	86	71	34.75	5.89
75	59	61	63	65	86	88	90	92	76	37.25	6.10
80	64	65	68	70	91	93	96	97	81	39.75	6.30
85	68	70	72	74	97	99	101	103	86	42.25	6.50
90	73	74	77	79	102	104	107	108	91	44.75	6.69
95	77	79	82	84	107	109	112	114	96	47.25	6.87
100	82	84	86	88	113	115	117	119	101	49.75	7.05

X.6 CRITICAL VALUES FOR THE KOLMOGOROV-SMIRNOV ONE-SAMPLE STATISTIC

A sample of size n is drawn from a population with cumulative distribution function (x). Define the empirical distribution function $F_n(x)$ to be the step function

$$F_n(x) = \frac{k}{n} \quad \text{for} \quad x_{(i)} \leq x \leq x_{(i+1)} \ ,$$

here k is the number of observations not greater than x. $x_{(1)} \ \ldots \ , x_{(n)}$ denote the sample alues arranged in ascending order. Under the null hypothesis that the sample has been rawn from the specified distribution, $F_n(x)$ should be fairly close to $F(x)$. Define

$$D = \max |F_n(x) - F(x)| \ .$$

'or a two-tailed test this table gives critical values of the sampling distribution of D under he null hypothesis. Reject the hypothetical distribution if D exceeds the tabulated value. f n is over 35, determine the critical values of D by the divisions indicated in the table.

A one-tailed test is provided by the statistic

$$D^+ = \max [F_n(x) - F(x)] \ .$$

Non-Parametric Statistics

CRITICAL VALUES FOR THE KOLMOGOROV-SMIRNOV TEST OF GOODNESS OF FIT

Sample Size (n)	Significance Level				
	.20	.15	.10	.05	.01
1	.900	.925	.950	.975	.995
2	.684	.726	.776	.842	.929
3	.565	.597	.642	.708	.829
4	.494	.525	.564	.624	.734
5	.446	.474	.510	.563	.669
6	.410	.436	.470	.521	.618
7	.381	.405	.438	.486	.577
8	.358	.381	.411	.457	.543
9	.339	.360	.388	.432	.514
10	.322	.342	.368	.409	.486
11	.307	.326	.352	.391	.468
12	.295	.313	.338	.375	.450
13	.284	.302	.325	.361	.433
14	.274	.292	.314	.349	.418
15	.266	.283	.304	.338	.404
16	.258	.274	.295	.328	.391
17	.250	.266	.286	.318	.380
18.	.244	.259	.278	.309	.370
19	.237	.252	.272	.301	.361
20	.231	.246	.264	.294	.352
25	.21	.22	.24	.264	.32
30	.19	.20	.22	.242	.29
35	.18	.19	.21	.23	.27
40				.21	.25
50				.19	.23
60				.17	.21
70				.16	.19
80				.15	.18
90				.14	
100				.14	
Asymptotic Formula:	$\dfrac{1.07}{\sqrt{n}}$	$\dfrac{1.14}{\sqrt{n}}$	$\dfrac{1.22}{\sqrt{n}}$	$\dfrac{1.36}{\sqrt{n}}$	$\dfrac{1.63}{\sqrt{n}}$

Reject the hypothetical distribution $F(x)$ if $D_n = \max |F_n(x) - F(x)|$ exceeds the tabulated value (For $\alpha = .01$ and $.05$, asymptotic formulas give values which are too high—by 1.5 per cent for $n = 80$.

X.7 CRITICAL VALUES FOR THE KOLMOGOROV-SMIRNOV TWO-SAMPLE STATISTIC

A sample of size n_1 is drawn from a population with cumulative distribution function $'(x)$. Define the empirical distribution function $F_{n_1}(x)$ to be the step function

$$F_{n_1}(x) = \frac{k}{n_1} \, ,$$

where k is the number of observations not greater than x. A second sample of size n_2 is drawn with empirical distribution function $F_{n_2}(x)$. Define

$$D_{n_1,n_2} = \max |F_{n_1}(x) - F_{n_2}(x)|$$

or a two-tailed test. This table gives critical values of the sampling distribution of D under the null hypothesis that two independent samples have been drawn from the same population or from populations with the same distribution. Reject the null hypothesis if D exceeds the tabulated value.

A one-tailed test is provided by the statistic

$$D^+ = \max [F_{n_1}(x) - F_{n_2}(x)] \ .$$

For large values of n_1 and n_2, approximate formulas to be used are given at the bottom of the table.

Non-Parametric Statistics

CRITICAL VALUES FOR THE KOLMOGOROV-SMIRNOV TEST OF H_0: $F_1(x) = F_2(x)$

Sample size n_1 across columns; Sample size n_2 down rows.

n_2 \ n_1	1	2	3	4	5	6	7	8	9	10	12	15
1	* *	* *	* *	* *	* *	* *	* *	* *	* *	* *		
2			* *	* *	* *	* *	* *	7/8 *	16/18 *	9/10 *		
3			* *	* *	12/15 *	5/6 *	18/21 *	18/24 *	7/9 8/9		9/12 11/12	
4				3/4 *	16/20 *	9/12 10/12	21/28 24/28	6/8 7/8	27/36 32/36	14/20 16/20	8/12 10/12	
5					4/5 4/5	20/30 25/30	25/35 30/35	27/40 32/40	31/45 36/45	7/10 8/10		10/15 11/15
6						4/6 5/6	29/42 35/42	16/24 18/24	12/18 14/18	19/30 22/30	7/12 9/12	
7							5/7 5/7	35/56 42/56	40/63 47/63	43/70 53/70		
8								5/8 6/8	45/72 54/72	23/40 28/40	14/24 16/24	
9									5/9 6/9	52/90 62/90	20/36 24/36	
10										6/10 7/10		15/30 19/30
12											6/12 7/12	30/60 35/60
15												7/15 8/15

Reject H_0 if

$$D = \max |F_{n_1}(x) - F_{n_2}(x)|$$

exceeds the tabulated value.
The upper value gives a level
at most .05 and the lower
value gives a level at most .01.

Note 1: Where * appears, do not reject H_0 at the given level.
Note 2: For large values of n_1 and n_2, the following approximate formulas may be used:

$$\alpha = .05: \quad 1.36 \sqrt{\frac{n_1 + n_2}{n_1 n_2}}$$

$$\alpha = .01: \quad 1.63 \sqrt{\frac{n_1 + n_2}{n_1 n_2}}$$

CRITICAL VALUES OF D IN THE KOLMOGOROV-SMIRNOV TWO-SAMPLE TEST

(Large samples: two-tailed test)

Level of significance	Value of D so large as to call for rejection of H_0 at the indicated level of significance, where $D = \text{maximum } \lvert F_{n_1}(X) - F_{n_2}(X) \rvert$
.10	$1.22 \sqrt{\dfrac{n_1 + n_2}{n_1 n_2}}$
.05	$1.36 \sqrt{\dfrac{n_1 + n_2}{n_1 n_2}}$
.025	$1.48 \sqrt{\dfrac{n_1 + n_2}{n_1 n_2}}$
.01	$1.63 \sqrt{\dfrac{n_1 + n_2}{n_1 n_2}}$
.005	$1.73 \sqrt{\dfrac{n_1 + n_2}{n_1 n_2}}$
.001	$1.95 \sqrt{\dfrac{n_1 + n_2}{n_1 n_2}}$

X.8   KRUSKAL-WALLIS ONE-WAY ANALYSIS OF VARIANCE BY RANKS. PROBABILITIES ASSOCIATED WITH VALUES AS LARGE AS OBSERVED VALUES OF H

Three samples of sizes n_1, n_2, and n_3 are combined and ranked in ascending order of magnitude: numbers from 1 up to $N = n_1 + n_2 + n_3$ are attached to the ranks. To test whether the three samples come from the same population, the test statistic is

$$H = \frac{12}{N(N + 1)} \sum_{j=1}^{3} \frac{R_j^2}{n_j} - 3(N + 1),$$

where n_j = number of observations in j^{th} sample, $j = 1, 2, 3$,

$$N = \sum_{j=1}^{3} n_j = \text{number of observations in all samples combined} ,$$

R_j = sum of ranks in j^{th} sample.

Large values of H lead to rejection of the null hypothesis. If the three samples are from identical populations and the sample sizes are not too small, then H is approximately distributed as chi-square with two degrees of freedom. The first column in the table gives the sizes of the three samples. The second gives various values of H. The third gives the probability associated with the occurrence under the null hypothesis of values as large as an observed H.

PROBABILITIES ASSOCIATED WITH VALUES AS LARGE AS OBSERVED VALUES OF *H* IN THE KRUSKAL-WALLIS ONE-WAY ANALYSIS OF VARIANCE BY RANKS

Sample sizes			*H*	*p*	Sample sizes			*H*	*p*
n_1	n_2	n_3			n_1	n_2	n_3		
2	1	1	2.7000	.500	4	3	2	6.4444	.008
								6.3000	.011
2	2	1	3.6000	.200				5.4444	.046
								5.4000	.051
2	2	2	4.5714	.067				4.5111	.098
			3.7143	.200				4.4444	.102
3	1	1	3.2000	.300	4	3	3	6.7455	.010
								6.7091	.013
3	2	1	4.2857	.100				5.7909	.046
			3.8571	.133				5.7273	.050
								4.7091	.092
3	2	2	5.3572	.029				4.7000	.101
			4.7143	.048					
			4.5000	.067	4	4	1	6.6667	.010
			4.4643	.105				6.1667	.022
								4.9667	.048
3	3	1	5.1429	.043				4.8667	.054
			4.5714	.100				4.1667	.082
			4.0000	.129				4.0667	.102
3	3	2	6.2500	.011	4	4	2	7.0364	.006
			5.3611	.032				6.8727	.011
			5.1389	.061				5.4545	.046
			4.5556	.100				5.2364	.052
			4.2500	.121				4.5545	.098
								4.4455	.103
3	3	3	7.2000	.004					
			6.4889	.011	4	4	3	7.1439	.010
			5.6889	.029				7.1364	.011
			5.6000	.050				5.5985	.049
			5.0667	.086				5.5758	.051
			4.6222	.100				4.5455	.099
								4.4773	.102
4	1	1	3.5714	.200					
					4	4	4	7.6538	.008
4	2	1	4.8214	.057				7.5385	.011
			4.5000	.076				5.6923	.049
			4.0179	.114				5.6538	.054
								4.6539	.097
4	2	2	6.0000	.014				4.5001	.104
			5.3333	.033					
			5.1250	.052	5	1	1	3.8571	.143
			4.4583	.100					
			4.1667	.105	5	2	1	5.2500	.036
								5.0000	.048
4	3	1	5.8333	.021				4.4500	.071
			5.2083	.050				4.2000	.095
			5.0000	.057				4.0500	.119
			4.0556	.093					
			3.8889	.129					

PROBABILITIES ASSOCIATED WITH VALUES AS LARGE AS OBSERVED VALUES OF H IN THE KRUSKAL-WALLIS ONE-WAY ANALYSIS OF VARIANCE BY RANKS

n_1	n_2	n_3	H	p	n_1	n_2	n_3	H	p
5	2	2	6.5333	.008				5.6308	.050
			6.1333	.013				4.5487	.099
			5.1600	.034				4.5231	.103
			5.0400	.056					
			4.3733	.090	5	4	4	7.7604	.009
			4.2933	.122				7.7440	.011
5	3	1	6.4000	.012				5.6571	.049
			4.9600	.048				5.6176	.050
			4.8711	.052				4.6187	.100
			4.0178	.095				4.5527	.102
			3.8400	.123	5	5	1	7.3091	.009
5	3	2	6.9091	.009				6.8364	.011
			6.8218	.010				5.1273	.046
			5.2509	.049				4.9091	.053
			5.1055	.052				4.1091	.086
			4.6509	.091				4.0364	.105
			4.4945	.101	5	5	2	7.3385	.010
5	3	3	7.0788	.009				7.2692	.010
			6.9818	.011				5.3385	.047
			5.6485	.049				5.2462	.051
			5.5152	.051				4.6231	.097
			4.5333	.097				4.5077	.100
			4.4121	.109	5	5	3	7.5780	.010
5	4	1	6.9545	.008				7.5429	.010
			6.8400	.011				5.7055	.046
			4.9855	.044				5.6264	.051
			4.8600	.056				4.5451	.100
			3.9873	.098				4.5363	.102
			3.9600	.102	5	5	4	7.8229	.010
5	4	2	7.2045	.009				7.7914	.010
			7.1182	.010				5.6657	.049
			5.2727	.049				5.6429	.050
			5.2682	.050				4.5229	.099
			4.5409	.098				4.5200	.101
			4.5182	.101	5	5	5	8.0000	.009
5	4	3	7.4449	.010				7.9800	.010
			7.3949	.011				5.7800	.049
			5.6564	.049				5.6600	.051
								4.5600	.100
								4.5000	.102

X.9   CRITICAL VALUES OF SPEARMAN'S RANK CORRELATION COEFFICIENT

Spearman's coefficient of rank correlation, denoted by the letter r_s, measures the correspondence between two rankings. If d_i is the difference between the ranks of the ith pair of a set of n pairs of elements, then Spearman's Rho is defined as

$$r_s = 1 - \frac{6 \sum\limits_{i=1}^{n} d_i^2}{n^3 - n}$$

$$= 1 - \frac{6S_r}{n^3 - n} \,, \qquad \text{where } S_r = \sum_{i=1}^{n} d_i^2 \,.$$

The exact distribution of S_r has been studied, and critical values when there is complete independence are given in this table.

Non-Parametric Statistics

CRITICAL VALUES OF SPEARMAN'S RANK CORRELATION COEFFICIENT

n	$\gamma = 0.10$	$\gamma = 0.05$	$\gamma = 0.02$	$\gamma = 0.01$
5	0.900	—	—	—
6	0.829	0.886	0.943	—
7	0.714	0.786	0.893	0.929
8	0.643	0.738	0.833	0.881
9	0.600	0.700	0.783	0.833
10	0.564	0.648	0.745	0.794
11	0.536	0.618	0.709	0.818
12	0.497	0.591	0.703	0.780
13	0.475	0.566	0.673	0.745
14	0.457	0.545	0.646	0.716
15	0.441	0.525	0.623	0.689
16	0.425	0.507	0.601	0.666
17	0.412	0.490	0.582	0.645
18	0.399	0.476	0.564	0.625
19	0.388	0.462	0.549	0.608
20	0.377	0.450	0.534	0.591
21	0.368	0.438	0.521	0.576
22	0.359	0.428	0.508	0.562
23	0.351	0.418	0.496	0.549
24	0.343	0.409	0.485	0.537
25	0.336	0.400	0.475	0.526
26	0.329	0.392	0.465	0.515
27	0.323	0.385	0.456	0.505
28	0.317	0.377	0.448	0.496
29	0.311	0.370	0.440	0.487
30	0.305	0.364	0.432	0.478

X.10 DISTRIBUTION OF KENDALL'S RANK CORRELATION COEFFICIENT

Consider any one of the $\frac{n}{2}(n-1)$ possible pairs for two sets of ranked elements. Associate with this pair (a) a score of $+1$ if the ranking for both sets is the same order or (b) a score of -1 if the ranking is in different order. Kendall's score S_t is then defined as the total of these $\frac{n}{2}(n-1)$ individual pairs. S_t will have a maximum value of $\frac{n}{2}(n-1)$ if the two rankings are identical and a minimum value of $-\frac{n}{2}(n-1)$ if the sets are ranked in exactly opposite order. Kendall's Tau is defined as

$$t_k = \frac{S_t}{\frac{n}{2}(n-1)} ,$$

and has the range $-1 \leq t_k \leq 1$. This table may be used to determine the exact probability associated with the occurrence (one-tailed) under the null hypothesis that the observed value of Kendall's Tau indicates the existence of an association between the two sets of any value as extreme as an observed S_t. The tabled value is the probability that S_t is equalled or exceeded.

Non-Parametric Statistics

DISTRIBUTION OF KENDALL'S RANK CORRELATION COEFFICIENT, t_k, IN RANDOM RANKINGS

S_t	Values of n				S_t	Values of n		
	4	5	8	9		6	7	10
0	0.625	0.592	0.548	0.540	1	0.500	0.500	0.500
2	.375	.408	.452	.460	3	.360	.386	.431
4	.167	.242	.360	.381	5	.235	.281	.364
6	.042	.117	.274	.306	7	.136	.191	.300
8		.042	.199	.238	9	.068	.119	.242
10		0.0083	0.138	0.179	11	0.028	0.068	0.190
12			.089	.130	13	.0083	.035	.146
14			.054	.090	15	.0014	.015	.108
16			.031	.060	17		.0054	.078
18			.016	.038	19		.0014	.054
20			0.0071	0.022	21		0.0002	0.036
22			.0028	.012	23			.023
24			.0009	.0063	25			.014
26			.0002	.0029	27			.0083
28				.0012	29			.0046
30				0.0004	31			0.0023
					33			.0011
					35			.0005

The distribution of S_t is symmetrical so that values of the probability for negative S_t can be obtaine by appropriate subtraction from unity; e.g. for $n = 9$,

$$\Pr\{S_t \geq -14\} = 1 - \Pr\{S_t \geq 16\} = 1 - 0.060 = 0.940 \ .$$

XI. Quality Control

XI.1 FACTORS FOR COMPUTING CONTROL LIMITS

Control Charts for Measurement

If the process mean and standard deviation, μ and σ, are known, and it is assumed that the underlying distribution is normal, it is possible to assert with probability $1 - \alpha$ that the mean of a random sample of size n will fall between $\bar{x} - z_{\alpha/2}\dfrac{\sigma}{\sqrt{n}}$ and $\bar{x} + z_{\alpha/2}\dfrac{\sigma}{\sqrt{n}}$.

These two limits on $\bar{x}$ provide upper and lower control limits. In actual practice, μ and σ are usually unknown and it is necessary to estimate their values from a large sample taken while the process is "in control." The central line of an $\bar{x}$-chart is given by μ and the lower and upper three-sigma control limits are given by $\mu - A\sigma$ and $\mu + A\sigma$, respectively, where $A = \dfrac{3}{\sqrt{n}}$ and n is the sample size. Where the population parameters are unknown, is necessary to estimate these parameters on the basis of preliminary samples. If k samples are used, each of size n, denote the mean of the ith sample by $\bar{x}_i$ and the grand mean of the k sample means by $\bar{\bar{x}}$, i.e.

$$\bar{\bar{x}} = \frac{1}{k} \sum_{i=1}^{k} \bar{x}_i .$$

Denote the range of the ith sample by R_i and by $\bar{R}$ the mean of the k sample ranges, i.e.

$$\bar{R} = \frac{1}{k} \sum_{i=1}^{k} R_i .$$

Since $\bar{\bar{x}}$ is an unbiased estimate of the population mean μ, the central line for the $\bar{x}$-chart is given by $\bar{\bar{x}}$. The statistic R does not provide an unbiased estimate of σ, but $A_2\bar{R}$ is an unbiased estimate of $\dfrac{3\sigma}{\sqrt{n}}$. The constant multiplier A_2 depends on the assumption of normality. Thus, the central line and the lower and upper three sigma limits, LCL and UCL, for an $\bar{x}$-chart (with μ and σ estimated from past date) are given by

$$\text{central line} = \bar{\bar{x}}$$
$$\text{LCL} = \bar{\bar{x}} - A_2\bar{R}$$
$$\text{UCL} = \bar{\bar{x}} + A_2\bar{R} .$$

The central line and control limits of an R chart are based on the distribution of the range of samples of size n from a normal population. The mean and standard deviation of the sampling distribution of R are given by $d_2\sigma$ and $d_3\sigma$, respectively, when σ is known. Here d_2 and d_3 are constants which depend on the size of the sample. The set of control chart values for an R chart (with σ known) is given by

$$\text{central line} = d_2\sigma$$
$$\text{LCL} = D_1\sigma$$
$$\text{UCL} = D_2\sigma,$$

where $D_1 = d_2 - 3d_3$ and $D_2 = d_2 + 3d_3$.

189

If σ is unknown, the control chart values for an R chart are given by

$$\text{central line} = \bar{R}$$
$$\text{LCL} = D_3\bar{R}$$
$$\text{UCL} = D_4\bar{R},$$

where $D_3 = \dfrac{D_1}{d_2}$ and $D_4 = \dfrac{D_2}{d_2}$.

The central line and control limits of an s-chart are based on estimates obtained fro▪ the samples. A pooled estimate of the population variance is obtained from the k sample▪ i.e.

$$s_p^2 = \frac{\sum_i (n_i - 1)s_i^2}{\sum_i (n_i - 1)} \, , \qquad i = 1, 2, \ldots, k \; .$$

If the sample sizes are all equal, the pooled estimate is

$$s_p^2 = \frac{1}{k} \sum_i s_i^2 \; .$$

The control chart values for an s-chart are given by

$$\text{central line} = C_2' s_p$$
$$\text{LCL} = B_2' s_p$$
$$\text{UCL} = B_4' s_p \; .$$

If one uses the biased estimator of the variance s_p', as is often done in quality contr▪ work, the control chart values are given by

$$\text{central line} = c_2 s_p'$$
$$\text{LCL} = B_2 s_p'$$
$$\text{UCL} = B_4 s_p' \; .$$

B. Control Charts for Attributes

Control limits for a fraction-defective chart are based on the sampling theory f▪ proportions, using the normal curve approximation to the binomial. If k samples are take▪ the estimator of p is given by

$$\bar{p} = \frac{\sum_i x_i}{\sum_i n_i} \, , \qquad i = 1, 2, \ldots, k$$

where x_i is the number of defectives in the ith sample of size n_i. The central line and con▪ trol limits of a fraction defective chart based on analysis of past data are given by

$$\text{central line} = \bar{p}$$
$$\text{LCL} = \bar{p} - 3\sqrt{\frac{\bar{p}(1 - \bar{p})}{n_i}}$$
$$\text{UCL} = \bar{p} + 3\sqrt{\frac{\bar{p}(1 - \bar{p})}{n_i}} \; .$$

When the sample sizes are approximately equal, n_i is replaced by $\bar{n} = \dfrac{1}{k} \sum_i n_i$.

Equivalent to the p chart for the fraction defective is the control chart for the number of defective. Here, if p is estimated by $\bar{p}$, the control chart values for a number-of-defectives chart are given by

$$\text{central line} = \bar{n}\bar{p}$$
$$\text{LCL} = \bar{n}\bar{p} - 3\sqrt{\bar{n}\bar{p}(1 - \bar{p})}$$
$$\text{UCL} = \bar{n}\bar{p} + 3\sqrt{\bar{n}\bar{p}(1 - \bar{p})} \ .$$

In many cases it is necessary to control the number of defects per unit C, where C is taken to be a value of a random variable having a Poisson distribution. If k is the number of units available for estimating λ, the parameter of the Poisson distribution, and if C_i the number of defects in the ith unit, then λ is estimated by

$$\bar{C} = \frac{1}{k}\sum_{i=1}^{k} C_i \ ,$$

and the control-chart values for the C-chart are

$$\text{central line} = \bar{C}$$
$$\text{LCL} = \bar{C} - 3\sqrt{\bar{C}}$$
$$\text{UCL} = \bar{C} + 3\sqrt{\bar{C}}$$

This table presents values of the factors for computing control limits for various sample sizes n.

Quality Control

FACTORS FOR COMPUTING CONTROL LIMITS

Number of observations in sample, n	$\bar{X}$ chart Factors for control limits		R chart Factor for central line	Factors for control limits		s chart Factor for central line	Factors for control limits		$\hat{\sigma}$ chart (biased) Factor for central line	Factors for control limits	
	A	A_2	d_2	D_3	D_4	c_2'	B_2'	B_4'	c_2	B_2	B_4
2	2.121	1.880	1.128	0	3.267	0.798	0	2.298	0.5642	0	3.2
3	1.732	1.023	1.693	0	2.575	0.886	0	2.111	0.7236	0	2.5
4	1.500	0.729	2.059	0	2.282	0.921	0	1.982	0.7979	0	2.2
5	1.342	0.577	2.326	0	2.115	0.940	0	1.889	0.8407	0	2.08
6	1.225	0.483	2.534	0	2.004	0.951	0.085	1.817	0.8686	0.030	1.97
7	1.134	0.419	2.704	0.076	1.924	0.960	0.158	1.762	0.8882	0.118	1.88
8	1.061	0.373	2.847	0.136	1.864	0.965	0.215	1.715	0.9027	0.185	1.81
9	1.000	0.337	2.970	0.184	1.816	0.969	0.262	1.676	0.9139	0.239	1.76
10	0.949	0.308	3.078	0.223	1.777	0.973	0.302	1.644	0.9227	0.284	1.71
11	0.905	0.285	3.173	0.256	1.744	0.976	0.336	1.616	0.9300	0.321	1.67
12	0.866	0.266	3.258	0.284	1.716	0.977	0.365	1.589	0.9359	0.354	1.64
13	0.832	0.249	3.336	0.308	1.692	0.980	0.392	1.568	0.9410	0.382	1.61
14	0.802	0.235	3.407	0.329	1.671	0.981	0.414	1.548	0.9453	0.406	1.59
15	0.775	0.223	3.472	0.348	1.652	0.982	0.434	1.530	0.9490	0.428	1.57
16	0.750	0.212	3.532	0.364	1.636	0.984	0.454	1.514	0.9523	0.448	1.55
17	0.728	0.203	3.588	0.379	1.621	0.984	0.469	1.499	0.9551	0.466	1.53
18	0.707	0.194	3.640	0.392	1.608	0.986	0.486	1.486	0.9576	0.482	1.51
19	0.688	0.187	3.689	0.404	1.596	0.986	0.500	1.472	0.9599	0.497	1.50
20	0.671	0.180	3.735	0.414	1.586	0.987	0.513	1.461	0.9619	0.510	1.49
21	0.655	0.173	3.778	0.425	1.575	0.988	0.525	1.451	0.9638	0.523	1.47
22	0.640	0.167	3.819	0.434	1.566	0.988	0.536	1.440	0.9655	0.534	1.46
23	0.626	0.162	3.858	0.443	1.557	0.989	0.546	1.432	0.9670	0.545	1.45
24	0.612	0.157	3.895	0.452	1.548	0.989	0.556	1.422	0.9684	0.555	1.44
25	0.600	0.153	3.931	0.459	1.541	0.990	0.566	1.414	0.9696	0.565	1.43

XI.2   PERCENTAGE POINTS OF THE DISTRIBUTION OF THE MEAN DEVIATION

If $x_1, x_2, \ldots, x_n$ is a random sample of n observations, the mean deviation is given

$$\text{M.D.} = \frac{1}{n} \sum_{i=1}^{n} |x_i - \bar{x}|$$

ere $\bar{x}$ is the sample mean. This table gives certain lower and upper percentage points of e standardized mean deviation $\dfrac{\text{M.D.}}{\sigma}$.

PERCENTAGE POINTS OF THE DISTRIBUTION OF THE MEAN DEVIATION

Size of sample n	\multicolumn{6}{c}{Lower percentage points}
	0.1	0.5	1.0	2.5	5.0	10.0
2	0.001	0.004	0.009	0.022	0.044	0.089
3	0.022	0.052	0.073	0.116	0.166	0.238
4	0.066	0.114	0.145	0.199	0.254	0.328
5	0.112	0.170	0.203	0.260	0.315	0.386
6	0.153	0.215	0.250	0.306	0.360	0.428
7	0.190	0.252	0.287	0.342	0.394	0.459
8	0.220	0.283	0.318	0.372	0.422	0.484
9	0.247	0.310	0.344	0.396	0.445	0.504
10	0.271	0.333	0.366	0.417	0.464	0.521
\multicolumn{7}{c}{Normal approximation}
10	0.171	0.269	0.316	0.386	0.445	0.514

Size of sample n	\multicolumn{6}{c}{Upper percentage points}
	10.0	5.0	2.5	1.0	0.5	0.1
2	1.163	1.386	1.585	1.821	1.985	2.327
3	1.117	1.276	1.417	1.586	1.703	1.949
4	1.089	1.224	1.344	1.489	1.590	1.806
5	1.069	1.187	1.292	1.419	1.507	1.693
6	1.052	1.158	1.253	1.366	1.445	1.613
7	1.038	1.135	1.222	1.325	1.397	1.550
8	1.026	1.116	1.196	1.292	1.358	1.499
9	1.016	1.100	1.175	1.264	1.326	1.457
10	1.007	1.086	1.156	1.240	1.299	1.422
\multicolumn{7}{c}{Normal approximation}
10	1.000	1.069	1.128	1.198	1.245	1.342

The unit is the population standard deviation.

XII. Miscellaneous Statistical Tables

XII.1 NUMBER OF PERMUTATIONS $P(n,m)$

This table contains the number of permutations of n distinct things taken m at a time, given by

$$P(n,m) = \frac{n!}{(n-m)!} = n(n-1) \cdots (n-m+1)$$

n \ m	0	1	2	3	4	5	6	7	8	9	10
0	1										
1	1	1									
2	1	2	2								
3	1	3	6	6							
4	1	4	12	24	24						
5	1	5	20	60	120	120					
6	1	6	30	120	360	720	720				
7	1	7	42	210	840	2520	5040	5040			
8	1	8	56	336	1680	6720	20160	40320	40320		
9	1	9	72	504	3024	15120	60480	1 81440	3 62880	3 62880	
10	1	10	90	720	5040	30240	1 51200	6 04800	18 14400	36 28800	36 288
11	1	11	110	990	7920	55440	3 32640	16 63200	66 52800	199 58400	399 168
12	1	12	132	1320	11880	95040	6 65280	39 91680	199 58400	798 33600	2395 008
13	1	13	156	1716	17160	1 54440	12 35520	86 48640	518 91840	2594 59200	10378 368
14	1	14	182	2184	24024	2 40240	21 62160	172 97280	1210 80960	7264 85760	36324 288
15	1	15	210	2730	32760	3 60360	36 03600	324 32400	2594 59200	18162 14400	1 08972 864

n \ m	11	12	13	14	15
8					
9					
10					
11	399 16800				
12	4790 01600	4790 01600			
13	31135 10400	62270 20800	62270 20800		
14	1 45297 15200	4 35891 45600	8 71782 91200	8 71782 91200	
15	5 44864 32000	21 79457 28000	65 38371 84000	130 76743 68000	130 76743 68000

XII.2 NUMBER OF COMBINATIONS $C(n,m)$

This table contains the number of combinations of n distinct things taken m at a time, given by

$$\binom{n}{m} = C(n,m) = \frac{n!}{m!(n-m)!} = \frac{P(n,r)}{m!} \ .$$

For values missing from the above table, use the relation $\binom{n}{m} = \binom{n}{n-m}$, e.g. $\binom{20}{12} = \binom{20}{8} = 125970$. $\binom{n}{m}$ is also referred to as a binomial coefficient. A recursion relation for the binomial coefficients is

$$\binom{n+1}{m+1} = \binom{n}{m} + \binom{n}{m+1} \ .$$

Miscellaneous Statistical Tables

NUMBER OF COMBINATIONS

$$\binom{n}{m} = C(n,m)$$

n \ m	0	1	2	3	4	5	6	7	8	9	10	11
0	1											
1	1	1										
2	1	2	1									
3	1	3	3	1								
4	1	4	6	4	1							
5	1	5	10	10	5	1						
6	1	6	15	20	15	6	1					
7	1	7	21	35	35	21	7	1				
8	1	8	28	56	70	56	28	8	1			
9	1	9	36	84	126	126	84	36	9	1		
10	1	10	45	120	210	252	210	120	45	10	1	
11	1	11	55	165	330	462	462	330	165	55	11	1
12	1	12	66	220	495	792	924	792	495	220	66	12
13	1	13	78	286	715	1287	1716	1716	1287	715	286	78
14	1	14	91	364	1001	2002	3003	3432	3003	2002	1001	364
15	1	15	105	455	1365	3003	5005	6435	6435	5005	3003	1365
16	1	16	120	560	1820	4368	8008	11440	12870	11440	8008	4368
17	1	17	136	680	2380	6188	12376	19448	24310	24310	19448	12376
18	1	18	153	816	3060	8568	18564	31824	43758	48620	43758	31824
19	1	19	171	969	3876	11628	27132	50388	75582	92378	92378	75582
20	1	20	190	1140	4845	15504	38760	77520	1 25970	1 67960	1 84756	1 67960
21	1	21	210	1330	5985	20349	54264	1 16280	2 03490	2 93930	3 52716	3 52716
22	1	22	231	1540	7315	26334	74613	1 70544	3 19770	4 97420	6 46646	7 05432
23	1	23	253	1771	8855	33649	1 00947	2 45157	4 90314	8 17190	11 44066	13 52078
24	1	24	276	2024	10626	42504	1 34596	3 46104	7 35471	13 07504	19 61256	24 96144
25	1	25	300	2300	12650	53130	1 77100	4 80700	10 81575	20 42975	32 68760	44 57400
26	1	26	325	2600	14950	65780	2 30230	6 57800	15 62275	31 24550	53 11735	77 26160
27	1	27	351	2925	17550	80730	2 96010	8 88030	22 20075	46 86825	84 36285	130 37895
28	1	28	378	3276	20475	98280	3 76740	11 84040	31 08105	69 06900	131 23110	214 74180
29	1	29	406	3654	23751	1 18755	4 75020	15 60780	42 92145	100 15005	200 30010	345 97290
30	1	30	435	4060	27405	1 42506	5 93775	20 35800	58 52925	143 07150	300 45015	546 27300
31	1	31	465	4495	31465	1 69911	7 36281	26 29575	78 88725	201 60075	443 52165	846 72315
32	1	32	496	4960	35960	2 01376	9 06192	33 65856	105 18300	280 48800	645 12240	1290 24480
33	1	33	528	5456	40920	2 37336	11 07568	42 72048	138 84156	385 67100	925 61040	1935 36720
34	1	34	561	5984	46376	2 78256	13 44904	53 79616	181 56204	524 51256	1311 28140	2860 97760
35	1	35	595	6545	52360	3 24632	16 23160	67 24520	235 35820	706 07460	1835 79396	4172 25900
36	1	36	630	7140	58905	3 76992	19 47792	83 47680	302 60340	941 43280	2541 86856	6008 05296
37	1	37	666	7770	66045	4 35897	23 24784	102 95472	386 08020	1244 03620	3483 30136	8549 92152
38	1	38	703	8436	73815	5 01942	27 60681	126 20256	489 03492	1630 11640	4727 33756	12033 22288
39	1	39	741	9139	82251	5 75757	32 62623	153 80937	615 23748	2119 15132	6357 45396	16760 56044
40	1	40	780	9880	91390	6 58008	38 38380	186 43560	769 04685	2734 38880	8476 60528	23118 01440
41	1	41	820	10660	1 01270	7 49398	44 96388	224 81940	955 48245	3503 43565	11210 99408	31594 61968
42	1	42	861	11480	1 11930	8 50668	52 45786	269 78328	1180 30185	4458 91810	14714 42973	42805 61376
43	1	43	903	12341	1 23410	9 62598	60 96454	322 24114	1450 08513	5639 21995	19173 34783	57520 04349
44	1	44	946	13244	1 35751	10 86008	70 59052	383 20568	1772 32627	7089 30508	24812 56778	76693 39132
45	1	45	990	14190	1 48995	12 21759	81 45060	453 79620	2155 53195	8861 63135	31901 87286	1 01505 95910
46	1	46	1035	15180	1 63185	13 70754	93 66819	535 24680	2609 32815	11017 16330	40763 50421	1 33407 83196
47	1	47	1081	16215	1 78365	15 33939	107 37573	628 91499	3144 57495	13626 49145	51780 66751	1 74171 33617
48	1	48	1128	17296	1 94580	17 12304	122 71512	736 29072	3773 48994	16771 06640	65407 15896	2 25952 00368
49	1	49	1176	18424	2 11876	19 06884	139 83816	859 00584	4509 78066	20544 55634	82178 22536	2 91359 16264
50	1	50	1225	19600	2 30300	21 18760	158 90700	998 84400	5368 78650	25054 33700	1 02722 78170	3 73537 38800

XII.3   RANDOM UNITS

Use of Table. If one wishes to select a random sample of N items from a universe of items, the following procedure may be applied. ($M > N$.)

1. Decide upon some arbitrary scheme of selecting entries from the table. For example, one may decide to use the entries in the first line, second column; second line, third column; third line, fourth column; etc.

2. Assign numbers to each of the items in the universe from 1 to M. Thus, if $M = 500$, the items would be numbered from 001 to 500, and therefore, each designated item is associated with a three digit number.

3. Decide upon some arbitrary scheme of selecting positional digits from each entry chosen according to Step 1. Thus, if $M = 500$, one may decide to use the first, third, and fourth digit of each entry selected, and as a consequence a three digit number is created for each entry choice.

4. If the number formed is $\leq M$, the correspondingly designated item in the universe is chosen for the random sample of N items. If a number formed is $> M$ or is a repeated number of one already chosen, it is passed over and the next desirable number is taken. This process is continued until the random sample of N items if selected.

Miscellaneous Statistical Tables

A TABLE OF 14,000 RANDOM UNITS

Line/Col.	(1)	(2)	(3)	(4)	(5)	(6)	(7)	(8)	(9)	(10)	(11)	(12)	(13)	(14)
1	10480	15011	01536	02011	81647	91646	69179	14194	62590	36207	20969	99570	91291	90700
2	22368	46573	25595	85393	30995	89198	27982	53402	93965	34095	52666	19174	39615	99505
3	24130	48360	22527	97265	76393	64809	15179	24830	49340	32081	30680	19655	63348	58629
4	42167	93093	06243	61680	07856	16376	39440	53537	71341	57004	00849	74917	97758	16379
5	37570	39975	81837	16656	06121	91782	60468	81305	49684	60672	14110	06927	01263	54613
6	77921	06907	11008	42751	27756	53498	18602	70659	90655	15053	21916	81825	44394	42880
7	99562	72905	56420	69994	98872	31016	71194	18738	44013	48840	63213	21069	10634	12952
8	96301	91977	05463	07972	18876	20922	94595	56869	69014	60045	18425	84903	42508	32307
9	89579	14342	63661	10281	17453	18103	57740	84378	25331	12566	58678	44947	05585	56941
10	85475	36857	43342	53988	53060	59533	38867	62300	08158	17983	16439	11458	18593	64952
11	28918	69578	88231	33276	70997	79936	56865	05859	90106	31595	01547	85590	91610	78188
12	63553	40961	48235	03427	49626	69445	18663	72695	52180	20847	12234	90511	33703	90322
13	09429	93969	52636	92737	88974	33488	36320	17617	30015	08272	84115	27156	30613	74952
14	10365	61129	87529	85689	48237	52267	67689	93394	01511	26358	85104	20285	29975	89868
15	07119	97336	71048	08178	77233	13916	47564	81056	97735	85977	29372	74461	28551	90707
16	51085	12765	51821	51259	77452	16308	60756	92144	49442	53900	70960	63990	75601	40719
17	02368	21382	52404	60268	89368	19885	55322	44819	01188	65255	64835	44919	05944	55157
18	01011	54092	33362	94904	31273	04146	18594	29852	71585	85030	51132	01915	92747	64951
19	52162	53916	46369	58586	23216	14513	83149	98736	23495	64350	94738	17752	35156	35749
20	07056	97628	33787	09998	42698	06691	76988	13602	51851	46104	88916	19509	25625	58104
21	48663	91245	85828	14346	09172	30168	90229	04734	59193	22178	30421	61666	99904	32812
22	54164	58492	22421	74103	47070	25306	76468	26384	58151	06646	21524	15227	96909	44592
23	32639	32363	05597	24200	13363	38005	94342	28728	35806	06912	17012	64161	18296	22851
24	29334	27001	87637	87308	58731	00256	45834	15398	46557	41135	10367	07684	36188	18510
25	02488	33062	28834	07351	19731	92420	60952	61280	50001	67658	32586	86679	50720	94953
26	81525	72295	04839	96423	24878	82651	66566	14778	76797	14780	13300	87074	79666	95725
27	29676	20591	68086	26432	46901	20849	89768	81536	86645	12659	92259	57102	80428	25280
28	00742	57392	39064	66432	84673	40027	32832	61362	98947	96067	64760	64584	96096	98253
29	05366	04213	25669	26422	44407	44048	37937	63904	45766	66134	75470	66520	34693	90449
30	91921	26418	64117	94305	26766	25940	39972	22209	71500	64568	91402	42416	07844	69618
31	00582	04711	87917	77341	42206	35126	74087	99547	81817	42607	43808	76655	62028	76630
32	00725	69884	62797	56170	86324	88072	76222	36086	84637	93161	76038	65855	77919	88006
33	69011	65797	95876	55293	18988	27354	26575	08625	40801	59920	29841	80150	12777	48501
34	25976	57948	29888	88604	67917	48708	18912	82271	65424	69774	33611	54262	85963	03547
35	09763	83473	73577	12908	30883	18317	28290	35797	05998	41688	34952	37888	38917	88050
36	91567	42595	27958	30134	04024	86385	29880	99730	55536	84855	29080	09250	79656	73211
37	17955	56349	90999	49127	20044	59931	06115	20542	18059	02008	73708	83517	36103	42791
38	46503	18584	18845	49618	02304	51038	20655	58727	28168	15475	56942	53389	20562	87338
39	92157	89634	94824	78171	84610	82834	09922	25417	44137	48413	25555	21246	35509	20468
40	14577	62765	35605	81263	39667	47358	56873	56307	61607	49518	89656	20103	77490	18062
41	98427	07523	33362	64270	01638	92477	66969	98420	04880	45585	46565	04102	46880	45709
42	34914	63976	88720	82765	34476	17032	87589	40836	32427	70002	70663	88863	77775	69348
43	70060	28277	39475	46473	23219	53416	94970	25832	69975	94884	19661	72828	00102	66794
44	53976	54914	06990	67245	68350	82948	11398	42878	80287	88267	47363	46634	06541	97809
45	76072	29515	40980	07391	58745	25774	22987	80059	39911	96189	41151	14222	60697	59583
46	90725	52210	83974	29992	65831	38857	50490	83765	55657	14361	31720	57375	56228	41546
47	64364	67412	33339	31926	14883	24413	59744	92351	97473	89286	35931	04110	23726	51900
48	08962	00358	31662	25388	61642	34072	81249	35648	56891	69352	48373	45578	78547	81788
49	95012	68379	93526	70765	10593	04542	76463	54328	02349	17247	28865	14777	62730	92277
50	15664	10493	20492	38391	91132	21999	59516	81652	27195	48223	46751	22923	32261	85653

A TABLE OF 14,000 RANDOM UNITS

Line/Col.	(1)	(2)	(3)	(4)	(5)	(6)	(7)	(8)	(9)	(10)	(11)	(12)	(13)	(14)
51	16408	81899	04153	53381	79401	21438	83035	92350	36693	31238	59649	91754	72772	02338
52	18629	81953	05520	91962	04739	13092	97662	24822	94730	06496	35090	04822	86772	98289
53	73115	35101	47498	87637	99016	71060	88824	71013	18735	20286	23153	72924	35165	43040
54	57491	16703	23167	49323	45021	33132	12544	41035	80780	45393	44812	12515	98931	91202
55	30405	83946	23792	14422	15059	45799	22716	19792	09983	74353	68668	30429	70735	25499
56	16631	35006	85900	98275	32388	52390	16815	69298	82732	38480	73817	32523	41961	44437
57	96773	20206	42559	78985	05300	22164	24369	54224	35083	19687	11052	91491	60383	19746
58	38935	64202	14349	82674	66523	44133	00697	35552	35970	19124	63318	29686	03387	59846
59	31624	76384	17403	53363	44167	64486	64758	75366	76554	31601	12614	33072	60332	92325
60	78919	19474	23632	27889	47914	02584	37680	20801	72152	39339	34806	08930	85001	87820
61	03931	33309	57047	74211	63445	17361	62825	39908	05607	91284	68833	25570	38818	46920
62	74426	33278	43972	10119	89917	15665	52872	73823	73144	88662	88970	74492	51805	99378
63	09066	00903	20795	95452	92648	45454	09552	88815	16553	51125	79375	97596	16296	66092
64	42238	12426	87025	14267	20979	04508	64535	31355	86064	29472	47689	05974	52468	16834
65	16153	08002	26504	41744	81959	65642	74240	56302	00033	67107	77510	70625	28725	34191
66	21457	40742	29820	96783	29400	21840	15035	34537	33310	06116	95240	15957	16572	06004
67	21581	57802	02050	89728	17937	37621	47075	42080	97403	48626	68995	43805	33386	21597
68	55612	78095	83197	33732	05810	24813	86902	60397	16489	03264	88525	42786	05269	92532
69	44657	66999	99324	51281	84463	60563	79312	93454	68876	25471	93911	25650	12682	73572
70	91340	84979	46949	81973	37949	61023	43997	15263	80644	43942	89203	71795	99533	50501
71	91227	21199	31935	27022	84067	05462	35216	14486	29891	68607	41867	14951	91696	85065
72	50001	38140	66321	19924	72163	09538	12151	06878	91903	18749	34405	56087	82790	70925
73	65390	05224	72958	28609	81406	39147	25549	48542	42627	45233	57202	94617	23772	07896
74	27504	96131	83944	41575	10573	08619	64482	73923	36152	05184	94142	25299	84387	34925
75	37169	94851	39117	89632	00959	16487	65536	49071	39782	17095	02330	74301	00275	48280
76	11508	70225	51111	38351	19444	66499	71945	05422	13442	78675	84081	66938	93654	59894
77	37449	30362	06694	54690	04052	53115	62757	95348	78662	11163	81651	50245	34971	52924
78	46515	70331	85922	38329	57015	15765	97161	17869	45349	61796	66345	81073	49106	79860
79	30986	81223	42416	58353	21532	30502	32305	86482	05174	07901	54339	58861	74818	46942
80	63798	64995	46583	09765	44160	78128	83991	42865	92520	83531	80377	35909	81250	54238
81	82486	84846	99254	67632	43218	50076	21361	64816	51202	88124	41870	52689	51275	83556
82	21885	32906	92431	09060	64297	51674	64126	62570	26123	05155	59194	52799	28225	85762
83	60336	98782	07408	53458	13564	59089	26445	29789	85205	41001	12535	12133	14645	23541
84	43937	46891	24010	25560	86355	33941	25786	54990	71899	15475	95434	98227	21824	19585
85	97656	63175	89303	16275	07100	92063	21942	18611	47348	20203	18534	03862	78095	50136
86	03299	01221	05418	38982	55758	92237	26759	86367	21216	98442	08303	56613	91511	75928
87	79626	06486	03574	17668	07785	76020	79924	25651	83325	88428	85076	72811	22717	50585
88	85636	68335	47539	03129	65651	11977	02510	26113	99447	68645	34327	15152	55230	93448
89	18039	14367	61337	06177	12143	46609	32989	74014	64708	00533	35398	58408	13261	47908
90	08362	15656	60627	36478	65648	16764	53412	09013	07832	41574	17639	82163	60859	75567
91	79556	29068	04142	16268	15387	12856	66227	38358	22478	73373	88732	09443	82558	05250
92	92608	82674	27072	32534	17075	27698	98204	63863	11951	34648	88022	56148	34925	57031
93	23982	25835	40055	67006	12293	02753	14827	22235	35071	99704	37543	11601	35503	85171
94	09915	96306	05908	97901	28395	14186	00821	80703	70426	75647	76310	88717	37890	40129
95	50937	33300	26695	62247	69927	76123	50842	43834	86654	70959	79725	93872	28117	19233
96	42488	78077	69882	61657	34136	79180	97526	43092	04098	73571	80799	76536	71255	64239
97	46764	86273	63003	93017	31204	36692	40202	35275	57306	55543	53203	18098	47625	88684
98	03237	45430	55417	63282	90816	17349	88298	90183	36600	78406	06216	95787	42579	90730
99	86591	81482	52667	61583	14972	90053	89534	76036	49199	43716	97548	04379	46370	28672
100	38534	01715	94964	87288	65680	43772	39560	12918	86537	62738	19636	51132	25739	56947

Miscellaneous Statistical Tables

A TABLE OF 14,000 RANDOM UNITS

Line/Col.	(1)	(2)	(3)	(4)	(5)	(6)	(7)	(8)	(9)	(10)	(11)	(12)	(13)	(14)
101	13284	16834	74151	92027	24670	36665	00770	22878	02179	51602	07270	76517	97275	45960
102	21224	00370	30420	03883	96648	89428	41583	17564	27395	63904	41548	49197	82277	24120
103	99052	47887	81085	64933	66279	80432	65793	83287	34142	13241	30590	97760	35848	91983
104	00199	50993	98603	38452	87890	94624	69721	57484	67501	77638	44331	11257	71131	11059
105	60578	06483	28733	37867	07936	98710	98539	27186	31237	80612	44488	97819	70401	95419
106	91240	18312	17441	01929	18163	69201	31211	54288	39296	37318	65724	90401	79017	62077
107	97458	14229	12063	59611	32249	90466	33216	19358	02591	54263	88449	01912	07436	50813
108	35249	38646	34475	72417	60514	69257	12489	51924	86871	92446	36607	11458	30440	52639
109	38980	46600	11759	11900	46743	27860	77940	39298	97838	95145	32378	68038	89351	37005
110	10750	52745	38749	87365	58959	53731	89295	59062	39404	13198	59960	70408	29812	83126
111	36247	27850	73958	20673	37800	63835	71051	84724	52492	22342	78071	17456	96104	18327
112	70994	66986	99744	72438	01174	42159	11392	20724	54322	36923	70009	23233	65438	59685
113	99638	94702	11463	18148	81386	80431	90628	52506	02016	85151	88598	47821	00265	82525
114	72055	15774	43857	99805	10419	76939	25993	03544	21560	83471	43989	90770	22965	44247
115	24038	65541	85788	55835	38835	59399	13790	35112	01324	39520	76210	22467	83275	32286
116	74976	14631	35908	28221	39470	91548	12854	30166	09073	75887	36782	00268	97121	57676
117	35553	71628	70189	26436	63407	91178	90348	55359	80392	41012	36270	77786	89578	21059
118	35676	12797	51434	82976	42010	26344	92920	92155	58807	54644	58581	95331	78629	73344
119	74815	67523	72985	23183	02446	63594	98924	20633	58842	85961	07648	70164	34994	67662
120	45246	88048	65173	50989	91060	89894	36063	32819	68559	99221	49475	50558	34698	71800
121	76509	47069	86378	41797	11910	49672	88575	97966	32466	10083	54728	81972	58975	30761
122	19689	90332	04315	21358	97248	11188	39062	63312	52496	07349	79178	33692	57352	72862
123	42751	35318	97513	61537	54955	08159	00337	80778	27507	95478	21252	12746	37554	97775
124	11946	22681	45045	13964	57517	59419	58045	44067	58716	58840	45557	96345	33271	53464
125	96518	48688	20996	11090	48396	57177	83867	86464	14342	21545	46717	72364	86954	55580
126	35726	58643	76869	84622	39098	36083	72505	92265	23107	60278	05822	46760	44294	07672
127	39737	42750	48968	70536	84864	64952	38404	94317	65402	13589	01055	79044	19308	83623
128	97025	66492	56177	04049	80312	48028	26408	43591	75528	65341	49044	95495	81256	53214
129	62814	08075	09788	56350	76787	51591	54509	49295	85830	59860	30883	89660	96142	18354
130	25578	22950	15227	83291	41737	79599	96191	71845	86899	70694	24290	01551	80092	82118
131	68763	69576	88991	49662	46704	63362	56625	00481	73323	91427	15264	06969	57048	54149
132	17900	00813	64361	60725	88974	61005	99709	30666	26451	11528	44323	34778	60342	60388
133	71944	60227	63551	71109	05624	43836	58254	26160	32116	63403	35404	57146	10909	07346
134	54684	93691	85132	64399	29182	44324	14491	55226	78793	34107	30374	48429	51376	09559
135	25946	27623	11258	65204	52832	50880	22273	05554	99521	73791	85744	29276	70326	60251
136	01353	39318	44961	44972	91766	90262	56073	06606	51826	18893	83448	31915	97764	75091
137	99083	88191	27662	99113	57174	35571	99884	13951	71057	53961	61448	74909	07322	80960
138	52021	45406	37945	75234	24327	86978	22644	87779	23753	99926	63898	54886	18051	96314
139	78755	47744	43776	83098	03225	14281	83637	55984	13300	52212	58781	14905	46502	04472
140	25282	69106	59180	16257	22810	43609	12224	25643	89884	31149	85423	32581	34374	70873
141	11959	94202	02743	86847	79725	51811	12998	76844	05320	54236	53891	70226	38632	84776
142	11644	13792	98190	01424	30078	28197	55583	05197	47714	68440	22016	79204	06862	94451
143	06307	97912	68110	59812	95448	43244	31262	88880	13040	16458	43813	89416	42482	33939
144	76285	75714	89585	99296	52640	46518	55486	90754	88932	19937	57119	23251	55619	23679
145	55322	07589	39600	60866	63007	20007	66819	84164	61131	81429	60676	42807	78286	29015
146	78017	90928	90220	92503	83375	26986	74399	30885	88567	29169	72816	53357	15428	86932
147	44768	43342	20696	26331	43140	69744	82928	24988	94237	46138	77426	39039	55596	12655
148	25100	19336	14605	86603	51680	97678	24261	02464	86563	74812	60069	71674	15478	47642
149	83612	46623	62876	85197	07824	91392	58317	37726	84628	42221	10268	20692	15699	29167
150	41347	81666	82961	60413	71020	83658	02415	33322	66036	98712	46795	16308	28413	05417

A TABLE OF 14,000 RANDOM UNITS

Line/Col.	(1)	(2)	(3)	(4)	(5)	(6)	(7)	(8)	(9)	(10)	(11)	(12)	(13)	(14)
151	38128	51178	75096	13609	16110	73533	42564	59870	29399	67834	91055	89917	51096	89011
152	60950	00455	73254	96067	50717	13878	03216	78274	65863	37011	91283	33914	91303	49326
153	90524	17320	29832	96118	75792	25326	22940	24904	80523	38928	91374	55597	97567	38914
154	49897	18278	67160	39408	97056	43517	84426	59650	20247	19293	02019	14790	02852	05819
155	18494	99209	81060	19488	65596	59787	47939	91225	98768	43688	00438	05548	09443	82897
156	65373	72984	30171	37741	70203	94094	87261	30056	58124	70133	18936	02138	59372	09075
157	40653	12843	04213	70925	95360	55774	76439	61768	52817	81151	52188	31940	54273	49032
158	51638	22238	56344	44587	83231	50317	74541	07719	25472	41602	77318	15145	57515	07633
159	69742	99303	62578	83575	30337	07488	51941	84316	42067	49692	28616	29101	03013	73449
160	58012	74072	67488	74580	47992	69482	58624	17106	47538	13452	22620	24260	40155	74716
161	18348	19855	42887	08279	43206	47077	42637	45606	00011	20662	14642	49984	94509	56380
162	59614	09193	58064	29086	44385	45740	70752	05663	49081	26960	57454	99264	24142	74648
163	75688	28630	39210	52897	62748	72658	98059	67202	72789	01869	13496	14663	87645	89713
164	13941	77802	69101	70061	35460	34576	15412	81304	58757	35498	94830	75521	00603	97701
165	96656	86420	96475	86458	54463	96419	55417	41375	76886	19008	66877	35934	59801	00497
166	03363	82042	15942	14549	38324	87094	19069	67590	11087	68570	22591	65232	85915	91499
167	70366	08390	69155	25496	13240	57407	91407	49160	07379	34444	94567	66035	38918	65708
168	47870	36605	12927	16043	53257	93796	52721	73120	48025	76074	95605	67422	41646	14557
169	79504	77606	22761	30518	28373	73898	30550	76684	77366	32276	04690	61667	64798	66276
170	46967	74841	50923	15339	37755	98995	40162	89561	69199	42257	11647	47603	48779	97907
171	14558	50769	35444	59030	87516	48193	02945	00922	48189	04724	21263	20892	92955	90251
172	12440	25057	01132	38611	28135	68089	10954	10097	54243	06460	50856	65435	79377	53890
173	32293	29938	68653	10497	98919	46587	77701	99119	93165	67788	17638	23097	21468	36992
174	10640	21875	72462	77981	56550	55999	87310	69643	45124	00349	25748	00844	96831	30651
175	47615	23169	39571	56972	20628	21788	51736	33133	72696	32605	41569	76148	91544	21121
176	16948	11128	71624	72754	49084	96303	27830	45817	67867	18062	87453	17226	72904	71474
177	21258	61092	66634	70335	92448	17354	83432	49608	66520	06442	59664	20420	39201	69549
178	15072	48853	15178	30730	47481	48490	41436	25015	49932	20474	53821	51015	79841	32405
179	99154	57412	09858	65671	70655	71479	63520	31357	56968	06729	34465	70685	04184	25250
180	08759	61089	23706	32994	35426	36666	63988	98844	37533	08269	27021	45886	22835	78451
181	67323	57839	61114	62192	47547	58203	64630	34886	98777	75442	95592	06141	45096	73117
182	09255	13986	84834	20764	72206	89393	34548	93438	88730	61805	78955	18952	46436	58740
183	36304	74712	00374	10107	85061	69228	81969	92216	03568	39630	81869	52824	50937	27954
184	15884	67429	86612	47367	10242	44880	12060	44309	46629	55105	66793	93173	00480	13311
185	18745	32031	35303	08134	33925	03044	59929	95418	04917	57596	24878	61733	92834	64454
186	72934	40086	88292	65728	38300	42323	64068	98373	48971	09049	59943	36538	05976	82118
187	17626	02944	20910	57662	80181	38579	24580	90529	52303	50436	29401	57824	86039	81062
188	27117	61399	50967	41399	81636	16663	15634	79717	94696	59240	25543	97989	63306	90946
189	93995	18678	90012	63645	85701	85269	62263	68331	00389	72571	15210	20769	44686	96176
190	67392	89421	09623	80725	62620	84162	87368	29560	00519	84545	08004	24526	41252	14521
191	04910	12261	37566	80016	21245	69377	50420	85658	55263	68667	78770	04533	14513	18099
192	81453	20283	79929	59839	23875	13245	46808	74124	74703	35769	95588	21014	37078	39170
193	19480	75790	48539	23703	15537	48885	02861	86587	74539	65227	90799	58789	96257	02708
194	21456	13162	74608	81011	55512	07481	93551	72189	76261	91206	89941	15132	37738	59284
195	89406	20912	46189	76376	25538	87212	20748	12831	57166	35026	16817	79121	18929	40628
196	09866	07414	55977	16419	01101	69343	13305	94302	80703	57910	36933	57771	42546	03003
197	86541	24681	23421	13521	28000	94917	07423	57523	97234	63951	42876	46829	09781	58160
198	10414	96941	06205	72222	57167	83902	07460	69507	10600	08858	07685	44472	64220	27040
199	49942	06683	41479	58982	56288	42853	92196	20632	62045	78812	35895	51851	83534	10689
200	23995	68882	42291	23374	24299	27024	67460	94783	40937	16961	26053	78749	46704	21983

Miscellaneous Statistical Tables

XII.4   RANDOM NORMAL NUMBERS, $\mu = 0$, $\sigma = 1$

01	02	03	04	05	06	07	08	09	10
0.464	0.137	2.455	−0.323	−0.068	0.296	−0.288	1.298	0.241	−0.957
0.060	−2.526	−0.531	−0.194	0.543	−1.558	0.187	−1.190	0.022	0.525
1.486	−0.354	−0.634	0.697	0.926	1.375	0.785	−0.963	−0.853	−1.865
1.022	−0.472	1.279	3.521	0.571	−1.851	0.194	1.192	−0.501	−0.273
1.394	−0.555	0.046	0.321	2.945	1.974	−0.258	0.412	0.439	−0.035
0.906	−0.513	−0.525	0.595	0.881	−0.934	1.579	0.161	−1.885	0.371
1.179	−1.055	0.007	0.769	0.971	0.712	1.090	−0.631	−0.255	−0.702
−1.501	−0.488	−0.162	−0.136	1.033	0.203	0.448	0.748	−0.423	−0.432
−0.690	0.756	−1.618	−0.345	−0.511	−2.051	−0.457	−0.218	0.857	−0.465
1.372	0.225	0.378	0.761	0.181	−0.736	0.960	−1.530	−0.260	0.120
−0.482	1.678	−0.057	−1.229	−0.486	0.856	−0.491	−1.983	−2.830	−0.238
−1.376	−0.150	1.356	−0.561	−0.256	−0.212	0.219	0.779	0.953	−0.869
−1.010	0.598	−0.918	1.598	0.065	0.415	−0.169	0.313	−0.973	−1.016
−0.005	−0.899	0.012	−0.725	1.147	−0.121	1.096	0.481	−1.691	0.417
1.393	−1.163	−0.911	1.231	−0.199	−0.246	1.239	−2.574	−0.558	0.056
−1.787	−0.261	1.237	1.046	−0.508	−1.630	−0.146	−0.392	−0.627	0.561
−0.105	−0.357	−1.384	0.360	−0.992	−0.116	−1.698	−2.832	−1.108	−2.357
−1.339	1.827	−0.959	0.424	0.969	−1.141	−1.041	0.362	−1.726	1.956
1.041	0.535	0.731	1.377	0.983	−1.330	1.620	−1.040	0.524	−0.281
0.279	−2.056	0.717	−0.873	−1.096	−1.396	1.047	0.089	−0.573	0.932
−1.805	−2.008	−1.633	0.542	0.250	−0.166	0.032	0.079	0.471	−1.029
−1.186	1.180	1.114	0.882	1.265	−0.202	0.151	−0.376	−0.310	0.479
0.658	−1.141	1.151	−1.210	−0.927	0.425	0.290	−0.902	0.610	2.709
−0.439	0.358	−1.939	0.891	−0.227	0.602	0.873	−0.437	−0.220	−0.057
−1.399	−0.230	0.385	−0.649	−0.577	0.237	−0.289	0.513	0.738	−0.300
0.199	0.208	−1.083	−0.219	−0.291	1.221	1.119	0.004	−2.015	−0.594
0.159	0.272	−0.313	0.084	−2.828	−0.439	−0.792	−1.275	−0.623	−1.047
2.273	0.606	0.606	−0.747	0.247	1.291	0.063	−1.793	−0.699	−1.347
0.041	−0.307	0.121	0.790	−0.584	0.541	0.484	−0.986	0.481	0.996
−1.132	−2.098	0.921	0.145	0.446	−1.661	1.045	−1.363	−0.586	−1.023
0.768	0.079	−1.473	0.034	−2.127	0.665	0.084	−0.880	−0.579	0.551
0.375	−1.658	−0.851	0.234	−0.656	0.340	−0.086	−0.158	−0.120	0.418
−0.513	−0.344	0.210	−0.735	1.041	0.008	0.427	−0.831	0.191	0.074
0.292	−0.521	1.266	−1.206	−0.899	0.110	−0.528	−0.813	0.071	0.524
1.026	2.990	−0.574	−0.491	−1.114	1.297	−1.433	−1.345	−3.001	0.479
−1.334	1.278	−0.568	−0.109	−0.515	−0.566	2.923	0.500	0.359	0.326
−0.287	−0.144	−0.254	0.574	−0.451	−1.181	−1.190	−0.318	−0.094	1.114
0.161	−0.886	−0.921	−0.509	1.410	−0.518	0.192	−0.432	1.501	1.068
−1.346	0.193	−1.202	0.394	−1.045	0.843	0.942	1.045	0.031	0.772
1.250	−0.199	−0.288	1.810	1.378	0.584	1.216	0.733	0.402	0.226
0.630	−0.537	0.782	0.060	0.499	−0.431	1.705	1.164	0.884	−0.298
0.375	−1.941	0.247	−0.491	−0.665	−0.135	−0.145	−0.498	0.457	1.064
−1.420	0.489	−1.711	−1.186	0.754	−0.732	−0.066	1.006	−0.798	0.162
−0.151	−0.243	−0.430	−0.762	0.298	1.049	1.810	2.885	−0.768	−0.129
−0.309	0.531	0.416	−1.541	1.456	2.040	−0.124	0.196	0.023	−1.204
0.424	−0.444	0.593	0.993	−0.106	0.116	0.484	−1.272	1.066	1.097
0.593	0.658	−1.127	−1.407	−1.579	−1.616	1.458	1.262	0.736	−0.916
0.862	−0.885	−0.142	−0.504	0.532	1.381	0.022	−0.281	−0.342	1.222
0.235	−0.628	−0.023	−0.463	−0.899	−0.394	−0.538	1.707	−0.188	−1.153
−0.853	0.402	0.777	0.833	0.410	−0.349	−1.094	0.580	1.395	1.298

RANDOM NORMAL NUMBERS, $\mu = 0$, $\sigma = 1$

11	12	13	14	15	16	17	18	19	20
1.329	−0.238	−0.838	−0.988	−0.445	0.964	−0.266	−0.322	−1.726	2.252
1.284	−0.229	1.058	0.090	0.050	0.523	0.016	0.277	1.639	0.554
0.619	0.628	0.005	0.973	−0.058	0.150	−0.635	−0.917	0.313	−1.203
0.699	−0.269	0.722	−0.994	−0.807	−1.203	1.163	1.244	1.306	−1.210
0.101	0.202	−0.150	0.731	0.420	0.116	−0.496	−0.037	−2.466	0.794
1.381	0.301	0.522	0.233	0.791	−1.017	−0.182	0.926	−1.096	1.001
0.574	1.366	−1.843	0.746	0.890	0.824	−1.249	−0.806	−0.240	0.217
0.096	0.210	1.091	0.990	0.900	−0.837	−1.097	−1.238	0.030	−0.311
1.389	−0.236	0.094	3.282	0.295	−0.416	0.313	0.720	0.007	0.354
1.249	0.706	1.453	0.366	−2.654	−1.400	0.212	0.307	−1.145	0.639
0.756	−0.397	−1.772	−0.257	1.120	1.188	−0.527	0.709	0.479	0.317
0.860	0.412	−0.327	0.178	0.524	−0.672	−0.831	0.758	0.131	0.771
0.778	−0.979	0.236	−1.033	1.497	−0.661	0.906	1.169	−1.582	1.303
0.037	0.062	0.426	1.220,	0.471	0.784	−0.719	0.465	1.559	−1.326
2.619	−0.440	0.477	1.063	0.320	1.406	−0.701	−0.128	0.518	−0.676
0.420	−0.287	−0.050	−0.481	1.521	−1.367	0.609	0.292	0.048	0.592
1.048	0.220	1.121	−1.789	−1.211	−0.871	−0.740	0.513	−0.558	−0.395
1.000	−0.638	1.261	0.510	−0.150	0.034	0.054	−0.055	0.639	−0.825
0.170	−1.131	−0.985	0.102	−0.939	−1.457	1.766	1.087	−1.275	2.362
0.389	−0.435	0.171	0.891	1.158	1.041	1.048	−0.324	−0.404	1.060
0.305	0.838	−2.019	−0.540	0.905	1.195	−1.190	0.106	0.571	0.298
0.321	−0.039	1.799	−1.032	−2.225	−0.148	0.758	−0.862	0.158	−0.726
1.900	1.572	−0.244	−1.721	1.130	0.495	−0.484	0.014	−0.778	−1.483
0.778	−0.288	−0.224	−1.324	−0.072	0.890	−0.410	0.752	0.376	−0.224
0.617	−1.718	−0.183	−0.100	1.719	0.696	−1.339	−0.614	1.071	−0.386
1.430	−0.953	0.770	−0.007	−1.872	1.075	−0.913	−1.168	1.775	0.238
0.267	−0.048	0.972	0.734	−1.408	−1.955	−0.848	2.002	0.232	−1.273
0.978	−0.520	−0.368	1.690	−1.479	0.985	1.475	−0.098	−1.633	2.399
1.235	−1.168	0.325	1.421	2.652	−0.486	−1.253	0.270	−1.103	0.118
0.258	0.638	2.309	0.741	−0.161	−0.679	0.336	1.973	0.370	−2.277
0.243	0.629	−1.516	−0.157	0.693	1.710	0.800	−0.265	1.218	0.655
0.292	−1.455	−1.451	1.492	−0.713	0.821	−0.031	−0.780	1.330	0.977
0.505	0.389	0.544	−0.042	1.615	−1.440	−0.989	−0.580	0.156	0.052
0.397	−0.287	1.712	0.289	−0.904	0.259	−0.600	−1.635	−0.009	−0.799
0.605	−0.470	0.007	0.721	−1.117	0.635	0.592	−1.362	−1.441	0.672
1.360	0.182	−1.476	−0.599	−0.875	0.292	−0.700	0.058	−0.340	−0.639
0.480	−0.699	1.615	−0.225	1.014	−1.370	−1.097	0.294	0.309	−1.389
0.027	−0.487	−1.000	−0.015	0.119	−1.990	−0.687	−1.964	−0.366	1.759
1.482	−0.815	−0.121	1.884	−0.185	0.601	0.793	0.430	−1.181	0.426
1.256	−0.567	−0.994	1.011	−1.071	−0.623	−0.420	−0.309	1.362	0.863
1.132	2.039	1.934	−0.222	0.386	1.100	0.284	1.597	−1.718	−0.560
0.780	−0.239	−0.497	−0.434	−0.284	−0.241	−0.333	1.348	−0.478	−0.169
0.859	−0.215	0.241	1.471	0.389	−0.952	0.245	0.781	1.093	−0.240
0.447	1.479	0.067	0.426	−0.370	−0.675	−0.972	0.225	0.815	0.389
0.269	0.735	−0.066	−0.271	−1.439	1.036	−0.306	−1.439	−0.122	−0.336
0.097	−1.883	−0.218	0.202	−0.357	0.019	1.631	1.400	0.223	−0.793
0.686	1.596	−0.286	0.722	0.655	−0.275	1.245	−1.504	0.066	−1.280
0.957	0.057	−1.153	0.701	−0.280	1.747	−0.745	1.338	−1.421	0.386
0.976	−1.789	−0.696	−1.799	−0.354	0.071	2.355	0.135	−0.598	1.883
0.274	0.226	−0.909	−0.572	0.181	1.115	0.406	0.453	−1.218	−0.115

Miscellaneous Statistical Tables

RANDOM NORMAL NUMBERS, $\mu = 0$, $\sigma = 1$

21	22	23	24	25	26	27	28	29	30
-1.752	-0.329	-1.256	0.318	1.531	0.349	-0.958	-0.059	0.415	-1.08
-0.291	0.085	1.701	-1.087	-0.443	-0.292	0.248	-0.539	-1.382	0.31
-0.933	0.130	0.634	0.899	1.409	-0.883	-0.095	0.229	0.129	0.36
-0.450	-0.244	0.072	1.028	1.730	-0.056	-1.488	-0.078	-2.361	-0.99
0.512	-0.882	0.490	-1.304	-0.266	0.757	-0.361	0.194	-1.078	0.52
-0.702	0.472	0.429	-0.664	-0.592	1.443	-1.515	-1.209	-1.043	0.27
0.284	0.039	-0.518	1.351	1.473	0.889	0.300	0.339	-0.206	1.39
-0.509	1.420	-0.782	-0.429	-1.266	0.627	-1.165	0.819	-0.261	0.40
-1.776	-1.033	1.977	0.014	0.702	-0.435	-0.816	1.131	0.656	0.06
-0.044	1.807	0.342	-2.510	1.071	-1.220	-0.060	-0.764	0.079	-0.96
0.263	-0.578	1.612	-0.148	-0.383	-1.007	-0.414	0.638	-0.186	0.50
0.986	0.439	-0.192	-0.132	0.167	0.883	-0.400	-1.440	-0.385	-1.41
-0.441	-0.852	-1.446	-0.605	-0.348	1.018	0.963	-0.004	2.504	-0.84
-0.866	0.489	0.097	0.379	0.192	-0.842	0.065	1.420	0.426	-1.19
-1.215	0.675	1.621	0.394	-1.447	2.199	-0.321	-0.540	-0.037	0.18
-0.475	-1.210	0.183	0.526	0.495	1.297	-1.613	1.241	-1.016	-0.09
1.200	0.131	2.502	0.344	-1.060	-0.909	-1.695	-0.666	-0.838	-0.86
-0.498	-1.202	-0.057	-1.354	-1.441	-1.590	0.987	0.441	0.637	-1.11
-0.743	0.894	-0.028	1.119	-0.598	0.279	2.241	0.830	0.267	-0.15
0.779	-0.780	-0.954	0.705	-0.361	-0.734	1.365	1.297	-0.142	-1.38
-0.206	-0.195	1.017	-1.167	-0.079	-0.452	0.058	-1.068	-0.394	-0.40
-0.092	-0.927	-0.439	0.256	0.503	0.338	1.511	-0.465	-0.118	-0.45
-1.222	-1.582	1.786	-0.517	-1.080	-0.409	-0.474	-1.890	0.247	0.57
0.068	0.075	-1.383	-0.084	0.159	1.276	1.141	0.186	-0.973	-0.26
0.183	1.600	-0.335	1.553	0.889	0.896	-0.035	0.461	0.486	1.24
-0.811	-2.904	0.618	0.588	0.533	0.803	-0.696	0.690	0.820	0.55
-1.010	1.149	1.033	0.336	1.306	0.835	1.523	0.296	-0.426	0.00
1.453	1.210	-0.043	0.220	-0.256	-1.161	-2.030	-0.046	0.243	1.08
0.759	-0.838	-0.877	-0.177	1.183	-0.218	-3.154	-0.963	-0.822	-1.11
0.287	0.278	-0.454	0.897	-0.122	0.013	0.346	0.921	0.238	-0.58
-0.669	0.035	-2.077	1.077	0.525	-0.154	-1.036	0.015	-0.220	0.88
0.392	0.106	-1.430	-0.204	-0.326	0.825	-0.432	-0.094	-1.566	0.67
-0.337	0.199	-0.160	0.625	-0.891	-1.464	-0.318	1.297	0.932	-0.03
0.369	-1.990	-1.190	0.666	-1.614	0.082	0.922	-0.139	-0.833	0.09
-1.694	0.710	-0.655	-0.546	1.654	0.134	0.466	0.033	-0.039	0.83
0.985	0.340	0.276	0.911	-0.170	-0.551	1.000	-0.838	0.275	-0.30
-1.063	-0.594	-1.526	-0.787	0.873	-0.405	-1.324	0.162	-0.163	-2.71
0.033	-1.527	1.422	0.308	0.845	-0.151	0.741	0.064	1.212	0.82
0.597	0.362	-3.760	1.159	0.874	-0.794	-0.915	1.215	1.627	-1.24
-1.601	-0.570	0.133	-0.660	1.485	0.682	-0.898	0.686	0.658	0.34
-0.266	-1.309	0.597	0.989	0.934	1.079	-0.656	-0.999	-0.036	-0.53
0.901	1.531	-0.889	-1.019	0.084	1.531	-0.144	-1.920	0.678	-0.40
-1.433	-1.008	-0.990	0.090	0.940	0.207	-0.745	0.638	1.469	1.21
1.327	0.763	-1.724	-0.709	-1.100	-1.346	-0.946	-0.157	0.522	-1.26
-0.248	0.788	-0.577	0.122	-0.536	0.293	1.207	-2.243	1.642	1.35
-0.401	-0.679	0.921	0.476	1.121	-0.864	0.128	-0.551	-0.872	1.51
0.344	-0.324	0.686	-1.487	-0.126	0.803	-0.961	0.183	-0.358	-0.18
0.441	-0.372	-1.336	0.062	1.506	-0.315	-0.112	-0.452	1.594	-0.26
0.824	0.040	-1.734	0.251	0.054	-0.379	1.298	-0.126	0.104	-0.52
1.385	1.320	-0.509	-0.381	-1.671	-0.524	-0.805	1.348	0.676	0.79

RANDOM NORMAL NUMBERS, $\mu = 0$, $\sigma = 1$

31	32	33	34	35	36	37	38	39	40
1.556	0.119	−0.078	0.164	−0.455	0.077	−0.043	−0.299	0.249	−0.182
0.647	1.029	1.186	0.887	1.204	−0.657	0.644	−0.410	−0.652	−0.165
0.329	0.407	1.169	−2.072	1.661	0.891	0.233	−1.628	−0.762	−0.717
−1.188	1.171	−1.170	−0.291	0.863	−0.045	−0.205	0.574	−0.926	1.407
−0.917	−0.616	−1.589	1.184	0.266	0.559	−1.833	−0.572	−0.648	−1.090
0.414	0.469	−0.182	0.397	1.649	1.198	0.067	−1.526	−0.081	−0.192
0.107	−0.187	1.343	0.472	−0.112	1.182	0.548	2.748	0.249	0.154
−0.497	1.907	0.191	0.136	−0.475	0.458	0.183	−1.640	−0.058	1.278
0.501	0.083	−0.321	1.133	1.126	−0.299	1.299	1.617	1.581	2.455
−1.382	−0.738	1.225	1.564	−0.363	−0.548	1.070	0.390	−1.398	0.524
−0.590	0.699	−0.162	−0.011	1.049	−0.689	1.225	0.339	−0.539	−0.445
−1.125	1.111	−1.065	0.534	0.102	0.425	−1.026	0.695	−0.057	0.795
0.849	0.169	−0.351	0.584	2.177	0.009	−0.696	−0.426	−0.692	−1.638
−1.233	−0.585	0.306	0.773	1.304	−1.304	0.282	−1.705	0.187	−0.880
0.104	−0.468	0.185	0.498	−0.624	−0.322	−0.875	1.478	−0.691	−0.281
0.261	−1.883	−0.181	1.675	−0.324	−1.029	−0.185	0.004	−0.101	−1.187
−0.007	1.280	0.568	−1.270	1.405	1.731	2.072	1.686	0.728	−0.417
0.794	−0.111	0.040	−0.536	−0.976	2.192	1.609	−0.190	−0.279	−1.611
0.431	−2.300	−1.081	−1.370	2.943	0.653	−2.523	0.756	0.886	−0.983
−0.149	1.294	−0.580	0.482	−1.449	−1.067	1.996	−0.274	0.721	0.490
−0.216	−1.647	1.043	0.481	−0.011	−0.587	−0.916	−1.016	−1.040	−1.117
1.604	−0.851	−0.317	−0.686	−0.008	1.939	0.078	−0.465	0.533	0.652
−0.212	0.005	0.535	0.837	0.362	1.103	0.219	0.488	1.332	−0.200
0.007	−0.076	1.484	0.455	−0.207	−0.554	1.120	0.913	−0.681	1.751
−0.217	0.937	0.860	0.323	1.321	−0.492	−1.386	−0.003	−0.230	0.539
−0.649	0.300	−0.698	0.900	0.569	0.842	0.804	1.025	0.603	−1.546
−1.541	0.193	2.047	−0.552	1.190	−0.087	2.062	−2.173	−0.791	−0.520
0.274	−0.530	0.112	0.385	0.656	0.436	0.882	0.312	−2.265	−0.218
0.876	−1.498	−0.128	−0.387	−1.259	−0.856	−0.353	0.714	0.863	1.169
−0.859	−1.083	1.288	−0.078	−0.081	0.210	0.572	1.194	−1.118	−1.543
−0.015	−0.567	0.113	2.127	−0.719	3.256	−0.721	−0.663	−0.779	−0.930
−1.529	−0.231	1.223	0.300	−0.995	−0.651	0.505	0.138	−0.064	1.341
0.278	−0.058	−2.740	−0.296	−1.180	0.574	1.452	0.846	−0.243	−1.208
1.428	0.322	2.302	−0.852	0.782	−1.322	−0.092	−0.546	0.560	−1.430
0.770	−1.874	0.347	0.994	−0.485	−1.179	0.048	−1.324	1.061	0.449
−0.303	−0.629	0.764	0.013	−1.192	−0.475	−1.085	−0.880	1.738	−1.225
−0.263	−2.105	0.509	−0.645	1.362	0.504	−0.755	1.274	1.448	0.604
0.997	−1.187	−0.242	0.121	2.510	−1.935	0.350	0.073	0.458	−0.446
−0.063	−0.475	−1.802	−0.476	0.193	−1.199	0.339	0.364	−0.684	1.353
−0.168	1.904	−0.485	−0.032	−0.554	0.056	−0.710	−0.778	0.722	−0.024
0.366	−0.491	0.301	−0.008	−0.894	−0.945	0.384	−1.748	−1.118	0.394
0.436	−0.464	0.539	0.942	−0.458	0.445	−1.883	1.228	1.113	−0.218
0.597	−1.471	−0.434	0.705	−0.788	0.575	0.086	0.504	1.445	−0.513
−0.805	−0.624	1.344	0.649	−1.124	0.680	−0.986	1.845	−1.152	−0.393
1.681	−1.910	0.440	0.067	−1.502	−0.755	−0.989	−0.054	−2.320	0.474
−0.007	−0.459	1.940	0.220	−1.259	−1.729	0.137	−0.520	−0.412	2.847
0.209	−0.633	0.299	0.174	1.975	−0.271	0.119	−0.199	0.007	2.315
1.254	1.672	−1.186	−1.310	0.474	0.878	−0.725	−0.191	0.642	−1.212
−1.016	−0.697	0.017	−0.263	−0.047	−1.294	−0.339	2.257	−0.078	−0.049
−1.169	−0.355	1.086	−0.199	0.031	0.396	−0.143	1.572	0.276	0.027

Miscellaneous Statistical Tables

RANDOM NORMAL NUMBERS, $\mu = 0$, $\sigma = 1$

41	42	43	44	45	46	47	48	49	50
−0.856	−0.063	0.787	−2.052	−1.192	−0.831	1.623	1.135	0.759	−0.18
−0.276	−1.110	0.752	−1.378	−0.583	0.360	0.365	1.587	0.621	1.34
0.379	−0.440	0.858	1.453	−1.356	0.503	−1.134	1.950	−1.816	−0.28
1.468	0.131	0.047	0.355	0.162	−1.491	−0.739	−1.182	−0.533	−0.49
−1.805	−0.772	1.286	−0.636	−1.312	−1.045	1.559	−0.871	−0.102	−0.12
2.285	0.554	0.418	−0.577	−1.489	−1.255	0.092	−0.597	−1.051	−0.98
−0.602	0.399	1.121	−1.026	0.087	1.018	−1.437	0.661	0.091	−0.63
0.229	−0.584	0.705	0.124	0.341	1.320	−0.824	−1.541	−0.163	2.32
1.382	−1.454	1.537	−1.299	0.363	−0.356	−0.025	0.294	2.194	−0.39
0.978	0.109	1.434	−1.094	−0.265	−0.857	−1.421	−1.773	0.570	−0.08
−0.678	−2.335	1.202	−1.697	0.547	−0.201	−0.373	−1.363	−0.081	0.98
−0.366	−1.084	−0.626	0.798	1.706	−1.160	−0.838	1.462	0.636	0.57
−1.074	−1.379	0.086	−0.331	−0.288	−0.309	−1.527	−0.408	0.183	0.88
−0.600	−0.096	0.696	0.446	1.417	−2.140	0.599	−0.157	1.485	1.38
0.918	1.163	−1.445	0.759	0.878	−1.781	−0.056	−2.141	−0.234	0.97
−0.791	−0.528	0.946	1.673	−0.680	−0.784	1.494	−0.086	−1.071	−1.11
0.598	−0.352	0.719	−0.341	0.056	−1.041	1.429	0.235	0.314	−1.69
0.567	−1.156	−0.125	−0.534	0.711	−0.511	0.187	−0.644	−1.090	−1.28
0.963	0.052	0.037	0.637	−1.335	0.055	0.010	−0.860	−0.621	0.71
0.489	−0.209	1.659	0.054	1.635	0.169	0.794	−1.550	1.845	−0.38
−1.627	−0.017	0.699	0.661	−0.073	0.188	1.183	−1.054	−1.615	−0.70
−1.096	1.215	0.320	0.738	1.865	−1.169	−0.667	−0.674	−0.062	1.37
−2.532	1.031	−0.799	1.665	−2.756	−0.151	−0.704	0.602	−0.672	1.20
0.024	−1.183	−0.927	−0.629	0.204	−0.825	0.496	2.543	0.262	−0.78
0.192	0.125	0.373	−0.931	−0.079	0.186	−0.306	0.621	−0.292	1.13
−1.324	−1.229	−0.648	−0.430	0.811	0.868	0.787	1.845	−0.374	−0.68
−0.726	−0.746	1.572	−1.420	1.509	−0.361	−0.310	−3.117	1.637	0.64
−1.618	1.082	−0.319	0.300	1.524	−0.418	−1.712	0.358	−1.032	0.53
1.695	0.843	2.049	0.388	−0.297	1.077	−0.462	0.655	0.940	−0.38
0.790	0.605	−3.077	1.009	−0.906	−1.004	0.693	−1.098	1.300	0.54
1.792	−0.895	−0.136	−1.765	1.077	0.418	−0.150	0.808	0.697	0.43
0.771	−0.741	−0.492	−0.770	−0.458	−0.021	1.385	−1.225	−0.066	−1.47
−1.438	0.423	−1.211	0.723	−0.731	0.883	−2.109	−2.455	−0.210	1.64
−0.294	1.266	−1.994	−0.730	0.545	0.397	1.069	−0.383	−0.097	−0.98
−1.966	0.909	0.400	0.685	−0.800	1.759	0.268	1.387	−0.414	1.61
0.999	1.587	1.423	0.937	−0.943	0.090	1.185	−1.204	0.300	−1.38
0.581	0.481	−2.400	0.000	0.231	0.079	−2.842	−0.846	−0.508	−0.51
0.370	−1.452	−0.580	−1.462	−0.972	1.116	−0.994	0.374	−3.336	−0.08
0.834	−1.227	−0.709	−1.039	−0.014	−0.383	−0.512	−0.347	0.881	−0.63
−0.376	−0.813	0.660	−1.029	−0.137	0.371	0.376	0.968	1.338	−0.78
−1.621	0.815	−0.544	−0.376	−0.852	0.436	1.562	0.815	−1.048	0.18
0.163	−0.161	2.501	−0.265	−0.285	1.934	1.070	0.215	−0.876	0.07
1.786	−0.538	−0.437	0.324	0.105	−0.421	−0.410	−0.947	0.700	−1.00
2.140	1.218	−0.351	−0.068	0.254	0.448	−1.461	0.784	0.317	1.01
0.064	0.410	0.368	0.419	−0.982	1.371	0.100	−0.505	0.856	0.89
0.789	−0.131	1.330	0.506	−0.645	−1.414	2.426	1.389	−0.169	−0.19
−0.011	−0.372	−0.699	2.382	−1.395	−0.467	1.256	−0.585	−1.359	−1.86
−0.463	0.003	−1.470	1.493	0.960	0.364	−1.267	−0.007	0.616	0.62
−1.210	−0.669	0.009	1.284	−0.617	0.355	−0.589	−0.243	−0.015	−0.71
−1.157	0.481	0.560	1.287	1.129	−0.126	0.006	1.532	1.328	0.98

RANDOM NORMAL NUMBERS, $\mu = 0$, $\sigma = 1$

51	52	53	54	55	56	57	58	59	60
0.240	1.774	0.210	−1.471	1.167	−1.114	0.182	−0.485	−0.318	1.156
0.627	−0.758	−0.930	1.641	0.162	−0.874	−0.235	0.203	−0.724	−0.155
−0.594	0.098	0.158	−0.722	1.385	−0.985	−1.707	0.175	0.449	0.654
1.082	−0.753	−1.944	−1.964	−2.131	−2.796	−1.286	0.807	−0.122	0.527
0.060	−0.014	1.577	−0.814	−0.633	0.275	−0.087	0.517	0.474	−1.432
−0.013	0.402	−0.086	−0.394	0.292	−2.862	−1.660	−1.658	1.610	−2.205
1.586	−0.833	1.444	−0.615	−1.157	−0.220	−0.517	−1.668	−2.036	−0.850
−0.405	−1.315	−1.355	−1.331	1.394	−0.381	−0.729	−0.447	−0.906	0.622
−0.329	1.701	0.427	0.627	−9.271	−0.971	−1.010	1.182	−0.143	0.844
0.992	0.708	−0.115	−1.630	0.596	0.499	−0.862	0.508	0.474	−0.974
0.296	−0.390	2.047	−0.363	0.724	0.788	−0.089	0.930	−0.497	0.058
2.069	−1.422	−0.948	−1.742	−1.173	0.215	0.661	0.842	−0.984	−0.577
−0.211	−1.727	−0.277	1.592	−0.707	0.327	−0.527	0.912	0.571	−0.525
−0.467	1.848	−0.263	−0.862	0.706	−0.533	0.626	−0.200	−2.221	0.368
1.284	0.412	1.512	0.328	0.203	−1.231	−1.480	−0.400	−0.491	0.913
0.821	−1.503	−1.066	1.624	1.345	0.440	−1.416	0.301	−0.355	0.106
1.056	1.224	0.281	−0.098	1.868	−0.395	0.610	−1.173	−1.449	1.171
1.090	−0.790	0.882	1.687	−0.009	−2.053	−0.030	−0.421	1.253	−0.081
0.574	0.129	1.203	0.280	1.438	−2.052	−0.443	0.522	0.468	−1.211
−0.531	2.155	0.334	0.898	−1.114	0.243	1.026	0.391	−0.011	−0.024
0.896	0.181	−0.941	−0.511	0.648	−0.710	−0.181	−1.417	−0.585	0.087
0.042	0.579	−0.316	0.394	1.133	−0.305	−0.683	−1.318	−0.050	0.993
2.328	−0.243	0.534	0.241	0.275	0.060	0.727	−1.459	0.174	−1.072
0.486	−0.558	0.426	0.728	−0.360	−0.068	0.058	1.471	−0.051	0.337
−0.304	−0.309	0.646	0.309	−1.320	0.311	−1.407	−0.011	0.387	0.128
−2.319	−0.129	0.866	−0.424	0.236	0.419	−1.359	−1.088	−0.045	1.096
1.098	−0.875	0.659	−1.086	−0.424	−1.462	0.743	−0.787	1.472	1.677
−0.038	−0.118	−1.285	−0.545	−0.140	1.244	−1.104	0.146	0.058	1.245
−0.207	−0.746	1.681	0.137	0.104	−0.491	−0.935	0.671	−0.448	−0.129
0.333	−1.386	1.840	1.089	0.837	−1.642	−0.273	−0.798	0.067	0.334
1.190	−0.547	−1.016	0.540	−0.993	0.443	−0.190	1.019	−1.021	−1.276
−1.416	−0.749	0.325	0.846	2.417	−0.479	−0.655	−1.326	−1.952	1.234
0.622	0.661	0.028	1.302	−0.032	−0.157	1.470	−0.766	0.697	−0.303
−1.134	0.499	0.538	0.564	−2.392	−1.398	0.010	1.874	1.386	0.000
0.725	−0.242	0.281	1.355	−0.036	0.204	−0.345	0.395	−0.753	1.645
−0.210	0.611	−0.219	0.450	0.308	0.993	−0.146	0.225	−1.496	0.246
0.219	0.302	0.000	−0.437	−2.127	0.883	−0.599	−1.516	0.826	1.242
−1.098	−0.252	−2.480	−0.973	0.712	−1.430	−0.167	−1.237	0.750	−0.763
0.144	0.489	−0.637	1.990	0.411	−0.563	0.027	1.278	2.105	−1.130
−1.738	−1.295	0.431	−0.503	2.327	−0.007	−1.293	−1.206	−0.066	1.370
−0.487	−0.097	−1.361	−0.340	0.204	0.938	−0.148	−1.099	−0.252	−0.384
−0.636	−0.626	1.967	1.677	−0.331	−0.440	−1.440	1.281	1.070	−1.167
−1.464	−1.493	0.945	0.180	−0.672	−0.035	−0.293	−0.905	0.196	−1.122
0.561	−0.375	−0.657	1.304	0.833	−1.159	1.501	1.265	0.438	−0.437
−0.525	−0.017	1.815	0.789	−1.908	−0.353	1.383	−1.208	−1.135	1.082
0.980	−0.111	−0.804	−1.078	−1.930	0.171	−1.318	2.377	−0.303	1.062
0.501	0.835	−0.518	−1.034	−1.493	0.712	0.421	−1.165	0.782	−1.484
1.081	−1.176	−0.542	0.321	0.688	0.670	−0.771	−0.090	−0.611	−0.813
−0.148	−1.203	−1.553	1.244	0.826	0.077	0.128	−0.772	1.683	0.318
0.096	−0.286	0.362	0.888	0.551	1.782	0.335	2.083	0.350	0.260

Miscellaneous Statistical Tables

RANDOM NORMAL NUMBERS, $\mu = 0$, $\sigma = 1$

61	62	63	64	65	66	67	68	69	70
0.052	1.504	−1.350	−1.124	−0.521	0.515	0.839	0.778	0.438	−0.55
−0.315	−0.865	0.851	0.127	−0.379	1.640	−0.441	0.717	0.670	−0.30
0.938	−0.055	0.947	1.275	1.557	−1.484	−1.137	0.398	1.333	1.98
0.497	0.502	0.385	−0.467	2.468	−1.810	−1.438	0.283	1.740	0.42
2.308	−0.399	−1.798	0.018	0.780	1.030	0.806	−0.408	−0.547	−0.28
1.815	0.101	−0.561	0.236	0.166	0.227	−0.309	0.056	0.610	0.73
−0.421	0.432	0.586	1.059	0.278	−1.672	1.859	1.433	−0.919	−1.77
0.008	0.555	−1.310	−1.440	−0.142	−0.295	−0.630	−0.911	0.133	−0.30
1.191	−0.114	1.039	1.083	0.185	−0.492	0.419	−0.433	−1.019	−2.26
1.299	1.918	0.318	1.348	0.935	1.250	−0.175	−0.828	−0.336	0.72
0.012	−0.739	−1.181	−0.645	−0.736	1.801	−0.209	−0.389	0.867	−0.55
−0.586	−0.044	−0.983	0.332	0.371	−0.072	−1.212	1.047	−1.930	0.81
−0.122	1.515	0.338	−1.040	−0.008	0.467	−0.600	0.923	1.126	−0.75
0.879	0.516	−0.920	2.121	0.674	1.481	0.660	−0.986	1.644	−2.15
0.435	1.149	−0.065	1.391	0.707	0.548	−0.490	−1.139	0.249	−0.93
0.645	0.878	−0.904	0.896	−1.284	0.237	−0.378	−0.510	−1.123	−0.12
−0.514	−1.017	0.529	0.973	−1.202	0.005	−0.644	−0.167	−0.664	0.16
0.242	−0.427	−0.727	−1.150	−1.092	−0.736	0.925	−0.050	−0.200	−0.77
0.443	0.445	−1.287	−1.463	−0.650	0.412	−2.714	−0.903	−0.341	0.95
0.273	0.203	0.423	1.423	0.508	1.058	−0.828	0.143	−1.059	0.34
0.255	1.036	1.471	0.476	0.592	−0.658	0.677	0.155	1.068	−0.75
0.858	−0.370	0.522	−1.890	−0.389	0.609	1.210	0.489	−0.006	0.83
0.097	−1.709	1.790	−0.929	0.405	0.024	−0.036	0.580	−0.642	−1.12
0.520	0.889	−0.540	0.266	−0.354	0.524	−0.788	−0.497	−0.973	1.48
−0.311	−1.772	−0.496	1.275	−0.904	0.147	1.497	0.657	−0.469	−0.78
−0.604	0.857	−0.695	0.397	0.296	−0.285	0.191	0.158	1.672	1.19
−0.001	0.287	−0.868	−0.013	−1.576	−0.168	0.047	−0.159	0.086	−1.07
1.160	0.989	0.205	0.937	−0.099	−1.281	−0.276	0.845	0.752	0.66
1.579	−0.303	−1.174	−0.960	−0.470	−0.556	−0.689	1.535	−0.711	−0.74
−0.615	−0.154	0.008	1.353	−0.381	1.137	0.022	0.175	0.586	2.94
1.578	1.529	−0.294	−1.301	0.614	0.099	−0.700	−0.003	1.052	1.64
0.626	−0.447	−1.261	−2.029	0.182	−1.176	0.083	1.868	0.872	0.96
−0.493	−0.020	0.920	1.473	1.873	−0.289	0.410	0.394	0.881	0.05
−0.217	0.342	1.423	0.364	−0.119	0.509	−2.266	0.189	0.149	1.04
−0.792	0.347	−1.367	−0.632	−1.238	−0.136	−0.352	−0.157	−1.163	1.30
0.568	−0.226	0.391	−0.074	−0.312	0.400	1.583	0.481	−1.048	0.75
0.051	0.549	−2.192	1.257	−1.460	0.363	0.127	−1.020	−1.192	0.44
−0.891	0.490	0.279	0.372	−0.578	−0.836	2.285	−0.448	0.720	0.51
0.622	−0.126	−0.637	1.255	−0.354	0.032	−1.076	0.352	0.103	−0.49
0.623	0.819	−0.489	0.354	−0.943	−0.694	0.248	0.092	−0.673	−1.42
−1.208	−1.038	0.140	−0.762	−0.854	−0.249	2.431	0.067	−0.317	−0.87
−0.487	−2.117	0.195	2.154	1.041	−1.314	−0.785	−0.414	−0.695	2.31
0.522	0.314	−1.003	0.134	−1.748	−0.107	0.459	1.550	1.118	−1.00
0.838	0.613	0.227	0.308	−0.757	0.912	2.272	0.556	−0.041	0.00
−1.534	−0.407	1.202	1.251	−0.891	−1.588	−2.380	0.059	0.682	−0.87
−0.099	2.391	1.067	−2.060	−0.464	−0.103	3.486	1.121	0.632	−1.62
0.070	1.465	−0.080	−0.526	−1.090	−1.002	0.132	1.504	0.050	−0.39
0.115	−0.601	1.751	1.956	−0.196	0.400	−0.522	0.571	−0.101	−2.16
0.252	−0.329	−0.586	−0.118	−0.242	−0.521	0.818	−0.167	−0.469	0.43
0.017	0.185	0.377	1.883	−0.443	−0.039	−1.244	−0.820	−1.171	0.10

RANDOM NORMAL NUMBERS, $\mu = 0$, $\sigma = 1$

71	72	73	74	75	76	77	78	79	80
2.988	0.423	−1.261	−1.893	0.187	−0.412	−0.228	0.002	−0.384	−1.032
0.760	0.995	−0.256	−0.505	0.750	−0.654	0.647	0.613	0.086	−0.118
0.650	−0.927	−1.071	−0.796	1.130	−1.042	−0.181	−1.020	1.648	−1.327
0.394	−0.452	0.893	1.410	1.133	0.319	0.537	−0.789	0.078	−0.062
1.168	1.902	0.206	0.303	1.413	2.012	0.278	−0.566	−0.900	0.200
1.343	−0.377	−0.131	−0.585	0.053	0.137	−1.371	−0.175	−0.878	0.118
0.733	−1.921	0.471	−1.394	−0.885	−0.523	0.553	0.344	−0.775	1.545
0.172	−0.575	0.066	−0.310	1.795	−1.148	0.772	−1.063	0.818	0.302
1.457	0.862	1.677	−0.507	−1.691	−0.034	0.270	0.075	−0.554	1.420
0.087	0.744	1.829	1.203	−0.436	−0.618	−0.200	−1.134	−1.352	−0.098
0.092	1.043	−0.255	0.189	0.270	−1.034	−0.571	−0.336	−0.742	2.141
0.441	−0.379	−1.757	0.608	0.527	−0.338	−1.995	0.573	−0.034	−0.056
0.073	−0.250	0.531	−0.695	1.402	−0.462	−0.938	1.130	1.453	−0.106
0.637	0.276	−0.013	1.968	−0.205	0.486	0.727	1.416	0.963	1.349
0.792	−1.778	1.284	−0.452	0.602	0.668	0.516	−0.210	0.040	−0.103
1.223	1.561	−2.099	1.419	0.223	−0.482	1.098	0.513	0.418	−1.686
0.407	1.587	0.335	−2.475	−0.284	1.567	−0.248	−0.759	1.792	−2.319
0.462	−0.193	−0.012	−1.208	2.151	1.336	−1.968	−1.767	−0.374	0.783
1.457	0.883	1.001	−0.169	0.836	−1.236	1.632	−0.142	−0.222	0.340
1.918	−1.246	−0.209	0.780	−0.330	−2.953	−0.447	−0.094	1.344	−0.196
0.126	1.094	−1.206	−1.426	1.474	−1.080	0.000	0.764	1.476	−0.016
0.306	−0.847	0.639	−0.262	−0.427	0.391	−1.298	−1.013	2.024	−0.539
0.477	1.595	−0.762	0.424	0.799	0.312	1.151	−1.095	1.199	−0.765
0.369	−0.709	1.283	−0.007	−1.440	−0.782	0.061	1.427	1.656	0.974
0.579	0.606	−0.866	−0.715	−0.301	−0.180	0.188	0.668	−1.091	1.476
0.418	−0.588	0.919	−0.083	1.084	0.944	0.253	−1.833	1.305	0.171
0.128	−0.834	0.009	0.742	0.539	−0.948	−1.055	−0.689	−0.338	1.091
0.291	0.235	−0.971	−1.696	1.119	0.272	0.635	−0.792	−1.355	1.291
1.024	1.212	−1.100	−0.348	1.741	0.035	1.268	0.192	0.729	−0.467
0.378	1.026	0.093	0.468	−0.967	0.675	0.807	−2.109	−1.214	0.559
1.232	−0.815	0.608	1.429	−0.748	0.201	0.400	−1.230	−0.398	−0.674
1.793	−0.581	−1.076	0.512	−0.442	−1.488	−0.580	0.172	−0.891	0.311
0.766	0.310	−0.070	0.624	−0.389	1.035	−0.101	−0.926	0.816	−1.048
0.606	−1.224	1.465	0.012	1.061	0.491	−1.023	1.948	0.866	−0.737
0.106	−2.715	0.363	0.343	−0.159	2.672	1.119	0.731	−1.012	−0.889
0.060	0.444	1.596	−0.630	0.362	−0.306	1.163	−0.974	0.486	−0.373
2.081	1.161	−1.167	0.021	0.053	−0.094	0.381	−0.628	−2.581	−1.243
1.727	−1.266	0.088	0.936	0.368	0.648	−0.799	1.115	−0.968	−2.588
0.001	1.364	1.677	0.644	1.505	0.440	−0.329	0.498	0.869	−0.965
1.114	−0.239	−0.409	−0.334	−0.605	0.501	−1.921	−0.470	2.354	−0.660
0.189	−0.547	−1.758	−0.295	−0.279	−0.515	−1.053	0.553	−0.297	0.496
0.065	−0.023	−0.267	−0.247	1.318	0.904	−0.712	−1.152	−0.543	0.176
1.742	−0.599	0.430	−0.615	1.165	0.084	2.017	−1.207	2.614	1.490
0.732	0.188	2.343	0.526	−0.812	0.389	1.036	−0.023	0.229	−2.262
1.490	0.014	0.167	1.422	0.015	0.069	0.133	0.897	−1.678	0.323
1.507	−0.571	−0.724	1.741	−0.152	−0.147	−0.158	−0.076	0.652	0.447
0.513	0.168	−0.076	−0.171	0.428	0.205	−0.865	0.107	1.023	0.077
0.834	−1.121	1.441	0.492	0.559	1.724	−1.659	0.245	1.354	−0.041
0.258	1.880	−0.536	1.246	−0.188	0.746	1.097	0.258	1.547	1.238
0.818	0.273	0.159	−0.765	0.526	1.281	1.154	−0.687	−0.793	0.795

Miscellaneous Statistical Tables

RANDOM NORMAL NUMBERS, $\mu = 0$, $\sigma = 1$

81	82	83	84	85	86	87	88	89	90
−0.713	−0.541	−0.571	−0.807	−1.560	1.000	0.140	−0.549	0.887	2.23
−0.117	0.530	−1.599	−1.602	0.412	−1.450	−1.217	1.074	−1.021	−0.42
1.187	−1.523	1.437	0.051	1.237	−0.798	1.616	−0.823	−1.207	1.25
−0.182	−0.186	0.517	1.438	0.831	−1.319	−0.539	−0.192	0.150	2.12
1.964	−0.629	−0.944	−0.028	0.948	1.005	0.242	−0.432	−0.329	0.11
0.230	1.523	1.658	0.753	0.724	0.183	−0.147	0.505	0.448	−0.05
0.839	−0.849	−0.145	−1.843	−1.276	0.481	−0.142	−0.534	0.403	0.37
−0.801	0.343	−1.822	0.447	−0.931	−0.824	−0.484	0.864	−1.069	0.86
−0.124	0.727	1.654	−0.182	−1.381	−1.146	−0.572	0.159	0.186	1.22
−0.088	0.032	−0.564	0.654	1.141	−0.056	−0.343	0.067	−0.267	−0.21
0.912	−1.114	−1.035	−1.070	−0.297	1.195	0.030	0.022	0.406	−0.41
1.397	−0.473	0.433	0.023	−1.204	1.254	0.551	−1.012	−0.789	0.90
−0.652	−0.029	0.064	0.511	1.117	−0.465	0.523	−0.083	0.386	0.25
1.236	−0.457	−1.354	−0.898	−0.270	−1.837	1.641	−0.657	−0.753	−1.68
−0.498	1.302	0.816	−0.936	1.404	0.555	2.450	−0.789	−0.120	0.50
−0.005	2.174	1.893	−1.361	−0.991	0.508	−0.823	0.918	0.524	0.48
0.115	−1.373	−0.900	−1.010	0.624	0.946	0.312	−1.384	0.224	2.34
0.167	0.254	1.219	1.153	−0.510	−0.007	−0.285	−0.631	−0.356	0.25
0.976	1.158	−0.469	1.099	0.509	−1.324	−0.102	−0.296	−0.907	0.44
0.653	−0.366	0.450	−2.653	−0.592	−0.510	0.983	0.023	−0.881	0.87
−0.150	−0.088	0.457	−0.448	0.605	0.668	−0.613	0.261	0.023	−0.05
0.060	0.276	0.229	−1.527	−0.316	−0.834	−1.652	−0.387	0.632	0.89
−0.678	0.547	0.243	−2.183	−0.368	1.158	−0.996	−0.705	−0.314	1.46
2.139	0.395	−0.376	−0.175	0.406	0.309	−1.021	−0.460	−0.217	0.30
0.091	1.793	0.822	0.054	0.573	−0.729	−0.517	0.589	1.927	0.94
−0.003	0.344	1.242	−1.105	0.234	−1.222	−0.474	1.831	0.124	−0.84
−0.965	0.268	−1.543	0.690	0.917	2.017	−0.297	1.087	0.371	1.49
−0.076	−0.495	−0.103	0.646	2.427	−2.172	0.660	−1.541	−0.852	0.58
−0.365	−3.305	0.805	−0.418	−1.201	0.623	−0.223	0.109	0.205	−0.66
0.578	0.145	−1.438	1.122	−1.406	1.172	0.272	−2.245	1.207	1.22
−0.398	−0.304	0.529	−0.514	−0.681	−0.366	0.338	0.801	−0.301	−0.79
−0.951	−1.483	−0.613	−0.171	−0.459	1.231	−1.232	−0.497	−0.779	0.24
1.025	−0.039	−0.721	0.813	1.203	0.245	0.402	1.541	0.691	−1.42
−0.958	0.791	0.948	0.222	−0.704	−0.375	−0.246	−0.682	−0.871	0.05
1.097	−1.428	1.402	−1.425	−0.877	0.536	0.988	2.529	0.768	−1.32
0.377	2.240	0.854	−1.158	0.066	−1.222	0.821	−1.602	−0.760	−0.87
1.729	0.073	1.022	0.891	0.659	−1.040	0.251	−0.710	−1.734	−0.03
−1.329	−0.381	−0.515	1.484	−0.430	−0.466	−0.167	−0.788	−0.660	0.00
−0.132	0.391	2.205	−1.165	0.200	0.415	−0.765	0.239	−1.182	1.13
0.336	0.657	−0.805	0.150	−0.938	1.057	−1.090	1.604	−0.598	−0.76
0.124	−1.812	1.750	0.270	−0.114	0.517	−0.226	0.127	0.129	−0.75
−0.036	0.365	0.766	0.877	−0.804	−0.140	0.182	−0.483	−0.376	−0.56
−0.609	−0.019	−0.992	−1.193	−0.516	0.517	1.677	0.839	−1.134	0.67
−0.894	0.318	0.607	−0.865	0.526	−0.971	1.365	0.319	1.804	1.74
−0.357	−0.802	0.635	−0.491	−1.110	0.785	−0.042	−1.042	−0.572	0.24
−0.258	−0.383	−1.013	0.001	−1.673	0.561	−1.054	−0.106	−0.760	−1.00
2.245	−0.431	−0.496	0.796	0.193	1.202	−0.429	−0.217	0.333	−0.64
1.956	0.477	0.812	−0.117	0.606	−0.330	0.425	−0.232	0.802	0.65
1.358	0.139	0.199	−0.475	−0.120	0.184	−0.020	−1.326	0.517	−1.70
0.656	1.081	0.180	0.145	0.376	−1.363	−0.491	0.352	−1.477	1.28

RANDOM NORMAL NUMBERS, $\mu = 0$, $\sigma = 1$

91	92	93	94	95	96	97	98	99	100
0.181	0.583	−1.478	−0.181	0.281	−0.559	1.985	−1.122	−1.106	1.441
1.549	−1.183	−2.089	−1.997	−0.343	1.275	0.676	−0.212	1.252	0.163
0.978	−1.067	−2.640	0.134	0.328	−0.052	−0.030	−0.273	−0.570	1.026
0.596	−0.420	−0.318	−0.057	−0.695	−1.148	0.333	−0.531	−2.037	−1.587
0.440	0.032	0.163	1.029	0.079	1.148	0.762	−1.961	−0.674	−0.486
0.443	−1.100	0.728	−2.397	−0.543	0.872	−0.568	0.980	−0.174	0.728
2.401	−1.375	−1.332	−2.177	−2.064	−0.245	−0.039	0.585	1.344	1.386
0.311	0.322	−0.158	0.359	0.103	0.371	0.735	0.011	2.091	0.490
1.209	0.241	−1.488	−0.667	−1.772	−0.197	0.741	−1.303	−1.149	2.251
0.575	−1.227	−1.674	1.400	0.289	0.005	0.185	−1.072	0.431	−1.096
0.190	0.272	1.216	0.227	1.358	0.215	−2.306	−1.301	−0.597	−1.401
0.817	−0.769	−0.470	−0.633	0.187	−0.517	−0.888	−1.712	1.774	−0.162
0.265	−0.676	0.244	1.897	−0.629	−0.206	−1.419	1.049	0.266	−0.438
0.221	0.678	2.149	1.486	−1.361	1.402	−0.028	0.493	0.744	0.195
0.436	0.358	−0.602	0.107	0.085	0.573	0.529	1.577	0.239	1.898
0.010	0.475	0.655	0.659	−0.029	−0.029	0.126	−1.335	−1.261	2.036
0.244	1.654	1.335	−0.610	0.617	0.642	0.371	0.241	0.001	−1.799
0.932	−1.275	−1.134	−1.246	−1.508	0.949	1.743	−0.271	−1.333	−1.875
0.199	−1.285	−0.387	0.191	0.726	−0.151	0.064	−0.803	−0.062	0.780
0.251	−0.431	−0.831	0.036	−0.464	−1.089	0.284	−0.451	1.693	1.004
1.074	−1.323	−1.659	−0.186	−0.612	1.612	−2.159	−1.210	0.596	−1.421
1.518	2.101	0.397	0.516	−1.169	−1.821	1.346	2.435	1.165	−0.428
0.935	−0.206	1.117	−0.241	−0.963	−0.099	0.412	−1.344	0.411	0.583
1.360	−0.380	0.031	1.066	0.893	0.431	−0.081	0.099	0.500	−2.441
0.115	−0.211	1.471	0.332	0.750	0.652	−0.812	1.383	−0.355	−0.638
0.082	−0.309	−0.355	−0.402	0.774	0.150	0.015	2.539	−0.756	−1.049
1.492	0.259	0.323	0.697	−0.509	0.968	−0.053	1.033	−0.220	−2.322
0.203	0.548	1.494	1.185	0.083	−1.196	−0.749	−1.105	1.324	0.689
1.857	−0.167	−1.531	1.551	0.848	0.120	0.415	−0.317	1.446	1.002
0.669	−1.017	−2.437	−0.558	−0.657	0.940	0.985	0.483	−0.361	0.095
0.128	1.463	−0.436	−0.239	−1.443	0.732	0.168	−0.144	−0.392	0.989
1.879	−2.456	0.029	0.429	0.618	−1.683	−2.262	0.034	−0.002	1.914
0.680	0.252	0.130	1.658	−1.023	0.407	−0.235	−0.224	−0.434	0.253
0.631	0.225	−0.951	1.072	−0.285	−1.731	−0.427	−1.446	−0.873	0.619
1.273	0.723	0.201	0.505	−0.370	−0.421	−0.015	−0.463	0.288	1.734
0.643	−1.485	0.403	0.003	−0.243	0.000	0.964	−0.703	0.844	−0.686
0.435	−2.162	−0.169	−1.311	−1.639	0.193	2.692	−1.994	0.326	0.562
1.706	0.119	−1.566	0.637	−1.948	−1.068	0.935	0.738	0.650	0.491
0.498	1.640	0.384	−0.945	−1.272	0.945	−1.013	−0.913	−0.469	2.250
0.065	−0.005	0.618	−0.523	−0.055	1.071	0.758	−0.736	−0.959	0.598
0.190	−1.020	−1.104	0.936	−0.029	−1.004	−0.657	1.270	−0.060	−0.809
0.879	−0.642	1.155	−0.523	−0.757	−1.027	0.985	−1.222	1.078	0.163
0.559	1.094	1.587	−0.384	−1.701	0.418	0.327	0.669	0.019	0.782
0.261	1.234	−0.505	−0.664	−0.446	−0.747	0.427	−0.369	0.089	−1.302
3.136	1.120	−0.591	2.515	−2.853	1.375	2.421	0.672	1.817	−0.067
1.307	−0.586	−0.311	−0.026	1.633	−1.340	−1.209	0.110	−0.126	−0.288
1.455	1.099	−1.225	−0.817	0.667	−0.212	0.684	0.349	−1.161	−2.432
0.443	−0.415	−0.660	0.098	0.435	−0.846	−0.375	−0.410	−1.747	−0.790
0.326	0.798	0.349	0.524	0.690	−0.520	−0.522	0.602	−0.193	−0.535
1.027	−1.459	−0.840	−1.637	−0.462	0.607	−0.760	1.342	−1.916	0.424

Miscellaneous Statistical Tables

XII.5 RANDOM NORMAL NUMBERS, $\mu = 2$, $\sigma = 1$

01	02	03	04	05	06	07	08	09	10
2.422	0.130	2.232	1.700	1.903	0.725	2.031	0.515	−0.684	2.7?
0.694	2.556	1.868	1.263	2.115	1.516	1.972	3.627	1.482	3.2?
1.875	2.273	0.655	2.299	0.055	1.955	−0.147	2.168	2.193	1.8?
1.017	0.757	1.288	1.322	2.080	2.170	1.502	2.953	0.171	1.9?
2.453	4.199	1.403	2.017	3.496	0.165	2.556	1.003	1.973	2.1?
2.274	1.767	1.564	2.412	2.207	0.475	2.656	1.579	0.394	1.2?
3.000	1.618	1.530	2.224	2.881	2.715	3.103	1.941	2.179	3.7
2.510	2.256	1.146	5.177	1.931	1.693	1.021	3.337	2.137	1.8?
1.233	2.085	2.251	1.578	3.796	3.017	2.863	2.514	1.615	1.5
3.075	1.730	2.427	2.990	1.680	3.250	3.050	3.243	1.846	1.7?
1.344	−0.095	2.166	4.116	2.500	1.939	1.567	3.047	1.385	−0.8?
1.246	3.860	1.253	1.876	4.373	1.993	1.262	2.319	2.488	2.4?
0.889	2.299	2.458	1.790	1.048	2.302	0.138	2.383	1.170	2.2?
1.154	1.401	1.935	3.106	1.548	−0.096	2.153	2.333	1.761	3.7?
3.031	1.048	0.719	1.474	2.779	0.292	2.341	2.707	1.741	2.3?
0.534	1.155	1.705	1.662	0.457	0.602	1.365	2.663	3.755	1.9?
2.230	3.096	0.045	3.639	0.680	0.970	1.593	2.117	2.395	1.9?
2.355	1.761	1.816	1.822	1.434	2.259	3.788	3.280	1.317	2.9
1.461	0.947	0.717	2.923	2.133	2.526	2.687	2.144	1.692	1.4?
3.034	1.778	2.122	2.025	3.008	1.447	−0.305	2.452	1.726	0.8?
2.761	0.473	3.726	1.893	2.455	1.633	1.654	3.006	3.523	2.3?
1.961	0.965	1.481	1.402	2.106	2.214	1.727	3.670	3.795	2.2?
2.639	4.010	1.915	1.713	1.484	1.443	1.444	2.394	1.688	0.7?
1.349	2.225	0.644	1.404	2.583	2.149	2.359	2.274	1.432	1.6?
2.959	2.797	4.635	3.268	2.889	2.349	0.933	3.403	2.206	−0.2?
2.440	2.919	1.455	0.695	1.466	1.124	1.257	1.265	0.096	3.4?
3.078	3.279	0.352	2.583	1.690	0.729	2.072	1.332	1.158	1.8?
1.736	1.968	0.011	2.418	1.026	1.342	2.103	1.792	2.175	1.6
3.275	3.147	2.800	2.172	0.004	1.763	3.801	2.510	2.517	−0.1?
2.579	2.297	2.030	2.725	3.721	2.545	1.631	−0.346	−0.011	1.9?
2.549	3.546	2.805	1.250	0.769	2.238	2.284	3.722	2.085	2.6
2.954	1.990	1.249	1.028	3.241	1.926	3.056	1.732	2.116	1.8?
1.442	2.542	2.557	1.741	0.630	2.117	1.662	2.237	−0.046	3.1?
4.039	2.030	2.859	3.538	2.424	2.169	3.643	3.290	2.742	1.3?
2.127	0.288	2.921	0.175	1.670	3.151	1.443	0.935	1.125	2.8
1.102	2.536	1.476	2.980	0.416	1.784	2.521	1.867	1.709	1.5
2.938	2.112	1.350	2.115	1.164	1.761	1.350	1.798	3.160	2.5?
2.975	2.681	0.721	1.291	2.276	2.131	2.187	2.752	1.380	0.6?
1.386	1.712	1.692	2.844	1.559	0.418	3.020	0.785	1.962	3.1
2.834	1.485	0.632	0.872	0.735	1.934	1.221	2.544	1.797	1.4
3.346	1.147	1.766	1.862	2.595	1.524	3.499	2.652	2.139	2.5
2.243	3.881	2.846	2.670	3.377	1.380	4.183	0.883	1.373	1.9?
2.705	2.661	1.521	1.290	2.280	1.638	0.884	2.636	2.077	1.0
2.760	1.182	1.152	3.074	1.073	2.917	2.150	2.866	1.688	1.6?
2.086	1.250	1.577	2.871	2.985	2.585	2.897	2.398	0.999	1.7
0.802	1.421	4.793	0.268	2.838	2.227	3.331	2.395	2.064	2.9
4.165	2.014	0.616	1.929	0.641	2.304	1.263	2.125	0.908	1.7
2.291	2.549	0.851	1.856	2.452	3.282	0.978	2.255	1.683	1.9
1.428	4.194	2.262	2.957	1.991	2.759	1.553	3.538	1.272	3.4
2.051	2.455	2.759	2.267	2.794	4.106	2.373	1.401	2.562	2.5?

RANDOM NORMAL NUMBERS, $\mu = 2$, $\sigma = 1$

11	12	13	14	15	16	17	18	19	20
1.911	0.626	2.289	1.628	1.638	2.676	0.900	1.685	1.605	1.366
3.196	2.979	2.447	2.099	1.273	2.733	2.653	2.219	1.318	3.129
0.398	2.304	1.019	0.363	1.286	2.428	0.677	1.684	1.267	0.651
1.228	2.134	0.300	1.785	2.547	1.566	2.545	2.428	1.702	2.276
1.190	3.020	0.954	2.907	2.916	1.279	3.403	2.698	1.629	1.448
0.953	2.127	1.723	2.302	1.474	0.826	1.644	2.035	2.359	2.930
1.479	1.956	1.280	1.722	0.938	0.922	2.734	3.484	1.659	2.789
1.509	0.952	1.258	−0.864	1.620	1.789	2.931	2.616	1.622	1.566
0.627	2.404	0.571	2.940	2.705	1.709	2.404	1.456	2.486	2.869
1.923	2.765	2.422	1.725	1.009	2.372	1.925	1.083	3.314	1.961
2.760	2.633	3.011	2.277	1.539	0.873	2.379	2.610	1.635	−0.625
2.009	3.204	1.114	2.269	0.912	0.831	2.485	2.076	1.230	3.607
0.876	1.124	2.137	1.448	1.236	1.699	1.408	1.454	2.018	1.514
1.430	1.920	2.969	1.518	1.543	1.509	4.071	3.444	0.907	2.478
3.422	2.307	2.919	1.833	1.792	3.090	2.212	0.814	1.661	0.865
3.304	1.292	1.863	2.785	1.666	0.323	2.384	3.133	3.393	2.814
2.329	2.671	3.353	1.166	1.016	3.036	2.024	1.439	2.203	1.128
1.402	1.964	1.505	1.746	1.912	1.202	0.595	0.527	1.881	3.456
2.274	1.209	1.450	2.241	1.678	1.565	2.746	2.149	1.829	1.520
1.205	0.531	2.975	3.024	3.357	2.558	1.450	2.192	1.665	3.373
2.462	1.328	1.301	3.312	1.959	2.010	2.482	1.530	1.909	3.171
0.227	3.166	1.989	2.976	2.188	1.399	1.407	2.610	1.903	0.624
2.142	2.926	1.634	1.940	0.785	2.331	1.663	0.847	2.533	1.166
2.558	0.903	0.082	1.299	2.366	2.554	1.948	1.055	1.559	1.787
0.818	3.174	0.123	−1.149	1.606	2.118	−0.044	0.022	0.866	2.336
3.083	2.287	2.379	2.909	2.520	0.708	1.600	0.790	1.751	2.480
2.517	1.470	2.621	0.880	1.931	1.495	1.943	1.868	2.048	3.879
2.594	1.571	1.218	2.346	2.267	0.946	2.840	1.753	2.237	0.687
0.411	0.760	1.114	1.842	1.756	3.951	2.110	2.251	2.116	1.042
2.853	3.054	2.421	2.418	1.542	2.070	0.641	0.753	1.040	0.702
1.262	−0.591	1.320	2.049	2.705	3.826	3.272	1.054	2.494	2.050
0.540	1.678	2.534	1.944	1.939	2.544	1.582	1.333	1.895	1.746
2.381	2.968	1.656	3.152	1.730	3.927	3.183	3.211	3.765	2.035
2.225	1.420	1.334	1.923	1.664	1.939	0.680	2.785	1.569	1.701
1.953	2.779	2.584	2.228	0.221	1.378	1.381	2.209	2.979	2.906
3.413	2.229	2.976	2.535	3.589	0.615	3.425	1.187	2.748	0.906
1.610	2.376	2.086	0.610	2.532	3.083	1.332	1.776	0.407	0.721
0.984	2.243	2.939	1.704	2.277	1.026	1.879	0.405	1.003	0.755
1.808	2.362	1.717	0.831	2.160	0.546	2.686	1.924	1.756	1.829
0.766	3.529	2.361	0.955	2.148	1.104	0.541	1.460	1.840	1.579
2.643	2.051	1.384	2.229	2.952	2.203	0.765	4.381	1.611	1.936
1.952	2.752	3.588	2.481	1.911	3.753	1.428	3.223	1.873	2.034
2.590	2.306	3.280	1.664	2.281	1.443	2.024	2.126	3.250	1.384
0.622	2.617	1.969	2.231	−0.079	0.768	2.547	1.365	1.163	1.280
0.433	−0.560	3.292	1.987	1.065	2.766	1.425	0.846	2.520	0.981
3.146	3.323	1.713	1.887	2.010	1.277	0.491	2.489	1.503	1.974
4.021	1.744	0.598	2.954	2.633	1.960	1.539	2.393	4.012	3.356
1.188	0.450	2.958	1.177	1.482	1.090	1.671	3.021	0.386	3.560
1.211	2.575	0.158	3.124	3.632	2.647	3.029	3.526	2.237	0.671
3.750	2.362	1.407	0.642	1.274	1.632	2.378	2.601	0.003	1.261

Miscellaneous Statistical Tables

RANDOM NORMAL NUMBERS, $\mu = 2$, $\sigma = 1$

21	22	23	24	25	26	27	28	29	30
1.707	2.089	1.315	0.278	3.045	2.968	1.396	1.534	2.365	2.74
1.113	1.779	1.935	0.971	4.024	0.847	1.382	2.342	2.110	0.31
1.847	0.547	3.697	1.250	1.586	2.036	2.924	0.585	0.456	2.85
2.713	2.761	1.664	2.461	2.158	3.453	2.078	1.113	1.769	1.26
0.676	0.432	2.667	2.515	1.369	3.196	2.979	2.447	2.099	1.27
2.167	1.828	2.867	1.178	2.078	1.500	2.622	2.341	1.504	2.40
3.445	3.323	2.558	1.789	1.595	1.191	1.175	2.872	1.257	1.06
1.284	3.180	3.315	1.210	1.842	3.384	2.942	2.550	0.727	1.73
2.135	2.590	2.533	1.635	1.983	0.614	0.377	−0.663	0.427	2.44
2.944	2.043	2.220	1.987	2.859	3.029	2.091	1.052	1.532	1.95
3.654	2.333	1.468	3.126	1.241	2.936	1.557	2.020	1.423	2.70
0.821	1.542	2.365	2.199	3.479	3.111	−0.107	1.644	1.337	1.44
2.483	2.583	2.075	1.026	0.668	2.281	1.566	1.255	2.020	1.13
0.715	1.384	2.080	2.542	2.368	0.019	2.906	2.325	2.175	5.19
4.638	2.662	1.012	2.941	1.336	0.574	3.034	2.937	2.553	0.17
2.327	2.152	3.057	2.077	2.321	0.861	2.892	1.394	−0.556	1.45
−0.082	0.676	3.038	2.470	1.394	2.131	1.262	3.207	1.810	0.32
2.051	1.576	2.087	3.030	2.030	2.827	2.183	1.182	1.507	−0.04
2.438	0.924	1.699	0.477	2.449	2.540	1.620	2.509	2.347	3.02
2.284	2.159	2.975	3.268	0.484	1.862	1.676	1.449	2.475	2.55
0.872	0.474	2.213	3.602	3.244	3.078	1.376	2.612	2.421	1.01
2.236	1.963	1.839	1.598	2.195	2.680	2.228	1.107	−0.661	1.04
2.425	2.412	1.500	2.278	2.328	2.102	2.087	3.098	2.697	0.76
1.511	2.431	1.434	1.558	1.020	2.864	0.871	2.523	1.878	1.37
1.600	2.040	2.993	0.873	0.568	2.703	2.578	1.515	3.627	2.09
3.076	1.939	0.682	3.085	2.877	2.696	−0.771	2.560	1.954	0.99
2.593	1.610	2.800	2.456	0.226	3.575	1.435	2.170	1.165	3.50
1.362	2.727	2.145	2.023	0.509	0.336	2.045	0.375	1.010	2.31
1.603	2.783	0.682	2.108	2.031	0.854	2.028	2.357	0.722	1.56
1.908	1.635	2.009	1.203	1.775	2.868	1.949	1.391	1.151	1.34
3.486	0.507	2.322	1.204	2.434	1.720	1.804	2.235	2.439	1.49
2.029	1.352	3.629	2.076	1.587	0.891	3.029	1.242	0.014	4.01
2.894	1.688	0.657	1.800	2.943	1.373	1.269	3.411	1.316	2.40
0.965	−0.028	1.904	2.241	2.563	1.149	2.375	1.386	1.562	2.85
2.191	2.133	2.676	0.229	2.319	1.114	3.197	2.588	3.163	2.42
2.115	2.418	2.741	1.839	2.416	1.452	0.319	0.853	2.774	0.92
1.120	2.126	0.773	0.798	3.436	2.374	2.173	−0.333	2.004	1.70
3.524	0.008	3.260	1.109	4.111	2.474	2.482	2.416	0.832	4.08
0.103	2.774	2.056	2.463	0.383	−0.962	2.458	2.388	1.556	1.08
1.573	2.519	2.153	3.188	1.618	2.477	2.185	1.851	0.498	2.00
1.138	3.032	2.390	2.436	2.655	1.484	2.378	3.166	2.531	2.08
2.665	2.960	2.518	1.940	0.026	2.570	2.703	2.592	3.094	1.80
1.397	1.859	2.208	2.559	1.749	0.624	0.074	1.398	0.996	2.91
1.875	2.250	0.183	2.214	1.356	4.282	2.370	3.006	1.413	2.41
3.195	2.671	1.918	3.305	3.722	1.372	2.564	2.106	1.871	1.79
1.464	2.055	3.045	2.367	1.992	0.919	3.006	2.713	4.049	4.61
3.328	1.781	2.565	1.304	2.041	1.597	0.225	2.309	−0.558	2.56
2.804	3.606	1.858	3.028	2.456	1.730	1.430	3.405	0.474	2.22
2.590	1.641	3.857	2.582	2.594	1.933	3.341	1.002	2.704	1.34
2.980	0.601	1.595	2.248	2.381	1.911	0.626	2.289	1.628	1.63

RANDOM NORMAL NUMBERS, $\mu = 2$, $\sigma = 1$

1	32	33	34	35	36	37	38	39	40
.355	0.073	3.139	2.472	1.825	0.296	1.685	3.401	1.820	1.428
.086	1.955	2.529	2.503	1.687	1.754	4.138	2.394	0.303	3.776
.367	1.525	2.625	1.789	0.991	4.127	0.915	3.023	1.377	3.435
.248	0.749	3.697	4.166	2.544	1.620	3.217	1.083	1.907	2.951
.694	0.258	1.836	1.953	1.853	3.590	3.604	1.907	1.995	2.468
.546	1.255	2.856	3.221	2.397	−0.010	2.169	2.781	3.001	2.536
.266	2.089	2.974	1.305	2.376	−0.475	1.792	1.546	0.583	1.214
.713	2.473	1.381	1.750	1.064	3.744	2.470	1.004	2.155	2.332
.001	1.600	2.166	0.561	0.898	2.587	0.580	−0.461	0.954	1.364
.406	2.207	2.110	1.522	3.923	1.379	3.613	3.379	2.716	2.796
.432	1.651	1.584	3.649	2.485	2.820	2.948	2.626	1.763	3.329
.541	1.154	4.311	2.354	2.257	1.262	2.304	2.178	1.657	2.126
.216	3.505	−0.056	1.332	0.980	1.675	1.850	2.487	2.051	1.433
.602	2.225	2.949	3.945	3.753	3.855	2.769	0.760	2.095	1.419
.211	1.804	2.642	0.975	1.646	2.552	2.291	1.277	2.341	−0.219
.006	2.279	1.097	3.473	0.919	2.535	2.459	3.934	1.826	1.587
.520	2.468	2.156	2.438	1.625	1.604	1.628	1.139	2.608	2.643
.666	4.058	2.805	3.069	0.945	2.533	2.761	1.140	2.604	1.627
.852	0.570	3.920	1.572	2.924	2.135	1.558	2.604	2.191	2.529
.014	2.825	3.502	2.006	1.879	3.304	1.538	0.906	3.125	1.009
.540	0.444	1.541	1.850	1.793	2.284	1.890	3.091	2.293	2.491
.190	2.087	2.159	1.157	2.314	1.753	0.722	2.447	2.124	2.927
.741	3.411	1.689	1.945	0.286	3.288	1.390	0.240	1.448	2.768
.169	1.937	2.261	0.766	2.075	0.457	2.031	0.831	−0.009	4.316
.979	1.935	2.232	1.812	3.290	2.031	3.222	2.520	4.105	0.705
.405	0.166	0.137	3.246	4.142	2.808	2.526	2.687	−0.627	2.023
.923	2.287	1.164	0.732	0.736	0.892	2.633	2.107	1.260	0.615
.529	3.188	2.153	3.828	3.610	1.654	2.596	0.957	1.479	1.497
.781	3.562	3.633	0.889	0.832	2.068	2.103	3.360	1.686	1.538
.153	2.125	1.930	3.161	2.931	1.941	3.108	1.732	4.296	1.830
.204	3.945	0.682	4.165	2.419	0.565	1.637	1.931	1.092	2.482
.154	1.889	1.391	1.690	1.356	2.560	1.784	1.041	2.808	0.576
.391	2.602	2.496	2.177	1.564	1.781	0.302	2.499	1.501	1.410
.266	2.051	1.958	0.979	2.454	1.438	2.098	3.208	2.374	2.710
.842	0.513	1.736	2.878	1.893	1.614	2.775	1.060	1.508	1.197
.799	0.757	0.625	3.336	2.268	1.418	1.616	2.363	0.751	2.138
.104	3.564	0.681	1.231	2.527	0.172	1.331	0.991	3.570	1.382
.232	3.514	−0.433	2.932	3.245	2.778	2.196	−0.326	1.034	1.889
.201	3.351	1.761	1.957	1.342	3.575	3.216	1.335	1.527	0.812
.046	1.646	1.363	1.051	4.600	3.209	3.041	3.234	2.034	0.682
.874	1.663	2.591	1.396	1.052	1.068	2.226	3.048	1.906	2.755
.389	2.966	2.846	2.410	1.663	3.620	2.151	2.036	3.733	1.462
.144	1.641	1.693	1.599	2.704	3.083	1.387	0.593	1.191	2.707
.177	0.829	2.094	1.737	1.625	1.766	1.415	2.238	0.549	1.887
.595	2.094	2.851	1.175	0.425	2.242	1.477	3.237	2.614	1.226
.655	3.804	0.607	1.958	4.251	1.457	3.369	2.077	1.511	1.458
.601	2.255	1.787	1.136	2.912	3.060	2.562	3.137	3.248	1.382
.308	2.422	3.081	2.185	1.963	3.855	2.389	4.057	2.428	3.054
.196	4.160	2.841	1.550	0.919	1.884	1.911	1.386	2.607	1.625
.843	1.330	1.678	2.198	1.398	0.709	1.810	2.269	4.242	0.777

Miscellaneous Statistical Tables

RANDOM NORMAL NUMBERS, $\mu = 2$, $\sigma = 1$

41	42	43	44	45	46	47	48	49	50
1.017	2.773	3.278	2.557	1.003	4.181	0.946	3.464	1.945	2.
0.723	0.781	1.546	1.649	2.723	4.542	1.819	0.511	2.580	3.
1.681	1.200	0.335	3.391	2.382	3.080	0.685	1.924	1.085	1.
0.622	0.742	2.495	1.860	1.145	2.040	2.103	−0.256	0.976	2.
1.815	2.061	2.092	2.089	2.281	2.377	1.821	1.760	2.515	1.
4.334	1.662	0.044	1.363	0.681	2.111	2.443	1.603	1.803	2.
0.863	2.642	5.436	0.332	2.847	1.466	3.031	1.571	3.024	1.
2.414	1.988	2.666	0.867	1.589	2.192	3.027	2.257	1.868	1.
1.505	2.364	0.762	1.955	1.888	1.845	3.180	2.618	1.730	1.
3.048	2.037	2.759	2.609	−0.042	1.215	1.292	1.582	1.522	2.
2.347	4.816	1.535	1.367	0.385	2.013	2.557	2.041	3.070	1.
2.637	2.563	1.892	2.131	0.191	2.484	1.788	2.762	2.166	1.
4.176	2.393	1.075	3.911	0.959	2.438	3.201	1.810	2.049	3.
0.814	1.055	0.395	2.185	1.741	2.742	2.228	2.151	1.997	3.
2.972	3.710	4.682	4.813	0.468	2.311	2.382	0.810	0.155	1.
3.210	2.294	1.751	2.719	3.103	2.459	3.656	4.862	3.724	3.
4.647	2.777	2.450	4.247	3.151	0.197	3.602	1.754	1.739	2.
2.398	2.318	1.071	4.416	1.063	1.568	3.057	0.985	3.425	1.
2.846	1.300	1.631	2.344	1.073	1.049	2.743	0.365	1.949	0.
2.654	1.044	4.907	3.688	2.752	2.365	2.083	1.669	2.538	2.
2.522	2.231	1.380	1.734	2.419	1.313	2.226	2.524	2.073	3.
0.711	1.460	1.175	2.244	0.929	−0.091	1.096	2.061	3.099	2.
3.372	3.769	0.942	3.646	2.481	1.554	3.715	1.193	1.956	1.
2.854	1.464	3.607	2.428	1.384	1.977	1.504	0.492	2.102	2.
1.851	0.855	2.913	2.684	3.043	2.595	1.803	1.303	0.233	−0.
0.851	0.943	2.635	1.671	0.778	2.899	2.145	2.747	1.342	2.
2.348	2.970	1.982	3.217	1.025	2.626	2.164	2.568	1.080	2.
2.284	2.458	3.307	0.374	1.370	2.631	−0.649	1.111	1.296	2.
0.983	2.360	1.880	4.331	3.672	−0.018	3.053	2.068	2.051	2.
3.603	1.047	1.433	3.600	2.465	2.472	1.190	3.504	2.205	0.
1.809	3.479	1.013	3.249	3.934	1.432	2.893	1.707	3.498	1.
1.277	2.925	2.783	1.597	2.619	2.000	1.513	2.888	1.421	0.
0.303	3.879	2.063	2.132	2.682	2.316	1.718	2.201	4.431	3.
2.498	3.072	3.567	2.302	3.157	1.860	2.802	2.098	0.902	1.
−0.542	0.666	3.987	2.668	2.360	2.762	1.351	2.835	2.972	1.
1.640	2.193	0.976	1.777	1.383	3.100	1.663	1.609	2.503	1.
2.248	1.911	0.620	2.295	1.884	3.421	1.086	2.085	1.861	−0.
1.900	0.623	3.047	1.127	−0.199	3.653	0.976	−0.088	−0.966	2.
1.536	0.718	−0.513	2.675	3.145	1.838	1.609	0.857	1.854	3.
2.503	3.434	2.290	2.397	1.162	1.932	2.626	0.816	1.229	1.
1.142	1.628	1.783	2.148	−0.105	3.072	2.312	3.666	2.784	2.
1.877	3.107	0.960	1.363	1.139	1.135	2.370	4.245	3.284	0.
3.632	2.586	1.531	1.613	1.645	0.963	4.596	1.979	2.649	1.
4.072	0.554	1.319	2.224	1.879	1.806	1.606	3.049	0.099	2.
1.564	1.624	1.014	1.414	1.796	1.244	1.712	2.319	2.166	2.
2.876	0.772	−0.646	1.254	3.797	1.827	2.039	4.280	2.208	1.
2.833	3.289	1.977	1.568	2.582	0.198	1.190	2.708	3.264	2.
1.108	2.332	1.546	0.872	4.085	2.583	3.384	3.934	3.073	1.
2.644	1.766	1.846	3.098	2.757	3.840	2.353	3.384	2.716	2.
2.105	1.828	1.899	0.854	2.878	1.569	4.151	0.821	2.818	2.

XII.6 RANDOM NORMAL NUMBERS, $\mu = 0$, $\sigma = 2$

01	02	03	04	05	06	07	08	09	10
.221	−0.540	−0.701	5.511	−2.404	−0.987	−0.158	−0.578	−1.893	0.854
.454	−2.816	0.580	−1.068	1.010	1.209	2.234	3.224	3.750	1.285
.089	0.418	−0.421	2.448	−0.279	1.916	−3.166	−0.773	−0.818	−1.411
.931	1.345	3.164	0.019	0.767	0.439	−3.412	−0.982	0.520	−0.473
.361	0.794	0.120	−0.347	2.785	0.980	1.003	−1.796	−1.778	−0.783
.559	−2.111	−3.396	4.236	2.764	−1.990	−0.060	−2.488	−0.503	−4.406
.816	−1.369	1.856	0.383	0.016	−2.144	−0.187	−1.561	1.441	−2.246
.784	0.607	0.663	−0.764	−1.395	1.738	−2.055	2.962	−1.616	1.326
.576	−3.024	0.191	1.084	−3.698	−3.031	−0.517	−0.179	0.681	−0.719
.232	1.234	−0.046	1.338	1.726	1.448	2.216	−1.662	2.188	2.308
.129	−1.936	3.381	1.319	−3.131	−1.037	1.191	1.449	0.690	−0.251
.753	1.049	1.616	1.232	2.910	0.389	−3.766	2.044	1.459	−0.002
.071	1.869	−5.827	0.866	−1.191	2.508	1.552	−1.052	1.914	−0.274
.507	0.595	−0.202	−0.775	−1.732	−2.771	0.049	5.221	3.059	0.015
.384	−1.574	1.414	−0.789	−1.263	−0.470	0.020	1.489	0.497	1.316
.688	4.311	2.305	−5.632	1.776	1.540	0.208	0.611	0.810	−1.241
.045	−1.563	3.687	−0.160	−0.101	1.838	1.590	1.222	0.377	2.069
.516	−1.339	0.956	−1.285	0.301	3.739	−3.320	0.183	0.993	−4.678
.536	1.965	−0.580	−0.307	1.564	0.163	−2.239	−2.460	−2.003	−1.609
.775	1.427	−0.626	−1.134	−3.109	1.652	2.331	−0.188	2.137	−1.316
.964	−3.740	1.995	−1.349	−1.068	−0.172	1.907	5.515	0.863	−1.018
.597	2.328	−0.722	3.057	1.632	0.655	0.972	1.401	1.840	4.508
.343	−2.859	0.903	−0.631	−2.810	3.345	1.997	0.356	1.215	0.501
.097	0.798	−1.057	3.880	2.321	−1.677	−3.746	−1.125	−1.090	−1.972
.977	0.225	−0.004	−0.513	3.613	1.030	2.349	−1.278	−1.301	−5.159
.421	−1.732	2.170	0.451	1.013	−0.912	0.615	−0.532	1.453	−2.155
.921	−0.932	−2.511	−0.164	0.154	−0.004	0.516	2.240	−0.020	4.432
.854	−3.192	−3.633	0.067	3.709	0.560	−0.156	0.964	−2.618	−0.718
.457	−0.566	−1.439	0.194	1.440	−1.568	−2.407	−1.356	0.849	0.801
.143	0.212	4.088	−0.832	−0.361	0.303	−2.984	1.378	−0.649	−2.399
.184	1.622	−1.896	0.026	−2.163	−1.683	3.778	3.585	−3.853	2.352
.842	1.179	−0.987	0.498	2.348	3.263	1.924	−4.421	−0.680	2.129
.930	0.114	−6.145	0.737	−0.353	2.478	−2.104	0.020	−2.250	−1.096
.174	−0.403	−1.539	1.740	−1.293	−1.922	1.228	1.433	−2.659	2.923
.017	2.409	1.876	4.534	−0.539	−1.534	−0.847	−0.107	0.796	−1.257
.632	−2.417	0.136	−1.155	4.277	−1.035	−0.968	−0.400	1.393	−0.858
.589	−2.199	0.776	0.821	3.237	4.810	1.012	−4.102	1.088	1.958
.308	0.561	−0.882	4.041	1.923	−0.717	0.599	1.705	0.241	2.677
.881	−1.808	−2.767	0.426	0.234	−4.060	1.036	−1.657	0.471	1.753
.869	4.397	1.986	−2.123	−0.065	−0.705	−0.968	−2.265	−1.979	−1.114
.252	−1.324	0.302	2.426	0.710	−1.454	−0.319	2.277	−0.971	−3.217
.833	1.653	−2.738	2.856	−0.789	−0.873	−0.809	−1.538	−1.334	2.289
.276	0.020	−0.162	−1.720	0.048	0.401	−2.073	2.430	2.776	1.174
.742	3.058	1.994	3.090	0.170	−0.789	−2.526	−0.980	−1.331	−1.834
.770	1.419	4.391	1.502	−2.856	−3.648	−1.179	1.556	3.176	2.613
.208	1.766	−0.282	3.051	−1.734	−0.032	1.234	−0.626	1.052	−3.146
.161	−0.803	5.530	2.219	−0.371	1.372	−1.649	−2.059	1.456	0.677
.276	−0.196	−1.456	0.139	0.094	2.367	−1.902	1.123	−1.222	0.323
.615	−0.140	0.697	−0.647	1.289	1.416	0.811	0.523	1.406	−1.022
.831	−0.105	−2.271	−3.207	0.539	−1.010	2.646	−1.985	0.347	0.712

Miscellaneous Statistical Tables

RANDOM NORMAL NUMBERS, $\mu = 0$, $\sigma = 2$

11	12	13	14	15	16	17	18	19	20
-0.686	0.678	-0.150	-0.334	5.096	0.708	0.403	1.538	1.217	-2.45
0.106	-0.018	2.558	-0.049	-1.061	-0.574	-0.510	-1.036	-0.168	2.51
-0.638	3.191	-4.587	-0.499	-0.510	-1.521	1.325	-1.355	-1.036	-3.30
-3.088	-0.998	0.883	2.230	0.603	0.906	-2.303	0.152	1.423	2.70
1.017	-2.496	-4.981	1.769	-0.252	3.018	2.764	1.105	-1.527	-1.18
0.132	-1.520	-1.748	-3.513	0.878	-0.483	-0.354	-1.103	-2.178	2.19
1.217	-2.380	-0.017	-0.072	-2.066	0.251	0.035	1.641	-1.298	2.07
0.540	-3.950	1.287	-0.771	-2.946	-1.181	-2.286	-1.561	0.420	-0.74
-0.395	-1.421	-2.163	0.270	-1.257	5.610	0.287	-1.138	-0.979	-2.22
-3.279	-2.813	1.125	-3.982	-0.102	0.576	-0.531	-0.695	-3.385	-1.04
3.510	3.235	2.623	-2.582	1.274	2.440	0.824	1.983	-0.770	2.48
-0.408	1.922	2.351	-0.146	-0.039	-0.648	-3.041	0.522	-1.659	-0.51
-2.434	1.297	2.813	-0.651	3.409	0.108	1.716	-1.801	0.344	0.92
0.541	0.820	-0.254	2.851	-3.027	1.553	2.218	1.758	1.003	-1.14
-0.813	1.562	2.116	-4.375	-2.289	1.593	0.163	-1.991	0.355	1.36
-0.238	-0.450	0.364	0.677	-0.711	-1.661	1.071	0.682	0.347	0.11
-1.214	-5.369	3.300	0.461	3.197	-0.368	-3.190	2.868	-0.943	0.93
3.521	1.655	2.373	1.993	-1.096	1.875	-1.143	-3.658	2.664	0.37
-1.075	-3.475	-2.069	-0.971	-0.909	-1.796	-0.760	-1.794	-1.576	3.86
-1.213	-1.760	0.397	-0.323	2.659	0.666	-4.368	2.704	1.160	-1.37
0.408	2.865	3.666	-0.433	-1.798	-1.434	-3.688	2.261	-1.084	0.72
-1.809	-2.890	-3.537	-2.701	0.656	0.684	0.905	1.953	2.720	-0.26
-1.633	0.283	-3.937	-0.224	-0.549	0.016	-1.265	-1.650	-1.506	0.50
-1.181	2.578	0.568	0.286	1.152	-0.929	-3.335	0.020	1.171	1.36
0.374	1.225	-0.213	-1.951	0.126	-1.551	-0.147	0.605	2.450	-1.5
-1.828	-3.459	2.624	2.605	0.698	-0.984	-2.289	-3.389	1.647	-2.59
-3.073	1.381	6.111	0.458	-0.792	-0.785	-1.254	2.784	-0.248	-0.96
-0.817	-1.048	-0.603	-0.647	-2.140	1.970	1.612	-2.050	1.962	-3.52
-0.778	-2.697	2.431	-2.011	2.810	0.010	1.830	-1.425	1.425	-2.50
1.866	0.259	-1.360	2.165	1.845	-0.326	2.054	-0.825	5.348	-0.38
0.715	-0.981	-0.126	0.263	-0.692	-3.790	-3.119	-0.547	-1.450	-0.79
0.802	-0.906	-0.726	0.071	3.693	-0.409	1.536	-0.907	3.216	-3.09
-0.363	0.030	-2.423	-0.517	-4.567	1.092	-1.194	0.253	-2.816	0.70
2.878	-2.689	0.797	0.820	0.764	0.366	-0.891	-0.122	0.196	1.05
-2.245	1.324	0.101	0.431	-2.152	0.779	-0.708	0.028	1.317	-1.28
-0.286	0.390	1.204	-4.414	-0.164	-3.724	0.207	1.835	0.334	0.66
-1.434	-0.736	-0.040	0.213	0.215	-0.565	0.915	-0.022	0.487	-0.48
0.004	-1.773	-0.480	0.768	-0.837	-0.513	0.828	4.563	1.298	2.83
1.299	-1.915	0.346	1.037	-2.953	-1.968	-1.704	1.639	2.802	-2.90
-1.694	1.993	1.021	-2.152	0.679	-0.763	0.577	2.860	0.329	-2.70
-0.506	0.328	-2.091	-0.238	2.582	-0.429	1.647	-1.048	-1.367	1.05
0.527	-3.033	-0.893	-0.776	-0.383	-0.708	-0.482	-2.686	1.369	1.0
3.815	-1.282	1.877	-1.177	0.000	-0.059	0.754	0.529	-0.522	2.07
1.838	-0.569	3.556	-0.956	-2.106	0.371	1.806	0.449	-0.867	2.6
0.810	3.394	1.224	-4.428	4.645	-0.644	-2.579	-2.198	2.683	1.33
4.048	-0.706	1.326	2.187	0.611	2.962	-2.137	-0.657	-3.539	-1.5
1.619	-1.214	-0.860	-0.625	-2.534	0.519	-2.539	0.737	-0.622	-0.1
1.732	-3.199	2.104	1.752	0.087	-0.333	3.943	0.037	0.135	1.7
-1.580	-3.456	2.934	0.580	1.665	2.331	-1.413	-1.558	2.144	1.8
1.176	-0.195	0.127	0.060	-3.145	-0.001	0.219	-1.803	-5.255	1.5

RANDOM NORMAL NUMBERS, $\mu = 0$, $\sigma = 2$

21	22	23	24	25	26	27	28	29	30
-0.625	-1.119	0.772	-1.479	0.164	3.051	0.297	1.904	2.864	3.093
1.375	1.994	1.004	-3.128	-1.517	-2.916	2.196	3.544	-1.858	-1.021
-1.835	0.393	-1.426	-0.469	-0.009	-2.366	0.408	-0.669	-0.266	-0.907
0.764	-2.796	-1.932	-1.144	-4.177	-2.150	4.163	1.003	-1.088	-0.346
-0.913	-4.834	2.310	-0.154	-2.007	-1.741	-3.570	1.361	-0.219	-1.424
-4.228	0.846	-0.794	-1.756	2.621	0.128	-1.369	2.090	-4.471	0.440
-2.724	2.694	-0.585	1.094	2.116	-1.176	0.180	1.438	1.260	-1.730
4.168	-0.764	-0.791	-2.517	-2.103	0.901	0.141	1.796	-4.435	1.711
3.286	2.374	1.605	-0.951	-1.308	-0.903	1.562	3.537	3.340	-1.417
0.197	0.212	1.303	-0.289	1.441	-2.913	-0.606	-1.302	1.281	0.147
0.713	-1.532	-4.409	-2.502	-1.488	1.696	-2.390	0.517	-2.406	-0.457
0.925	-2.267	2.010	-1.381	-2.057	0.988	-0.024	-2.096	0.116	1.383
-3.312	1.604	0.955	-0.184	0.074	-0.714	2.059	-2.293	0.899	-0.837
0.320	-2.893	-1.005	1.527	-0.990	1.930	-1.512	1.333	3.188	-1.555
0.619	-1.545	1.543	-0.207	-0.586	2.409	-2.454	-0.738	-0.060	-1.533
0.119	-0.542	-2.461	-2.475	-1.265	-3.598	0.983	-1.702	-1.735	-4.773
1.814	-0.053	-0.063	-2.921	2.076	-0.535	2.585	-3.066	-0.771	1.553
-2.068	0.648	2.066	0.610	-0.681	0.845	1.349	0.515	-1.106	-3.860
-1.881	-2.033	1.704	1.161	0.316	1.623	-3.370	-0.261	-3.559	-0.647
-2.125	0.620	-0.838	2.278	0.230	2.962	1.925	-2.209	-0.676	0.859
2.054	-2.290	2.264	1.598	2.064	-1.129	-1.381	-1.149	-0.488	0.568
-2.516	-2.190	-0.629	2.361	1.734	0.607	0.935	1.275	3.125	-0.224
-0.143	-1.222	1.061	-2.668	-4.419	0.569	0.259	-0.027	1.989	4.602
-0.007	0.017	-0.811	-0.166	0.850	0.565	0.184	-2.887	1.101	0.192
-1.749	0.231	-2.380	-3.177	-1.077	4.460	0.494	1.941	-0.106	0.015
1.300	-0.289	-2.657	-0.160	-0.490	-0.329	1.602	-1.110	4.204	2.552
0.588	-1.072	0.935	-0.164	0.113	1.139	-0.923	-0.953	0.001	-0.033
1.719	-1.183	-1.051	-0.944	0.734	1.965	2.121	2.213	3.826	-2.004
0.726	1.867	0.624	2.066	-2.792	-2.507	-0.816	-0.569	0.002	-1.934
2.369	-0.361	2.216	-1.500	-0.350	-1.063	-3.979	-3.626	-1.326	-0.488
1.790	-0.290	2.601	6.261	-0.622	-0.534	0.477	0.075	0.167	-2.351
1.801	-2.408	0.408	-2.039	0.175	3.839	3.096	-0.001	2.912	-0.560
0.392	1.600	-0.940	-0.160	-0.885	-1.083	-3.503	1.814	-0.563	-2.682
-4.113	-3.018	0.523	-1.915	-0.722	-2.769	0.210	-0.381	-0.724	-2.013
-0.393	-0.828	-0.102	-2.457	1.702	2.257	-2.473	-1.459	-1.385	-3.669
0.688	-0.214	2.741	2.906	-0.778	1.158	0.713	0.815	-0.670	0.144
1.957	1.104	3.540	2.726	-0.028	-0.181	-1.477	-4.434	0.457	0.057
1.823	-1.371	-4.951	3.333	0.248	1.691	2.311	-2.996	1.573	2.319
0.277	0.346	-1.354	3.170	0.268	0.773	1.242	3.542	0.940	-0.535
0.484	1.447	-0.512	-1.379	-0.808	1.014	2.103	0.005	-2.122	0.843
3.209	-1.924	-0.833	1.158	3.203	-0.040	-0.880	-2.217	0.007	0.022
1.834	2.064	-0.319	2.672	1.281	4.921	0.819	0.634	-4.961	-0.739
-1.618	0.705	0.220	-0.177	-0.117	-4.699	2.210	0.035	2.403	-0.816
2.553	1.710	-2.844	-4.619	-4.328	0.459	-2.373	-1.069	2.792	1.942
-1.665	1.627	1.072	-0.902	1.336	3.850	-0.804	1.254	-2.493	-1.101
-0.468	2.177	1.703	1.897	1.139	-1.606	-1.139	-1.435	5.162	-0.146
-1.296	-0.646	0.193	0.534	-0.863	-3.178	2.461	-1.275	0.731	2.983
-3.086	-0.115	2.325	0.088	4.652	2.833	-0.054	-0.670	-2.313	-1.956
-1.324	0.102	0.665	0.878	-1.760	-1.038	0.685	-1.034	0.380	-0.463
1.936	2.264	0.379	4.480	-3.841	-3.992	-3.565	2.558	-0.906	-0.432

Miscellaneous Statistical Tables

RANDOM NORMAL NUMBERS, $\mu = 0$, $\sigma = 2$

31	32	33	34	35	36	37	38	39	40
−0.660	−0.996	−0.264	−1.823	0.818	−0.410	−1.786	2.399	1.986	0.24
1.785	−0.471	0.082	−3.006	−2.286	−0.222	0.388	−0.110	−0.358	−0.33
0.880	0.224	2.561	2.165	2.974	2.516	−4.148	−0.241	−1.318	−0.67
1.021	3.100	−1.783	−2.063	−2.176	1.959	0.248	0.597	1.394	0.61
−2.420	5.579	2.351	1.601	1.045	−2.857	1.400	3.411	−0.239	−2.32
−2.542	−3.145	−2.432	−0.444	−1.276	−3.342	3.479	2.630	−0.405	1.84
−2.437	0.104	1.110	0.008	−2.173	2.294	−0.529	1.723	−1.609	2.11
3.280	−1.213	−3.063	3.637	−0.038	0.217	−0.790	−1.412	−0.055	−0.61
0.502	−0.767	−1.569	0.386	−1.990	−0.062	3.319	−2.448	−0.445	0.05
−1.684	0.290	3.179	0.158	3.562	−1.929	−1.170	1.179	0.110	3.65
0.023	1.652	3.049	−1.410	−1.447	−0.638	−2.483	2.386	−0.331	1.21
−2.741	−2.313	−2.069	−1.305	−0.934	−7.769	2.654	−1.032	1.489	0.67
−0.746	2.099	−3.225	1.533	−1.741	−1.922	0.895	−2.974	−0.828	1.7
−0.934	−4.158	−3.297	−2.859	−4.026	2.722	−1.268	0.991	−1.196	−0.4
−1.574	0.097	2.122	−3.279	−0.820	0.483	2.196	0.642	−1.488	0.37
1.261	−0.663	0.616	−2.801	1.065	4.845	0.418	−0.226	1.897	3.5
2.030	1.692	0.265	0.511	−1.959	0.247	−1.381	−2.625	0.695	−2.24
−4.452	0.900	−1.646	0.573	0.973	−0.350	2.649	4.114	2.497	0.2
−0.075	−2.069	−0.574	0.001	−0.784	−1.235	−3.191	2.128	1.168	−0.74
0.369	0.919	−2.760	1.878	−5.001	−1.670	0.913	−2.853	0.002	1.88
1.360	−2.214	−2.175	0.193	3.298	−0.103	2.226	0.164	−2.429	−0.5
−3.271	2.845	−0.102	−0.822	−3.646	0.361	−3.188	−1.031	1.846	−1.6
−0.908	−3.907	1.407	0.078	1.324	0.276	−2.805	0.604	1.632	2.4
−1.323	2.717	−0.083	−1.645	1.103	−1.539	−0.173	2.429	−0.343	0.0
1.095	−0.871	1.636	2.345	−3.127	0.500	1.250	−0.072	−3.248	2.6
−1.907	−0.869	−4.388	−0.114	1.890	0.218	0.510	−0.768	−2.610	3.6
1.508	−0.333	−2.433	1.237	−1.733	−2.826	−3.761	−1.125	0.720	−0.8
2.937	1.887	−0.430	−5.194	4.716	−2.950	−0.393	−1.111	0.008	−0.1
−4.706	−1.302	−2.011	−0.124	−2.037	0.140	1.392	−1.869	−2.249	−0.0
5.019	−3.900	1.300	−0.034	−1.679	−0.621	0.285	1.197	−0.871	−1.2
−0.910	−0.495	0.074	3.144	−2.631	−3.152	0.192	−1.073	0.646	4.3
1.304	−1.010	−0.739	−1.028	2.886	−1.418	1.314	0.779	−2.139	0.1
0.371	−5.663	−0.017	−1.551	−3.508	1.305	−0.819	−0.199	−0.331	−0.3
−2.039	−3.961	0.679	2.451	−2.802	1.449	−0.964	−1.170	0.891	1.5
−0.911	1.904	0.062	−2.375	−1.548	−0.361	2.692	3.772	2.005	3.7
−1.720	0.871	3.594	0.889	0.162	0.112	−0.053	−2.597	−1.310	−2.2
−4.091	0.430	2.222	−0.141	0.506	2.751	−0.472	−1.141	1.671	−0.9
−1.797	−1.272	1.847	0.039	0.689	−0.080	1.457	−3.856	1.332	−2.8
−0.719	0.829	2.570	1.107	−0.314	−3.750	1.041	−1.657	−0.233	1.4
−1.890	3.240	1.877	2.552	3.389	0.215	1.979	−0.895	−1.996	0.6
−1.952	−1.276	−2.754	−0.049	−2.916	3.820	0.381	1.337	2.211	3.4
0.502	1.812	−0.577	0.551	−0.257	0.883	4.377	−4.180	−2.266	−0.1
1.971	2.333	−0.945	2.618	2.953	−1.997	1.491	−0.082	2.617	0.7
0.561	−1.506	4.127	0.933	−1.930	0.460	−0.008	0.352	−1.274	0.2
−1.409	−0.638	2.757	0.461	1.331	2.030	−0.846	−1.035	−1.580	−0.7
−2.066	0.218	0.070	−3.420	0.089	−0.084	4.944	−4.285	0.200	0.2
−2.734	3.622	−0.300	1.648	1.328	0.479	−0.498	1.997	2.203	2.7
−1.434	1.441	0.258	−1.893	−2.925	−1.753	0.272	0.747	−0.999	−0.1
−0.071	−4.344	−2.763	4.371	1.547	2.588	2.914	0.261	3.381	5.4
4.574	1.751	3.420	−1.383	0.966	−2.731	3.444	1.410	2.740	−2.0

RANDOM NORMAL NUMBERS, $\mu = 0$, $\sigma = 2$

41	42	43	44	45	46	47	48	49	50
-1.739	0.276	1.761	0.092	0.820	1.772	-3.258	0.707	-0.578	-1.611
-1.776	-1.482	1.399	1.031	-0.546	-0.204	2.591	2.129	1.615	0.919
-1.894	0.388	1.023	-1.493	1.513	1.003	2.547	-2.443	-1.855	2.898
-2.042	1.064	-2.399	-0.333	-2.141	-1.022	-2.976	-0.485	0.073	-0.891
0.287	0.120	-2.013	0.598	0.001	3.454	2.077	-1.966	-4.187	0.452
0.739	-4.324	0.088	1.124	0.610	0.368	0.953	-0.141	-0.441	3.163
0.749	0.597	2.194	-0.771	1.063	0.246	0.465	-3.122	-1.995	3.015
-1.111	-2.558	0.146	-0.590	-3.278	2.649	-1.299	-1.809	3.938	-1.766
-0.347	2.324	0.083	-0.152	-3.563	-1.062	0.901	0.882	0.865	0.581
-0.105	1.781	-0.775	-0.726	-3.211	-1.200	-2.688	1.639	-0.945	-1.022
1.968	2.056	-4.124	-1.126	-2.798	-1.150	-1.632	-3.405	1.182	1.985
1.614	-1.436	-4.649	-1.168	2.549	0.522	-0.616	2.009	-0.465	1.362
-1.671	-0.907	-0.459	2.880	2.640	-0.751	-2.414	-1.195	-2.334	-0.240
-1.328	0.335	-0.049	-1.903	0.225	-0.140	-1.121	0.820	0.282	0.635
-0.623	-0.823	1.655	1.997	-3.841	3.318	1.035	1.056	2.112	2.166
-2.292	-1.662	2.136	-0.223	1.372	-3.612	-0.276	-4.097	-0.419	-0.017
3.146	1.248	0.090	-1.069	-0.022	1.017	-1.157	1.803	-1.585	-1.526
1.553	-1.369	0.044	0.606	-1.734	-1.443	-3.016	-0.977	3.150	0.264
-3.179	3.510	-2.299	0.371	1.071	-1.044	-1.352	1.740	1.936	0.242
1.701	-0.455	2.119	-1.716	-2.857	-0.991	-1.621	2.934	3.487	0.754
-2.088	1.495	2.961	-2.029	-0.072	-0.664	0.992	1.659	0.834	-2.175
-3.757	0.316	0.763	-3.035	0.907	-3.804	-3.403	3.689	0.901	-2.386
-1.177	-1.422	-4.712	0.235	-1.048	-2.627	0.794	-1.473	2.598	-0.364
-1.380	-1.661	-1.714	-1.396	0.477	1.750	-2.458	-5.077	-0.194	1.093
-0.852	0.562	-0.199	0.802	0.494	-0.294	0.205	0.260	-2.616	4.117
2.591	1.323	0.458	4.020	-1.907	-0.065	-2.786	0.137	0.446	4.368
-2.240	2.744	0.551	-3.005	-2.677	4.492	2.928	0.061	-0.216	2.566
-1.488	-0.163	-0.187	2.081	-0.993	1.160	1.301	-2.236	1.586	0.011
0.622	-0.988	-0.956	-0.484	-0.648	-3.467	-3.778	1.181	1.740	0.092
-0.949	-2.527	-1.934	1.318	0.422	3.848	0.050	-1.448	0.278	3.041
-4.952	0.019	1.793	0.881	0.282	0.621	1.202	-0.373	3.665	3.386
-0.751	3.342	0.969	0.821	1.983	-0.533	-1.273	-2.214	-0.774	-1.210
0.618	-0.688	-2.960	-5.252	-0.543	0.104	-0.468	-3.139	0.594	-1.302
2.371	0.160	1.715	0.319	1.387	5.138	3.883	-1.869	-0.899	-1.019
-1.184	0.047	1.453	-0.889	-1.292	0.197	-0.302	-1.497	-1.838	-0.940
-0.287	2.329	2.028	-1.765	1.669	-1.024	1.600	0.454	3.098	2.275
1.764	-2.839	-1.942	0.008	4.001	0.083	-1.631	2.968	-0.146	-2.079
1.149	-1.571	1.296	1.510	-0.599	0.083	-0.688	6.017	0.012	-1.451
2.984	-1.432	-0.960	-2.124	1.353	0.934	0.666	3.096	2.905	-1.472
-1.701	-0.004	2.710	0.573	2.424	-0.119	-1.410	3.413	-3.588	0.047
-2.333	0.912	-0.773	-2.016	2.253	2.784	3.764	0.559	4.791	1.288
-0.214	2.787	0.095	-3.174	1.460	0.411	0.922	-0.474	3.113	-1.067
1.214	0.785	-2.686	1.909	-1.747	-4.551	0.589	-0.573	-1.364	-2.583
0.878	0.097	1.650	1.437	-1.643	-2.608	1.122	0.538	0.664	-0.323
-0.105	-0.297	3.821	2.105	2.021	-1.922	1.472	0.042	1.403	1.465
-0.593	0.136	0.910	-0.549	-1.472	3.214	-2.273	3.458	1.436	0.500
2.198	2.325	-1.229	-0.276	1.560	-0.482	-0.482	0.455	-0.181	1.417
1.160	0.139	0.997	-0.082	-0.689	0.995	-5.301	0.998	3.413	-1.797
3.024	-1.561	0.982	-1.244	1.407	-0.063	-1.176	2.355	2.006	-4.833
0.955	0.174	-0.401	2.472	0.584	3.811	1.115	0.951	-2.136	-2.324

XII.7 ORTHOGONAL POLYNOMIALS

In fitting a curvilinear model of the form

$$y_i = \beta_0 + \beta_1 x_i + \beta_2 x_i^2 + \cdots + \beta_p x_i^p + e_i ,$$
$$i = 1, 2, \ldots , n,$$

it is convenient computationally to fit the model using orthogonal polynomials. Here (fits the model

$$y_i = \alpha_0 \xi_0(x) + \alpha_1 \xi_1(x) + \cdots + \alpha_p \xi_p(x) + e_i$$

where the $\xi_j(x)$ are orthogonal polynomials in x of the jth degree, namely, the Tchebych polynomials. The least-squares estimators $\hat{\alpha}_j$ of the α_j are given by

$$\hat{\alpha}_j = \frac{\sum_i y_i \xi_j(x_i)}{\sum_i [\xi_j(x_i)]^2} .$$

The estimators $\hat{\alpha}_j$ for any $j \le n - 1$ are independent normal variates with means α_j i $j \le p$ and 0 for $j > p$ and with variances $\dfrac{\sigma^2}{\sum_i [\xi_j(x_i)]^2}$. A mean-square estimate of σ^2 is p

vided by the error sum of squares

$$s^2 = \left\{ \sum_{i=1}^{n} y_i^2 - \sum_{j=0}^{p} \hat{\alpha}_j^2 \left(\sum_{i=1}^{n} \left[\xi_j(x_i) \right]^2 \right) \right\} \Big/ (n - p - 1) .$$

Thus the ratios $\dfrac{(\alpha_j - \hat{\alpha}_j) \sqrt{\sum_i [\xi_j(x_i)]^2}}{s}$ with $\hat{\alpha}_j = 0$ for $j > p$ is distributed as Studen t-distribution.

This table provides values of $\xi_j(x_i)$ for various values of n and j. To avoid fraction values and to reduce the size of the integers, the table is arranged so that the highest pow of x_i in $\xi_j(x_i)$ has a coefficient λ_j. The two values at the bottom of each column are th values $D_j = \displaystyle\sum_{i=1}^{n} [\xi_j(x_i)]^2$ and λ_j.

ORTHOGONAL POLYNOMIALS

n = 3, 4, 5, 6, 7

ξ_1'	ξ_2'	ξ_1'	ξ_2'	ξ_3'	ξ_1'	ξ_2'	ξ_3'	ξ_4'	ξ_1'	ξ_2'	ξ_3'	ξ_4'	ξ_5'	ξ_1'	ξ_2'	ξ_3'	ξ_4'	ξ_5'
−1	+1	−3	+1	−1	−2	+2	−1	+1	−5	+5	−5	+1	−1	−3	+5	−1	+3	−1
0	−2	−1	−1	+3	−1	−1	+2	−4	−3	−1	+7	−3	+5	−2	0	+1	−7	+4
+1	+1	+1	−1	−3	0	−2	0	+6	−1	−4	+4	+2	−10	−1	−3	+1	+1	−5
		+3	+1	+1	+1	−1	−2	−4	+1	−4	−4	+2	+10	0	−4	0	+6	0
					+2	+2	+1	+1	+3	−1	−7	−3	−5	+1	−3	−1	+1	+5
									+5	+5	+5	+1	+1	+2	0	−1	−7	−4
														+3	+5	+1	+3	+1
2	6	20	4	20	10	14	10	70	70	84	180	28	252	28	84	6	154	84
1	3	2	1	$\frac{10}{3}$	1	1	$\frac{5}{6}$	$\frac{35}{12}$	2	$\frac{3}{2}$	$\frac{5}{3}$	$\frac{7}{12}$	$\frac{21}{10}$	1	1	$\frac{1}{6}$	$\frac{7}{12}$	$\frac{7}{20}$

n = 8, 9, 10, 11

ξ_1'	ξ_2'	ξ_3'	ξ_4'	ξ_5'	ξ_1'	ξ_2'	ξ_3'	ξ_4'	ξ_5'	ξ_1'	ξ_2'	ξ_3'	ξ_4'	ξ_5'	ξ_1'	ξ_2'	ξ_3'	ξ_4'	ξ_5'
−7	+7	−7	+7	−7	0	−20	0	+18	0	+1	−4	−12	+18	+6	0	−10	0	+6	0
−5	+1	+5	−13	+23	+1	−17	−9	+9	+9	+3	−3	−31	+3	+11	+1	−9	−14	+4	+4
−3	−3	+7	−3	−17	+2	−8	−13	−11	+4	+5	−1	−35	−17	+1	+2	−6	−23	−1	+4
−1	−5	+3	+9	−15	+3	+7	−7	−21	−11	+7	+2	−14	−22	−14	+3	−1	−22	−6	−1
+1	−5	−3	+9	+15	+4	+28	+14	+14	+4	+9	+6	+42	+18	+6	+4	+6	−6	−6	−6
+3	−3	−7	+3	+17											+5	+15	+30	+6	+3
+5	+1	−5	−13	−23															
+7	+7	+7	+7	+7															
168	168	264	616	2184	60	2,772	990	2,002	468	330	132	8,580	2,860	780	110	858	4,290	286	156
2	1	$\frac{2}{3}$	$\frac{7}{12}$	$\frac{7}{10}$	1	3	$\frac{5}{6}$	$\frac{7}{12}$	$\frac{3}{20}$	2	$\frac{1}{2}$	$\frac{5}{3}$	$\frac{5}{12}$	$\frac{1}{10}$	1	1	$\frac{5}{6}$	$\frac{1}{12}$	$\frac{1}{40}$

n = 12, 13, 14

ξ_1'	ξ_2'	ξ_3'	ξ_4'	ξ_5'	ξ_1'	ξ_2'	ξ_3'	ξ_4'	ξ_5'	ξ_1'	ξ_2'	ξ_3'	ξ_4'	ξ_5'
+1	−35	−7	+28	+20	0	−14	0	+84	0	+1	−8	−24	+108	+60
+3	−29	−19	+12	+44	+1	−13	−4	+64	+20	+3	−7	−67	+63	+145
+5	−17	−25	−13	+29	+2	−10	−7	+11	+26	+5	−5	−95	−13	+139
+7	+1	−21	−33	−21	+3	−5	−8	−54	+11	+7	−2	−98	−92	+28
+9	+25	−3	−27	−57	+4	+2	−6	−96	−18	+9	+2	−66	−132	−132
+11	+55	+33	+33	+33	+5	+11	0	−66	−33	+11	+7	+11	−77	−187
					+6	+22	+11	+99	+22	+13	+13	+143	+143	+143
572	12,012	5,148	8,008	15,912	182	2,002	572	68,068	6,188	910	728	97,240	136,136	235,144
2	3	$\frac{2}{3}$	$\frac{7}{24}$	$\frac{3}{20}$	1	1	$\frac{1}{6}$	$\frac{7}{12}$	$\frac{7}{120}$	2	$\frac{1}{2}$	$\frac{5}{3}$	$\frac{7}{12}$	$\frac{7}{30}$

n = 15, 16

ξ_1'	ξ_2'	ξ_3'	ξ_4'	ξ_5'	ξ_1'	ξ_2'	ξ_3'	ξ_4'	ξ_5'
0	−56	0	+756	0	+1	−21	−63	+189	+45
+1	−53	−27	+621	+675	+3	−19	−179	+129	+115
+2	−44	−49	+251	+1000	+5	−15	−265	+23	+131
+3	−29	−61	−249	+751	+7	−9	−301	−101	+77
+4	−8	−58	−704	−44	+9	−1	−267	−201	−33
+5	+19	−35	−869	−979	+11	+9	−143	−221	−143
+6	+52	+13	−429	−1144	+13	+21	+91	−91	−143
+7	+91	+91	+1001	+1001	+15	+35	+455	+273	+143
280	37,128	39,780	6,466,460	10,581,480	1,360	5,712	1,007,760	470,288	201,552
1	3	$\frac{5}{6}$	$\frac{35}{12}$	$\frac{21}{20}$	2	1	$\frac{10}{3}$	$\frac{7}{12}$	$\frac{1}{10}$

ORTHOGONAL POLYNOMIALS

	17					18			
ξ_1'	ξ_2'	ξ_3'	ξ_4'	ξ_5'	ξ_1'	ξ_2'	ξ_3'	ξ_4'	ξ_5'
0	−24	0	+36	0	+1	−40	−8	+44	+2
+1	−23	−7	+31	+55	+3	−37	−23	+33	+5
+2	−20	−13	+17	+88	+5	−31	−35	+13	+7
+3	−15	−17	−3	+83	+7	−22	−42	−12	+5
+4	−8	−18	−24	+36	+9	−10	−42	−36	+1
+5	+1	−15	−39	−39	+11	+5	−33	−51	−4
+6	+12	−7	−39	−104	+13	+23	−13	−47	−8
+7	+25	+7	−13	−91	+15	+44	+20	−12	−6
+8	+40	+28	+52	+104	+17	+68	+68	+68	+8
D 408	7,752	3,876	16,796	100,776	1,938	23,256	23,256	28,424	6,953,5
λ 1	1	$\frac{1}{6}$	$\frac{1}{12}$	$\frac{1}{20}$	2	$\frac{3}{2}$	$\frac{1}{3}$	$\frac{1}{12}$	$\frac{3}{10}$

	19					20			
ξ_1'	ξ_2'	ξ_3'	ξ_4'	ξ_5'	ξ_1'	ξ_2'	ξ_3'	ξ_4'	ξ_5'
0	−30	0	+396	0	+1	−33	−99	+1188	+3
+1	−29	−44	+352	+44	+3	−31	−287	+948	+10
+2	−26	−83	+227	+74	+5	−27	−445	+503	+14
+3	−21	−112	+42	+79	+7	−21	−553	−77	+13
+4	−14	−126	−168	+54	+9	−13	−591	−687	+7
+5	−5	−120	−354	+3	+11	−3	−539	−1187	−1
+6	+6	−89	−453	−58	+13	+9	−377	−1402	−12
+7	+19	−28	−388	−98	+15	+23	−85	−1122	−18
+8	+34	+68	−68	−68	+17	+39	+357	−102	−11
+9	+51	+204	+612	+102	+19	+57	+969	+1938	+19
D 570	13,566	213,180	2,288,132	89,148	2,660	17,556	4,903,140	22,881,320	31,201,80
λ 1	1	$\frac{5}{6}$	$\frac{7}{12}$	$\frac{1}{40}$	2	1	$\frac{10}{3}$	$\frac{35}{24}$	$\frac{7}{20}$

	21					22			
ξ_1'	ξ_2'	ξ_3'	ξ_4'	ξ_5'	ξ_1'	ξ_2'	ξ_3'	ξ_4'	ξ_5'
0	−110	0	+594	0	+1	−20	−12	+702	+3
+1	−107	−54	+540	+1404	+3	−19	−35	+585	+10
+2	−98	−103	+385	+2444	+5	−17	−55	+365	+15
+3	−83	−142	+150	+2819	+7	−14	−70	+70	+15
+4	−62	−166	−130	+2354	+9	−10	−78	−258	+11
+5	−35	−170	−406	+1063	+11	−5	−77	−563	+3
+6	−2	−149	−615	−788	+13	+1	−65	−775	−6
+7	+37	−98	−680	−2618	+15	+8	−40	−810	−15
+8	+82	−12	−510	−3468	+17	+16	0	−570	−19
+9	+133	+114	0	−1938	+19	+25	+57	+57	−9
+10	+190	+285	+969	+3876	+21	+35	+133	+1197	+22
D 770	201,894	432,630	5,720,330	121,687,020	3,542	7,084	96,140	8,748,740	40,562,34
λ 1	3	$\frac{5}{6}$	$\frac{7}{12}$	$\frac{21}{40}$	2	$\frac{1}{2}$	$\frac{1}{3}$	$\frac{7}{12}$	$\frac{7}{30}$

ORTHOGONAL POLYNOMIALS

		23					24		
ξ_1'	ξ_2'	ξ_3'	ξ_4'	ξ_5'	ξ_1'	ξ_2'	ξ_3'	ξ_4'	ξ_5'
0	−44	0	+858	0	+1	−143	−143	+143	+715
+1	−43	−13	+793	+65	+3	−137	−419	+123	+2005
+2	−40	−25	+605	+116	+5	−125	−665	+85	+2893
+3	−35	−35	+315	+141	+7	−107	−861	+33	+3171
+4	−28	−42	−42	+132	+9	−83	−987	−27	+2721
+5	−19	−45	−417	+87	+11	−53	−1023	−87	+1551
+6	−8	−43	−747	+12	+13	−17	−949	−137	−169
+7	+5	−35	−955	−77	+15	+25	−745	−165	−2071
+8	+20	−20	−950	−152	+17	+73	−391	−157	−3553
+9	+37	+3	−627	−171	+19	+127	+133	−97	−3743
+10	+56	+35	+133	−76	+21	+187	+847	+33	−1463
+11	+77	+77	+1463	+209	+23	+253	+1771	+253	+4807
1,012	35,420	32,890	13,123,110	340,860	4,600	394,680	17,760,600	394,680	177,928,920
1	1	$\frac{1}{6}$	$\frac{7}{12}$	$\frac{1}{60}$	2	3	$\frac{10}{3}$	$\frac{1}{12}$	$\frac{3}{10}$

		25					26		
ξ_1'	ξ_2'	ξ_3'	ξ_4'	ξ_5'	ξ_1'	ξ_2'	ξ_3'	ξ_4'	ξ_5'
0	−52	0	+858	0	+1	−28	−84	+1386	+330
+1	−51	−77	+803	+275	+3	−27	−247	+1221	+935
+2	−48	−149	+643	+500	+5	−25	−395	+905	+1381
+3	−43	−211	+393	+631	+7	−22	−518	+466	+1582
+4	−36	−258	+78	+636	+9	−18	−606	−54	+1482
+5	−27	−285	−267	+501	+11	−13	−649	−599	+1067
+6	−16	−287	−597	+236	+13	−7	−637	−1099	+377
+7	−3	−259	−857	−119	+15	0	−560	−1470	−482
+8	+12	−196	−982	−488	+17	+8	−408	−1614	−1326
+9	+29	−93	−897	−753	+19	+17	−171	−1419	−1881
+10	+48	+55	−517	−748	+21	+27	+161	−759	−1771
+11	+69	+253	+253	−253	+23	+38	+598	+506	−506
+12	+92	+506	+1518	+1012	+25	+50	+1150	+2530	+2530
1,300	53,820	1,480,050	14,307,150	7,803,900	5,850	16,380	7,803,900	40,060,020	48,384,180
1	1	$\frac{5}{6}$	$\frac{5}{12}$	$\frac{1}{20}$	2	$\frac{1}{2}$	$\frac{5}{3}$	$\frac{7}{12}$	$\frac{1}{10}$

		27					28		
ξ_1'	ξ_2'	ξ_3'	ξ_4'	ξ_5'	ξ_1'	ξ_2'	ξ_3'	ξ_4'	ξ_5'
0	−182	0	+1638	0	+1	−65	−39	+936	+1560
+1	−179	−18	+1548	+3960	+3	−63	−115	+840	+4456
+2	−170	−35	+1285	+7304	+5	−59	−185	+655	+6701
+3	−155	−50	+870	+9479	+7	−53	−245	+395	+7931
+4	−134	−62	+338	+10058	+9	−45	−291	+81	+7887
+5	−107	−70	−262	+8803	+11	−35	−319	−259	+6457
+6	−74	−73	−867	+5728	+13	−23	−325	−590	+3718
+7	−35	−70	−1400	+1162	+15	−9	−305	−870	−22
+8	+10	−60	−1770	−4188	+17	+7	−255	−1050	−4182
+9	+61	−42	−1872	−9174	+19	+25	−171	−1074	−7866
+10	+118	−15	−1587	−12144	+21	+45	−49	−879	−9821
+11	+181	+22	−782	−10879	+23	+67	+115	−395	−8395
+12	+250	+70	+690	−2530	+25	+91	+325	+455	−1495
+13	+325	+130	+2990	+16445	+27	+117	+585	+1755	+13455
1,638	712,530	101,790	56,448,210	2,032,135,560	7,308	95,004	2,103,660	19,634,160	1,354,757,040
1	3	$\frac{1}{6}$	$\frac{7}{12}$	$\frac{21}{40}$	2	1	$\frac{2}{3}$	$\frac{7}{24}$	$\frac{7}{20}$

ORTHOGONAL POLYNOMIALS

		29					30		
ξ_1'	ξ_2'	ξ_3'	ξ_4'	ξ_5'	ξ_1'	ξ_2'	ξ_3'	ξ_4'	ξ_5'
0	−70	0	+2184	0	+1	−112	−112	+12376	+176
+1	−69	−104	+2080	+1768	+3	−109	−331	+11271	+508
+2	−66	−203	+1775	+3298	+5	−103	−535	+9131	+775
+3	−61	−292	+1290	+4373	+7	−94	−714	+6096	+940
+4	−54	−366	+660	+4818	+9	−82	−858	+2376	+976
+5	−45	−420	−66	+4521	+11	−67	−957	−1749	+867
+6	−34	−449	−825	+3454	+13	−49	−1001	−5929	+614
+7	−21	−448	−1540	+1695	+15	−28	−980	−9744	+238
+8	−6	−412	−2120	−556	+17	−4	−884	−12704	−217
+9	+11	−336	+2460	−2946	+19	+23	−703	−14249	−682
+10	+30	−215	−2441	−4958	+21	+53	−427	−13749	−1053
+11	+51	−44	−1930	−5885	+23	+86	−46	−10504	−1196
+12	+74	+182	−780	−4810	+25	+122	+450	−3744	−936
+13	+99	+468	+1170	−585	+27	+161	+1071	+7371	−58
+14	+126	+819	+4095	+8190	+29	+203	+1827	+23751	+1696

D 2,030　113,274　4,207,320　107,987,880　500,671,080 | 8,990　302,064　21,360,240　3,671,587,920　2,145,733,20

λ　1　1　$\frac{5}{6}$　$\frac{7}{12}$　$\frac{7}{40}$ | 2　$\frac{3}{2}$　$\frac{5}{3}$　$\frac{35}{12}$　$\frac{3}{10}$

		31					32		
ξ_1'	ξ_2'	ξ_3'	ξ_4'	ξ_5'	ξ_1'	ξ_2'	ξ_3'	ξ_4'	ξ_5'
+0	−80	0	+408	0	+1	−85	−51	+459	+25
+1	−79	−119	+391	+221	+3	−83	−151	+423	+73
+2	−76	−233	+341	+416	+5	−79	−245	+353	+113
+3	−71	−337	+261	+561	+7	−73	−329	+253	+140
+4	−64	−426	+156	+636	+9	−65	−399	+129	+150
+5	−55	−495	+33	+627	+11	−55	−451	−11	+141
+6	−44	−539	−99	+528	+13	−43	−481	−157	+113
+7	−31	−553	−229	+343	+15	−29	−485	−297	+66
+8	−16	−532	−344	+88	+17	−13	−459	−417	+5
+9	+1	−471	−429	−207	+19	+5	−399	−501	−62
+10	+20	−365	−467	−496	+21	+25	−301	−531	−126
+11	+41	−209	−439	−715	+23	+47	−161	−487	−172
+12	+64	+2	−324	−780	+25	+71	+25	−347	−181
+13	+89	+273	−99	−585	+27	+97	+261	−87	−130
+14	+116	+609	+261	0	+29	+125	+551	+319	+87
+15	+145	+1015	+783	+1131	+31	+155	+899	+899	+2697

D 2,480　158,224　6,724,520　4,034,712　9,536,592 | 10,912　185,504　5,379,616　5,379,616　54,285,216

λ　1　1　$\frac{5}{6}$　$\frac{1}{12}$　$\frac{1}{60}$ | 2　1　$\frac{2}{3}$　$\frac{1}{12}$　$\frac{1}{30}$

ORTHOGONAL POLYNOMIALS

		33					34		
ξ'_1	ξ'_2	ξ'_3	ξ'_4	ξ'_5	ξ'_1	ξ'_2	ξ'_3	ξ'_4	ξ'_5
0	−272	0	+3672	0	1	−48	−144	+4104	+6840
+1	−269	−27	+3537	+2565	3	−47	−427	+3819	+19855
+2	−260	−53	+3139	+4864	5	−45	−695	+3263	+30917
+3	−245	−77	+2499	+6649	7	−42	−938	+2464	+38864
+4	−224	−98	+1652	+7708	9	−38	−1146	+1464	+42744
+5	−197	−115	+647	+7883	11	−33	−1309	+319	+41899
+6	−164	−127	−453	+7088	13	−27	−1417	−901	+36049
+7	−125	−133	−1571	+5327	15	−20	−1460	−2112	+25376
+8	−80	−132	−2616	+2712	17	−12	−1428	−3216	+10608
+9	−29	−123	−3483	−519	19	−3	−1311	−4101	−6897
+10	+28	−105	−4053	−3984	21	+7	−1099	−4641	−25067
+11	+91	−77	−4193	−7139	23	+18	−782	−4696	−41032
+12	+160	−38	−3756	−9260	25	+30	−350	−4112	−51040
+13	+235	+13	−2581	−9425	27	+43	+207	−2721	−50373
+14	+316	+77	−493	−6496	29	+57	+899	−341	−33263
+15	+403	+155	+2697	+899	31	+72	+1736	+3224	+7192
+16	+496	+248	+7192	+14384	33	+88	+2728	+8184	+79112
2,992	1,947,792	417,384	348,330,136	1,547,128,656	13,090	62,832	51,477,360	456,432,592	46,929,569,232
1	3	$\frac{1}{6}$	$\frac{7}{12}$	$\frac{3}{20}$	2	$\frac{1}{2}$	$\frac{5}{3}$	$\frac{7}{12}$	$\frac{7}{10}$

		35		
ξ'_1	ξ'_2	ξ'_3	ξ'_4	ξ'_5
0	−102	0	+23256	0
1	−101	−152	+22496	+3800
2	−98	−299	+20251	+7250
3	−93	−436	+16626	+10021
4	−86	−558	+11796	+11826
5	−77	−660	+6006	+12441
6	−66	−737	−429	+11726
7	−53	−784	−7124	+9646
8	−38	−796	−13624	+6292
9	−21	−768	−19404	+1902
10	−2	−695	−23869	−3118
11	+19	−572	−26354	−8173
12	+42	−394	−26124	−12458
13	+67	−156	−22374	−14937
14	+94	+147	−14229	−14322
15	+123	+520	−744	−9052
16	+154	+968	+19096	+2728
17	+187	+1496	+46376	+23188
D	3,570	290,598	15,775,320	14,834,059,240	4,045,652,520
λ	1	1	$\frac{5}{6}$	$\frac{35}{12}$	$\frac{7}{40}$

Miscellaneous Statistical Tables

ORTHOGONAL POLYNOMIALS

36

ξ_1'	ξ_2'	ξ_3'	ξ_4'	ξ_5'
1	−323	−323	+2584	+12920
3	−317	−959	+2424	+37640
5	−305	−1565	+2111	+59063
7	−287	−2121	+1659	+75201
9	−263	−2607	+1089	+84381
11	−233	−3003	+429	+85371
13	−197	−3289	−286	+77506
15	−155	−3445	−1014	+60814
17	−107	−3451	−1706	+36142
19	−53	−3287	−2306	+5282
21	+7	−2933	−2751	−28903
23	+73	−2369	−2971	−62353
25	+145	−1575	−2889	−89685
27	+223	−531	−2421	−104067
29	+307	+783	−1476	−97092
31	+397	+2387	+44	−58652
33	+493	+4301	+2244	+23188
35	+595	+6545	+5236	+162316
D	15,540	3,011,652	307,618,740	191,407,216	199,046,103,984
λ	2	3	$\frac{10}{3}$	$\frac{7}{24}$	$\frac{21}{20}$

		37					38		
ξ_1'	ξ_2'	ξ_3'	ξ_4'	ξ_5'	ξ_1'	ξ_2'	ξ_3'	ξ_4'	ξ_5'
0	−114	0	+5814	0	1	−60	−36	+918	+15?
1	−113	−34	+5644	+680	3	−59	−107	+867	+44?
2	−110	−67	+5141	+1304	5	−57	−175	+767	+70(
3	−105	−98	+4326	+1819	7	−54	−238	+622	+90?
4	−98	−126	+3234	+2178	9	−50	−294	+438	+103(
5	−89	−150	+1914	+2343	11	−45	−341	+223	+107?
6	−78	−169	+429	+2288	13	−39	−377	−13	+101?
7	−65	−182	−1144	+2002	15	−32	−400	−258	+85?
8	−50	−188	−2714	+1492	17	−24	−408	−498	+60?
9	−33	−186	−4176	+786	19	−15	−399	−717	+26?
10	−14	−175	−5411	−64	21	−5	−371	−897	−12?
11	+7	−154	−6286	−979	23	+6	−322	−1018	−52?
12	+30	−122	−6654	−1850	25	+18	−250	−1058	−90?
13	+55	−78	−6354	−2535	27	+31	−153	−993	−119?
14	+82	−21	−5211	−2856	29	+45	−29	−797	−130?
15	+111	+50	−3036	−2596	31	+60	+124	−442	−115?
16	+142	+136	+374	−1496	33	+76	+308	+102	−63?
17	+175	+238	+5236	+748	35	+93	+525	+867	+39?
18	+210	+357	+11781	+4488	37	+111	+777	+1887	+207?
D	4,218	383,838	932,178	980,961,982	152,877,192	18,278	109,668	4,496,388	25,479,532	3,286,859,6?
λ	1	1	$\frac{1}{6}$	$\frac{7}{12}$	$\frac{1}{40}$	2	$\frac{1}{2}$	$\frac{1}{3}$	$\frac{1}{12}$	$\frac{1}{10}$

ORTHOGONAL POLYNOMIALS

	39			
ξ_1'	ξ_2'	ξ_3'	ξ_4'	ξ_5'
0	-380	0	$+1026$	0
1	-377	-189	$+999$	$+5049$
2	-368	-373	$+919$	$+9724$
3	-353	-547	$+789$	$+13669$
4	-332	-706	$+614$	$+16564$
5	-305	-845	$+401$	$+18143$
6	-272	-959	$+159$	$+18212$
7	-233	-1043	-101	$+16667$
8	-188	-1092	-366	$+13512$
9	-137	-1101	-621	$+8877$
10	-80	-1065	-849	$+3036$
11	-17	-979	-1031	-3575
12	$+52$	-838	-1146	-10340
13	$+127$	-637	-1171	-16445
14	$+208$	-371	-1081	-20860
15	$+295$	-35	-849	-22321
16	$+388$	$+376$	-446	-19312
17	$+487$	$+867$	$+159$	-10047
18	$+592$	$+1443$	$+999$	$+7548$
19	$+703$	$+2109$	$+2109$	$+35853$
D	4,940	4,496,388	33,722,910	32,224,114	9,860,578,884
λ	1	3	$\frac{5}{6}$	$\frac{1}{12}$	$\frac{3}{20}$

	40			
ξ_1'	ξ_2'	ξ_3'	ξ_4'	ξ_5'
1	-133	-399	$+39501$	$+627$
3	-131	-1187	$+37521$	$+1837$
5	-127	-1945	$+33631$	$+2917$
7	-121	-2653	$+27971$	$+3787$
9	-113	-3291	$+20751$	$+4377$
11	-103	-3839	$+12251$	$+4631$
13	-91	-4277	$+2821$	$+4511$
15	-77	-4585	-7119	$+4001$
17	-61	-4743	-17079	$+3111$
19	-43	-4731	-26499	$+1881$
21	-23	-4529	-34749	$+385$
23	-1	-4117	-41129	-1265
25	$+23$	-3475	-44869	-2915
27	$+49$	-2583	-45129	-4365
29	$+77$	-1421	-40999	-5365
31	$+107$	$+31$	-31499	-5611
33	$+139$	$+1793$	-15579	-4741
35	$+173$	$+3885$	$+7881$	-2331
37	$+209$	$+6327$	$+40071$	$+2109$
39	$+247$	$+9139$	$+82251$	$+9139$
D	21,320	567,112	644,482,280	49,625,135,560	644,482,280
λ	2	1	$\frac{10}{3}$	$\frac{35}{12}$	$\frac{1}{30}$

Miscellaneous Statistical Tables

ORTHOGONAL POLYNOMIALS

		41		
ξ_1'	ξ_2'	ξ_3'	ξ_4'	ξ_5'
0	−140	0	+8778	0
1	−139	−209	+8569	+4807
2	−136	−413	+7949	+9292
3	−131	−607	+6939	+13147
4	−124	−786	+5574	+16092
5	−115	−945	+3903	+17889
6	−104	−1079	+1989	+18356
7	−91	−1183	−91	+17381
8	−76	−1252	−2246	+14936
9	−59	−1281	−4371	+11091
10	−40	−1265	−6347	+6028
11	−19	−1199	−8041	+55
12	+4	−1078	−9306	−6380
13	+29	−897	−9981	−12675
14	+56	−651	−9891	−18060
15	+85	−335	−8847	−21583
16	+116	+56	−6646	−22096
17	+149	+527	−3071	−18241
18	+184	+1083	+2109	−8436
19	+221	+1729	+9139	+9139
20	+260	+2470	+18278	+36556
D	5,740	641,732	47,900,710	2,481,256,778	10,376,164,708
λ	1	1	$\frac{5}{6}$	$\frac{7}{12}$	$\frac{7}{60}$

		42		
ξ_1'	ξ_2'	ξ_3'	ξ_4'	ξ_5'
1	−220	−44	+9614	+48070
3	−217	−131	+9177	+141151
5	−211	−215	+8317	+225181
7	−202	−294	+7062	+294546
9	−190	−366	+5454	+344262
11	−175	−429	+3549	+370227
13	−157	−481	+1417	+369473
15	−136	−520	−858	+340418
17	−112	−544	−3178	+283118
19	−85	−551	−5431	+199519
21	−55	−539	−7491	+93709
23	−22	−506	−9218	−27830
25	+14	−450	−10458	−155970
27	+53	−369	−11043	−278685
29	+95	−261	−10791	−380799
31	+140	−124	−9506	−443734
33	+188	+44	−6978	−445258
35	+239	+245	−2983	−359233
37	+293	+481	+2717	−155363
39	+350	+754	+10374	+201058
41	+410	+1066	+20254	+749398
D	24,682	1,629,012	9,075,924	3,084,805,724	4,389,117,671,484
λ	2	$\frac{3}{2}$	$\frac{1}{3}$	$\frac{7}{12}$	$\frac{21}{10}$

ORTHOGONAL POLYNOMIALS

		43		
ξ_1'	ξ_2'	ξ_3'	ξ_4'	ξ_5'
0	−154	0	+10626	0
1	−153	−46	+10396	+8740
2	−150	−91	+9713	+16948
3	−145	−134	+8598	+24113
4	−138	−174	+7086	+29766
5	−129	−210	+5226	+33501
6	−118	−241	+3081	+34996
7	−105	−266	+728	+34034
8	−90	−284	−1742	+30524
9	−73	−294	−4224	+24522
10	−54	−295	−6599	+16252
11	−33	−286	−8734	+6127
12	−10	−266	−10482	−5230
13	+15	−234	−11682	−16965
14	+42	−189	−12159	−27972
15	+71	−130	−11724	−36872
16	+102	−56	−10174	−41992
17	+135	+34	−7292	−41344
18	+170	+141	−2847	−32604
19	+207	+266	+3406	−13091
20	+246	+410	+11726	+20254
21	+287	+574	+22386	+70889
D	6,622	814,506	2,676,234	3,815,417,606	39,541,600,644
λ	1	1	$\frac{1}{6}$	$\frac{7}{12}$	$\frac{7}{40}$

		44		
ξ_1'	ξ_2'	ξ_3'	ξ_4'	ξ_5'
1	−161	−483	+5796	+1380
3	−159	−1439	+5556	+4060
5	−155	−2365	+5083	+6503
7	−149	−3241	+4391	+8561
9	−141	−4047	+3501	+10101
11	−131	−4763	+2441	+11011
13	−119	−5369	+1246	+11206
15	−105	−5845	−42	+10634
17	−89	−6171	−1374	+9282
19	−71	−6327	−2694	+7182
21	−51	−6293	−3939	+4417
23	−29	−6049	−5039	+1127
25	−5	−5575	−5917	−2485
27	+21	−4851	−6489	−6147
29	+49	−3857	−6664	−9512
31	+79	−2573	−6344	−12152
33	+111	−979	−5424	−13552
35	+145	+945	−3792	−13104
37	+181	+3219	−1329	−10101
39	+219	+5863	+2091	−3731
41	+259	+8897	+6601	+6929
43	+301	+12341	+12341	+22919
D	28,380	913,836	1,257,829,980	1,173,974,648	4,162,273,752
λ	2	1	$\frac{10}{3}$	$\frac{7}{24}$	$\frac{1}{20}$

Miscellaneous Statistical Tables

ORTHOGONAL POLYNOMIALS

45

ξ_1'	ξ_2'	ξ_3'	ξ_4'	ξ_5'
0	-506	0	$+9108$	0
1	-503	-252	$+8928$	$+4500$
2	-494	-499	$+8393$	$+8750$
3	-479	-736	$+7518$	$+12509$
4	-458	-958	$+6328$	$+15554$
5	-431	-1160	$+4858$	$+17689$
6	-398	-1337	$+3153$	$+18754$
7	-359	-1484	$+1268$	$+18634$
8	-314	-1596	-732	$+17268$
9	-263	-1668	-2772	$+14658$
10	-206	-1695	-4767	$+10878$
11	-143	-1672	-6622	$+6083$
12	-74	-1594	-8232	$+518$
13	$+1$	-1456	-9482	-5473
14	$+82$	-1253	-10247	-11438
15	$+169$	-980	-10392	-16808
16	$+262$	-632	-9772	-20888
17	$+361$	-204	-8232	-22848
18	$+466$	$+309$	-5607	-21714
19	$+577$	$+912$	-1722	-16359
20	$+694$	$+1610$	$+3608$	-5494
21	$+817$	$+2408$	$+10578$	-12341
22	$+946$	$+3311$	$+19393$	$+38786$
D	7,590	9,203,634	92,036,340	2,934,936,620	12,006,558,900
λ	1	3	$\frac{5}{6}$	$\frac{5}{12}$	$\frac{3}{40}$

46

ξ_2'	ξ_3'	ξ_4'	ξ_5'
-88	-264	$+1980$	$+3300$
-87	-787	$+1905$	$+9725$
-85	-1295	$+1757$	$+15631$
-82	-1778	$+1540$	$+20692$
-78	-2226	$+1260$	$+24612$
-73	-2629	$+925$	$+27137$
-67	-2977	$+545$	$+28067$
-60	-3260	$+132$	-27268
-52	-3468	-300	$+24684$
-43	-3591	-735	$+20349$
-33	-3619	-1155	$+14399$
-22	-3542	-1540	$+7084$
-10	-3350	-1868	-1220
$+3$	-3033	-2115	-9999
$+17$	-2581	-2255	-18589
$+32$	-1984	-2260	-26164
$+48$	-1232	-2100	-31724
$+65$	-315	-1743	-34083
$+83$	$+777$	-1155	-31857
$+102$	$+2054$	-300	-23452
$+122$	$+3526$	$+860$	-7052
$+143$	$+5203$	$+2365$	$+19393$
$+165$	$+7095$	$+4257$	$+58179$
D	285,384	429,502,920	143,167,640	27,214,866,840
λ	$\frac{1}{2}$	$\frac{5}{3}$	$\frac{1}{12}$	$\frac{1}{10}$

ORTHOGONAL POLYNOMIALS

	47			
	ξ_2'	ξ_3'	ξ_4'	ξ_5'
	−184	0	+15180	0
	−183	−55	+14905	+3575
	−180	−109	+14087	+6968
	−175	−161	+12747	+10003
	−168	−210	+10920	+12516
	−159	−255	+8655	+14361
	−148	−295	+6015	+15416
	−135	−329	+3077	+15589
	−120	−356	−68	+14824
	−103	−375	−3315	+13107
	−84	−385	−6545	+10472
	−63	−385	−9625	+7007
	−40	−374	−12408	+2860
	−15	−351	−14733	−1755
	+12	−315	−16425	−6552
	+41	−265	−17295	−11167
	+72	−200	−17140	−15152
	+105	−119	−15743	−17969
	+140	−21	−12873	−18984
	+177	+95	−8285	−17461
	+216	+230	−1720	−12556
	+257	+385	+7095	−3311
	+300	+561	+18447	+11352
	+345	+759	+32637	+32637
D	1,271,256	4,994,220	8,518,474,580	8,629,104,120
λ	1	$\frac{1}{6}$	$\frac{7}{12}$	$\frac{1}{20}$

	48			
	ξ_2'	ξ_3'	ξ_4'	ξ_5'
	−575	−115	+16445	+82225
	−569	−343	+15873	+242671
	−557	−565	+14743	+391231
	−539	−777	+13083	+520401
	−515	−975	+10935	+623307
	−485	−1155	+8355	+693957
	−449	−1313	+5413	+727493
	−407	−1445	+2193	+720443
	−359	−1547	−1207	+670973
	−305	−1615	−4675	+579139
	−245	−1645	−8085	+447139
	−179	−1633	−11297	+279565
	−107	−1575	−14157	+83655
	−29	−1467	−16497	−130455
	+55	−1305	−18135	−349479
	+145	−1085	−18875	−556729
	+241	−803	−18507	−731863
	+343	−455	−16807	−850633
	+451	−37	−13537	−884633
	+565	+455	−8445	−801047
	+685	+1025	−1265	−562397
	+811	+1677	+8283	−126291
	+943	+2415	+20493	+554829
	+1081	+3243	+35673	+1533939
D	12,712,560	92,620,080	10,301,411,120	19,208,385,771,120
λ	3	$\frac{2}{3}$	$\frac{7}{12}$	$\frac{21}{10}$

Miscellaneous Statistical Tables

ORTHOGONAL POLYNOMIALS

		49	
ξ_2'	ξ_3'	ξ_4'	ξ_5'
−200	0	+17940	0
−199	−299	+17641	+9867
−196	−593	+16751	+19272
−191	−877	+15291	+27767
−184	−1146	+13296	+34932
−175	−1395	+10815	+40389
−164	−1619	+7911	+43816
−151	−1813	+4661	+44961
−136	−1972	+1156	+43656
−119	−2091	−2499	+39831
−100	−2165	−6185	+33528
−79	−2189	−9769	+24915
−56	−2158	−13104	+14300
−31	−2067	−16029	+2145
−4	−1911	−18369	−10920
+25	−1685	−19935	−24083
+56	−1384	−20524	−36336
+89	−1003	−19919	−46461
+124	−537	−17889	−53016
+161	+19	−14189	−54321
+200	+670	−8560	−48444
+241	+1421	−729	−33187
+284	+2277	+9591	−6072
+329	+3243	+22701	+35673
+376	+4324	+38916	+95128
D 1,566,040	167,230,700	12,408,517,940	74,451,107,640
λ 1	$\frac{5}{6}$	$\frac{7}{12}$	$\frac{7}{60}$

ORTHOGONAL POLYNOMIALS

		50	
ξ_2'	ξ_3'	ξ_4'	ξ_5'
−104	−312	+96876	+10764
−103	−931	+93771	+31809
−101	−1535	+87631	+51419
−98	−2114	+78596	+68684
−94	−2658	+66876	+82764
−89	−3157	+52751	+92917
−83	−3601	+36571	−98527
−76	−3980	+18756	−99132
−68	−4284	−204	+94452
−59	−4503	−19749	+84417
−49	−4627	−39249	+69195
−38	−4646	−58004	+49220
−26	−4550	−75244	+25220
−13	−4329	−90129	−1755
+1	−3973	−101749	−30305
+16	−3472	−109124	−58652
+32	−2816	−111204	−84612
+49	−1995	−106869	−105567
+67	−999	−94929	−118437
+86	+182	−74124	−119652
+106	+1558	−43124	−105124
+127	+3139	−529	−70219
+149	+4935	+55131	−9729
+172	+6956	+125396	+82156
+196	+9212	+211876	+211876
D 433,160	770,715,400	372,255,538,200	372,255,538,200
λ $\frac{1}{2}$	$\frac{5}{3}$	$\frac{35}{12}$	$\frac{7}{30}$

XII.8   PERCENTAGE POINTS OF PEARSON CURVES

A system of frequency curves, defined by the solutions of the differential equation

$$\frac{dy}{dx} = \frac{y(x + a)}{b_0 + b_1 x + b_2 x^2}$$

was presented by Karl Pearson, and hence has been known as the Pearson distributions, or Pearson curves.

The parameters a, b_0, b_1, and b_2 determine and are determined by the first four moments of a Pearson curve; thus the shape of such a curve is specified completely by $\sqrt{\beta_1}$, β_2, and the information that it is a Pearson curve, where

$$\beta_1 = \frac{\mu_3^2}{\mu_2^3},$$ where μ_i represents the ith population moment around the mean,

and

$$\beta_2 = \frac{\mu_4}{\mu_2^2}.$$

This table gives upper and lower percentage points for the Pearson curves. Generally to obtain the percentage points, a solution for x_α must be obtained from the integral equation

$$\alpha = \int_{l_2}^{x} f(x)\, dx ,$$

where $f(x)$ is a particular Pearson distribution, l_2 is the lower limit of x, and α is a given probability level. Then in the body of the tables are the values of

$$X_\alpha = \frac{x_\alpha - \mu}{\sigma}$$

for $\alpha = 0.005, 0.01, 0.025, 0.05, 0.95, 0.975, 0.99,$ and 0.995. The ranges of β_1 and β_2 have been extended so as to have $0 \le \beta_1 \le 1.8$ and $1.2 \le \beta_2 \le 6.6$. This extension also allows coverage of the J and U-shaped curves.

The tables are presented assuming that $\mu_3 > 0$, i.e., the distributions are assumed to be positively skewed (long tail at right). The upper percentage points ($\alpha > 0.50$) are positive and the lower percentage points ($\alpha < 0.50$) are negative.

If $\mu_3 < 0$, the roles of the tables must be interchanged. That is to say, if $\mu_3 < 0$ and the lower percentage points are desired, i.e., $\alpha < 0.50$, obtain the value desired from the tabled upper percentage points, attaching a negative sign; and if $\mu_3 < 0$ and the upper percentage points are desired, i.e., $\alpha > 0.50$, then read the desired value from the tabled lower percentage points, attaching a positive sign.

PERCENTAGE POINTS OF PEARSON CURVES

Upper 5% points of the standardized deviate $(x_\alpha - \mu)/\sigma$, $(\alpha = 0.95)$.

β_2 \ β_1	0.00	0.01	0.03	0.05	0.10	0.15	0.20	0.30	0.40	0.50	0.60	0.70
1.2	1.1547	1.2056	1.2326	1.2458	1.2579							
1.4	1.3191	1.3781	1.4106	1.4271	1.4438	1.4436	1.4348	1.4042				
1.6	1.4638	1.5249	1.5618	1.5832	1.6128	1.6249	1.6258	1.6031	1.5604			
1.8	1.5588	1.6151	1.6517	1.6751	1.7138	1.7390	1.7558	1.7687	1.7546	1.7153	1.6598	1.6005
2.0	1.6108	1.6602	1.6941	1.7168	1.7576	1.7881	1.8129	1.8503	1.8721	1.8748	1.8538	1.8078
2.2	1.6361	1.6793	1.7100	1.7310	1.7702	1.8011	1.8279	1.8741	1.9127	1.9426	1.9606	1.9609
2.4	1.6467	1.6849	1.7126	1.7318	1.7682	1.7977	1.8238	1.8709	1.9138	1.9535	1.9888	2.0174
2.6	1.6495	1.6837	1.7088	1.7263	1.7600	1.7874	1.8119	1.8569	1.8991	1.9400	1.9799	2.0181
2.8	1.6483	1.6792	1.7021	1.7183	1.7493	1.7746	1.7975	1.8394	1.8792	1.9183	1.9574	1.9968
3.0	1.6449	1.6733	1.6944	1.7093	1.7380	1.7616	1.7827	1.8216	1.8585	1.8949	1.9317	1.9690
3.2						1.7488	1.7686	1.8046	1.8388	1.8724	1.9064	1.9410
3.4							1.7890	1.8207	1.8517	1.8830	1.9148	
3.6								1.8041	1.8330	1.8618	1.8911	
3.8									1.8160	1.8428	1.8698	
4.0										1.8258	1.8508	
4.2											1.8338	

β_2 \ β_1	0.80	0.90	1.00	1.10	1.20	1.30	1.40	1.50	1.60	1.70	1.80
2.0	1.7453	1.6803									
2.2	1.9377	1.8886	1.8221	1.7532							
2.4	2.0350	2.0354	2.0119	1.9614	1.8924	1.8210					
2.6	2.0531	2.0822	2.1007	2.1022	2.0792	2.0285	1.9580	1.8847			
2.8	2.0360	2.0743	2.1100	2.1402	2.1602	2.1632	2.1415	2.0910	2.0197	1.9451	
3.0	2.0073	2.0464	2.0859	2.1249	2.1618	2.1935	2.2152	2.2200	2.1997	2.1499	2.0782
3.2	1.9767	2.0136	2.0518	2.0912	2.1314	2.1715	2.2096	2.2429	2.2665	2.2732	2.2547
3.4	1.9476	1.9816	2.0170	2.0540	2.0926	2.1325	2.1736	2.2147	2.2543	2.2893	2.3148
3.6	1.9211	1.9523	1.9847	2.0185	2.0541	2.0913	2.1304	2.1711	2.2130	2.2554	2.2965
3.8	1.8975	1.9259	1.9555	1.9864	2.0187	2.0526	2.0884	2.1261	2.1658	2.2073	2.2503
4.0	1.8763	1.9026	1.9296	1.9577	1.9871	2.0179	2.0503	2.0846	2.1207	2.1589	2.1993
4.2	1.8575	1.8817	1.9067	1.9324	1.9592	1.9872	2.0166	2.0475	2.0801	2.1146	2.1511
4.4	1.8407	1.8631	1.8861	1.9100	1.9345	1.9601	1.9868	2.0148	2.0443	2.0753	2.1082
4.6		1.8466	1.8679	1.8899	1.9126	1.9361	1.9606	1.9861	2.0128	2.0409	2.0704
4.8			1.8517	1.8722	1.8932	1.9148	1.9374	1.9608	1.9852	2.0107	2.0374
5.0				1.8562	1.8758	1.8959	1.9166	1.9383	1.9607	1.9840	2.0084
5.2					1.8602	1.8790	1.8983	1.9184	1.9389	1.9606	1.9828
5.4						1.8637	1.8817	1.9003	1.9197	1.9396	1.9602
5.6							1.8670	1.8844	1.9023	1.9203	1.9399
5.8							1.8531	1.8699	1.8867	1.9040	1.9217
6.0								1.8567	1.8725	1.8887	1.9054
6.2									1.8596	1.8749	1.8906
6.4										1.8623	1.8771
6.6											1.8647

Miscellaneous Statistical Tables

PERCENTAGE POINTS OF PEARSON CURVES

Lower 5% points of the standardized deviate $(x_\alpha - \mu)/\sigma$, $(\alpha = 0.05)$. Note that for positive skewness, i.e., $\mu_3 > 0$, the deviates in this table are negative.

β_2 \ β_1	0.00	0.01	0.03	0.05	0.10	0.15	0.20	0.30	0.40	0.50	0.60	0.70
1.2	1.1547	1.0899	1.0355	0.9954								
1.4	1.3191	1.2450	1.1828	1.1368	1.0477	0.9771	0.9170					
1.6	1.4639	1.3899	1.3270	1.2794	1.1839	1.1055	1.0380	0.9254	0.8331			
1.8	1.5588	1.4936	1.4384	1.3960	1.3078	1.2312	1.1614	1.0389	0.9365	0.8497	0.7746	
2.0	1.6108	1.5556	1.5097	1.4746	1.4007	1.3342	1.2710	1.1516	1.0433	0.9483	0.8656	0.7934
2.2	1.6361	1.5893	1.5513	1.5226	1.4622	1.4074	1.3544	1.2494	1.1463	1.0485	0.9595	0.8806
2.4	1.6467	1.6064	1.5743	1.5504	1.5006	1.4559	1.4124	1.3247	1.2342	1.1427	1.0534	0.9699
2.6	1.6495	1.6141	1.5864	1.5659	1.5241	1.4870	1.4511	1.3786	1.3025	1.2223	1.1397	1.0576
2.8	1.6483	1.6165	1.5921	1.5742	1.5382	1.5067	1.4766	1.4162	1.3526	1.2845	1.2121	1.1367
3.0	1.6449	1.6160	1.5940	1.5781	1.5464	1.5192	1.4933	1.4422	1.3887	1.3313	1.2693	1.2030
3.2						1.5264	1.5043	1.4602	1.4145	1.3658	1.3130	1.2558
3.4								1.4727	1.4331	1.3912	1.3460	1.2969
3.6									1.4466	1.4100	1.3709	1.3285
3.8										1.4241	1.3900	1.3528
4.0											1.4042	1.3716
4.2												1.3862

β_2 \ β_1	0.80	0.90	1.00	1.10	1.20	1.30	1.40	1.50	1.60	1.70	1.80
2.2	0.8107	0.7484									
2.4	0.8944	0.8267	0.7659	0.7109							
2.6	0.9792	0.9068	0.8412	0.7820	0.7282	0.6788					
2.8	1.0608	0.9872	0.9179	0.8545	0.7966	0.7440	0.6956	0.6509			
3.0	1.1337	1.0632	0.9938	0.9277	0.8663	0.8100	0.7584	0.7110	0.6672		
3.2	1.1947	1.1304	1.0646	0.9993	0.9362	0.8769	0.8222	0.7716	0.7251	0.6820	0.6419
3.4	1.2437	1.1868	1.1269	1.0653	1.0036	0.9435	0.8863	0.8332	0.7836	0.7380	0.6957
3.6	1.2824	1.2326	1.1793	1.1232	1.0652	1.0069	0.9496	0.8946	0.8427	0.7946	0.7498
3.8	1.3127	1.2692	1.2222	1.1720	1.1192	1.0646	1.0093	0.9546	0.9017	0.8514	0.8046
4.0	1.3363	1.2982	1.2569	1.2124	1.1650	1.1151	1.0634	1.0109	0.9587	0.9078	0.8591
4.2	1.3550	1.3212	1.2848	1.2454	1.2031	1.1581	1.1108	1.0617	1.0117	0.9618	0.9129
4.4	1.3698	1.3400	1.3072	1.2722	1.2346	1.1943	1.1514	1.1064	1.0596	1.0120	0.9642
4.6		1.3545	1.3254	1.2938	1.2605	1.2244	1.1857	1.1448	1.1018	1.0572	1.0117
4.8			1.3403	1.3121	1.2819	1.2494	1.2145	1.1775	1.1383	1.0971	1.0545
5.0				1.3268	1.2995	1.2701	1.2387	1.2052	1.1695	1.1318	1.0924
5.2					1.3142	1.2876	1.2591	1.2286	1.1962	1.1618	1.1255
5.4						1.3022	1.2762	1.2484	1.2189	1.1875	1.1542
5.6							1.2907	1.2653	1.2383	1.2092	1.1790
5.8							1.3030	1.2797	1.2549	1.2285	1.2005
6.0								1.2920	1.2691	1.2448	1.2190
6.2									1.2814	1.2589	1.2351
6.4										1.2711	1.2490
6.6											1.2612

PERCENTAGE POINTS OF PEARSON CURVES

Upper 2.5% points of the standardized variate $(x_\alpha - \mu)/\sigma$, $(\alpha = 0.975)$.

β_2 \ β_1	0.00	0.01	0.03	0.05	0.10	0.15	0.20	0.30	0.40	0.50	0.60	0.70
1.2	1.1547	1.2056	1.2326	1.2458	1.2579							
1.4	1.3223	1.3823	1.4144	1.4303	1.4453	1.4440	1.4348	1.4042				
1.6	1.4955	1.5631	1.6012	1.6214	1.6439	1.6472	1.6397	1.6060	1.5604			
1.8	1.6454	1.7149	1.7567	1.7809	1.8146	1.8295	1.8330	1.8157	1.7746	1.7195	1.6598	1.6005
2.0	1.7567	1.8233	1.8657	1.8918	1.9331	1.9581	1.9735	1.9833	1.9691	1.9330	1.8785	1.8131
2.2	1.8332	1.8953	1.9363	1.9626	2.0071	2.0377	2.0605	2.0902	2.1024	2.0968	2.0723	2.0279
2.4	1.8847	1.9422	1.9811	2.0066	2.0516	2.0842	2.1106	2.1511	2.1793	2.1959	2.1997	2.1881
2.6	1.9197	1.9727	2.0094	2.0338	2.0778	2.1108	2.1383	2.1834	2.2197	2.2485	2.2695	2.2813
2.8	1.9434	1.9928	2.0273	2.0507	2.0930	2.1254	2.1529	2.1995	2.2391	2.2735	2.3032	2.3277
3.0	1.9600	2.0060	2.0387	2.0609	2.1016	2.1330	2.1600	2.2062	2.2466	2.2832	2.3167	2.3471
3.2						2.1362	2.1623	2.2076	2.2476	2.2845	2.3192	2.3420
3.4								2.2060	2.2450	2.2813	2.3159	2.3492
3.6									2.2404	2.2758	2.3096	2.3424
3.8										2.2690	2.3018	2.3338
4.0											2.2935	2.3243
4.2												2.3147

β_2 \ β_1	0.80	0.90	1.00	1.10	1.20	1.30	1.40	1.50	1.60	1.70	1.80
2.0	1.7455	1.6803									
2.2	1.9664	1.8949	1.8223	1.7532							
2.4	2.1586	2.1098	2.0441	1.9687	1.8928	1.8210					
2.6	2.2813	2.2667	2.2344	2.1829	2.1145	2.0366	1.9584	1.8847			
2.8	2.3456	2.3551	2.3532	2.3368	2.3028	2.2496	2.1795	2.0998	2.0201	1.9451	
3.0	2.3741	2.3966	2.4130	2.4212	2.4183	2.4009	2.3658	2.3115	2.2402	2.1594	2.0787
3.2	2.3829	2.4115	2.4371	2.4586	2.4743	2.4819	2.4783	2.4603	2.4245	2.3695	2.2975
3.4	2.3814	2.4124	2.4421	2.4698	2.4947	2.5158	2.5311	2.5383	2.5345	2.5161	2.4799
3.6	2.3745	2.4061	2.4371	2.4672	2.4963	2.5236	2.5483	2.5692	2.5843	2.5915	2.5874
3.8	2.3653	2.3964	2.4273	2.4580	2.4883	2.5181	2.5468	2.5740	2.5987	2.6195	2.6347
4.0	2.3548	2.3851	2.4153	2.4456	2.4759	2.5061	2.5363	2.5657	2.5946	2.6218	2.6465
4.2	2.3441	2.3734	2.4027	2.4321	2.4616	2.4914	2.5214	2.5514	2.5814	2.6111	2.6400
4.4	2.3336	2.3617	2.3900	2.4183	2.4468	2.4757	2.5050	2.5345	2.5644	2.5945	2.6244
4.6			2.3776	2.4047	2.4323	2.4600	2.4882	2.5168	2.5460	2.5755	2.6054
4.8			2.3657	2.3918	2.4182	2.4448	2.4720	2.4996	2.5276	2.5561	2.5851
5.0				2.3797	2.4050	2.4304	2.4565	2.4829	2.5098	2.5371	2.5651
5.2					2.3925	2.4170	2.4418	2.4673	2.4928	2.5190	2.5458
5.4						2.4043	2.4281	2.4523	2.4769	2.5019	2.5275
5.6							2.4152	2.4385	2.4621	2.4859	2.5100
5.8							2.4033	2.4256	2.4482	2.4711	2.4941
6.0								2.4136	2.4352	2.4573	2.4795
6.2									2.4232	2.4444	2.4659
6.4										2.4324	2.4530
6.6											2.4410

Miscellaneous Statistical Tables

PERCENTAGE POINTS OF PEARSON CURVES

Lower 2.5% points of the standardized deviate $(x_\alpha - \mu)/\sigma$, $(\alpha = 0.025)$. Note: If $\mu_3 > 0$, the variates in this table are negative.

β_1 / β_2	0.00	0.01	0.03	0.05	0.10	0.15	0.20	0.30	0.40	0.50	0.60	0.70
1.2	1.1547	1.0899	1.0355	0.9954								
1.4	1.3223	1.2461	1.1835	1.1371	1.0477	0.9771	0.9170					
1.6	1.4955	1.4115	1.3409	1.2886	1.1870	1.1064	1.0381	0.9254	0.8331			
1.8	1.6454	1.5615	1.4906	1.4372	1.3304	1.2426	1.1665	1.0396	0.9365	0.8497		
2.0	1.7567	1.6785	1.6129	1.5631	1.4613	1.3744	1.2962	1.1595	1.0448	0.9483	0.8656	0.7934
2.2	1.8332	1.7625	1.7037	1.6593	1.5677	1.4879	1.4139	1.2782	1.1572	1.0514	0.9600	0.8806
2.4	1.8847	1.8210	1.7688	1.7295	1.6488	1.5779	1.5114	1.3852	1.2665	1.1568	1.0581	0.9710
2.6	1.9197	1.8616	1.8149	1.7801	1.7089	1.6465	1.5878	1.4746	1.3645	1.2579	1.1571	1.0644
2.8	1.9434	1.8903	1.8481	1.8167	1.7533	1.6983	1.6464	1.5463	1.4471	1.3482	1.2508	1.1573
3.0	1.9600	1.9109	1.8722	1.8437	1.7866	1.7374	1.6914	1.6027	1.5143	1.4248	1.3344	1.2445
3.2						1.7674	1.7260	1.6469	1.5682	1.4880	1.4057	1.3222
3.4								1.6818	1.6112	1.5394	1.4653	1.3890
3.6									1.6458	1.5811	1.5144	1.4452
3.8										1.6152	1.5548	1.4922
4.0											1.5883	1.5313
4.2												1.5640

β_1 / β_2	0.80	0.90	1.00	1.10	1.20	1.30	1.40	1.50	1.60	1.70	1.80
2.2	0.8107	0.7484									
2.4	0.8944	0.8267	0.7659	0.7109							
2.6	0.9811	0.9072	0.8412	0.7820	0.7282	0.6788					
2.8	1.0700	0.9903	0.9189	0.8545	0.7966	0.7440	0.6956				
3.0	1.1573	1.0747	0.9985	0.9292	0.8666	0.8100	0.7584	0.7110	0.6672		
3.2	1.2386	1.1568	1.0786	1.0055	0.9385	0.8776	0.8222	0.7716	0.7251	0.6820	0.6419
3.4	1.3112	1.2329	1.1558	1.0815	1.0115	0.9467	0.8874	0.8332	0.7836	0.7380	0.6957
3.6	1.3738	1.3009	1.2272	1.1544	1.0837	1.0166	0.9540	0.8962	0.8433	0.7946	0.7498
3.8	1.4272	1.3600	1.2911	1.2215	1.1524	1.0851	1.0207	0.9602	0.9040	0.8524	0.8046
4.0	1.4722	1.4106	1.3470	1.2818	1.2158	1.1501	1.0858	1.0240	0.9655	0.9109	0.8603
4.2	1.5100	1.4537	1.3952	1.3348	1.2728	1.2100	1.1474	1.0858	1.0264	0.9699	0.9169
4.4	1.5420	1.4900	1.4366	1.3806	1.3231	1.2640	1.2041	1.1443	1.0853	1.0282	0.9736
4.6			1.4721	1.4200	1.3671	1.3118	1.2555	1.1982	1.1409	1.0844	1.0293
4.8			1.5026	1.4549	1.4055	1.3541	1.3012	1.2470	1.1922	1.1372	1.0829
5.0				1.4847	1.4389	1.3905	1.3417	1.2908	1.2387	1.1861	1.1333
5.2					1.4679	1.4235	1.3773	1.3297	1.2808	1.2306	1.1800
5.4						1.4519	1.4089	1.3643	1.3183	1.2710	1.2227
5.6							1.4366	1.3948	1.3517	1.3073	1.2612
5.8							1.4611	1.4220	1.3814	1.3395	1.2959
6.0								1.4460	1.4079	1.3685	1.3277
6.2									1.4315	1.3943	1.3559
6.4										1.4176	1.3812
6.6											1.4041

PERCENTAGE POINTS OF PEARSON CURVES

Upper 1% points of the standardized deviate $(x_\alpha - \mu)/\sigma$, $(\alpha = 0.99)$.

β_2 \ β_1	0.00	0.01	0.03	0.05	0.10	0.15	0.20	0.30	0.40	0.50	0.60	0.70
1.2	1.1547	1.2056	1.2326	1.2458								
1.4	1.3229	1.3831	1.4151	1.4308	1.4453	1.4440	1.4348					
1.6	1.5079	1.5786	1.6169	1.6359	1.6543	1.6535	1.6428	1.6063	1.5604			
1.8	1.6974	1.7764	1.8208	1.8444	1.8713	1.8762	1.8688	1.8320	1.7791	1.7200	1.6598	
2.0	1.8687	1.9511	1.9999	2.0274	2.0644	2.0794	2.0815	2.0601	2.0145	1.9532	1.8841	1.8137
2.2	2.0097	2.0918	2.1425	2.1726	2.2175	2.2418	2.2541	2.2548	2.2304	2.1855	2.1238	2.0507
2.4	2.1207	2.2004	2.2512	2.2826	2.3323	2.3632	2.3835	2.4030	2.4009	2.3798	2.3408	2.2848
2.6	2.2067	2.2833	2.3333	2.3649	2.4172	2.4521	2.4775	2.5103	2.5253	2.5249	2.5096	2.4785
2.8	2.2737	2.3469	2.3957	2.4270	2.4803	2.5174	2.5459	2.5872	2.6136	2.6280	2.6308	2.6215
3.0	2.3263	2.3963	2.4436	2.4744	2.5278	2.5659	2.5961	2.6424	2.6763	2.7003	2.7155	2.7217
3.2						2.6025	2.6336	2.6829	2.7211	2.7513	2.7745	2.7911
3.4								2.7129	2.7536	2.7875	2.8158	2.8390
3.6									2.7775	2.8137	2.8450	2.8723
3.8										2.8327	2.8659	2.8957
4.0											2.8809	2.9122
4.2												2.9237

β_2 \ β_1	0.80	0.90	1.00	1.10	1.20	1.30	1.40	1.50	1.60	1.70	1.80
2.0	1.7455										
2.2	1.9727	1.8956	1.8223								
2.4	2.2144	2.1344	2.0508	1.9694	1.8928						
2.6	2.4315	2.3690	2.2932	2.2087	2.1215	2.0373	1.9584				
2.8	2.5991	2.5621	2.5101	2.4432	2.3637	2.2762	2.1867	2.1006	2.0201		
3.0	2.7183	2.7037	2.6767	2.6358	2.5801	2.5101	2.4281	2.3386	2.2476	2.1602	2.0787
3.2	2.8005	2.8021	2.7945	2.7763	2.7461	2.7022	2.6439	2.5716	2.4877	2.9369	2.3049
3.4	2.8569	2.8692	2.8751	2.8734	2.8630	2.8421	2.8094	2.7631	2.7028	2.6287	2.5435
3.6	2.8957	2.9149	2.9295	2.9388	2.9421	2.9383	2.9257	2.9028	2.8679	2.8198	2.7577
3.8	2.9225	2.9459	2.9662	2.9827	2.9952	3.0026	3.0040	2.9983	2.9838	2.9592	2.9227
4.0	2.9409	2.9672	2.9910	3.0121	3.0303	3.0449	3.0559	3.0616	3.0616	3.0543	3.0384
4.2	2.9537	2.9818	3.0077	3.0318	3.0536	3.0731	3.0898	3.1032	3.1126	3.1171	3.1159
4.4	2.9623	2.9912	3.0186	3.0444	3.0686	3.0911	3.1118	3.1299	3.1455	3.1579	3.1661
4.6			3.0255	3.0525	3.0783	3.1023	3.1255	3.1468	3.1662	3.1836	3.1982
4.8			3.0301	3.0575	3.0837	3.1090	3.1334	3.1568	3.1789	3.1995	3.2179
5.0				3.0601	3.0869	3.1123	3.1377	3.1623	3.1859	3.2084	3.2296
5.2					3.0876	3.1140	3.1400	3.1647	3.1894	3.2128	3.2356
5.4						3.1138	3.1395	3.1647	3.1896	3.2138	3.2379
5.6							3.1381	3.1636	3.1885	3.2131	3.2372
5.8							3.1356	3.1611	3.1860	3.2110	3.2349
6.0								3.1583	3.1830	3.2073	3.2320
6.2									3.1789	3.2033	3.2278
6.4										3.1991	3.2231
6.6											3.2179

Miscellaneous Statistical Tables

PERCENTAGE POINTS OF PEARSON CURVES

Lower 1% points of the standardized variate $(x_\alpha - \mu)/\sigma$, $(\alpha = 0.01)$. Note: If $\mu_3 > 0$, the deviates in this table are negative.

β_2 \ β_1	0.00	0.01	0.03	0.05	0.10	0.15	0.20	0.30	0.40	0.50	0.60	0.70
1.2	1.1547	1.0899	1.0355	0.9954								
1.4	1.3229	1.2468	1.1835	1.1371	1.0477	0.9771	0.9170					
1.6	1.5079	1.4192	1.3453	1.2912	1.1876	1.1064	1.0381	0.9254	0.8331			
1.8	1.6974	1.5996	1.5176	1.4569	1.3393	1.2462	1.1678	1.0396	0.9365	0.8497		
2.0	1.8687	1.7685	1.6842	1.6212	1.4963	1.3946	1.3070	1.1617	1.0451	0.9483	0.8656	0.7934
2.2	2.0097	1.9121	1.8304	1.7689	1.6450	1.5413	1.4494	1.2915	1.1609	1.0521	0.9600	0.8806
2.4	2.1207	2.0279	1.9509	1.8929	1.7753	1.6751	1.5844	1.4226	1.2825	1.1621	1.0594	0.9712
2.6	2.2067	2.1193	2.0475	1.9936	1.8842	1.7904	1.7042	1.5466	1.4042	1.2766	1.1642	1.0665
2.8	2.2737	2.1915	2.1244	2.0745	1.9734	1.8866	1.8065	1.6576	1.5190	1.3901	1.2722	1.1665
3.0	2.3263	2.2488	2.1861	2.1397	2.0461	1.9660	1.8920	1.7535	1.6223	1.4968	1.3783	1.2684
3.2						2.0314	1.9631	1.8350	1.7124	1.5934	1.4782	1.3681
3.4								1.9037	1.7900	1.6785	1.5688	1.4617
3.6									1.8562	1.7522	1.6492	1.5470
3.8										1.8159	1.7196	1.6232
4.0											1.7809	1.6899
4.2												1.7495

β_2 \ β_1	0.80	0.90	1.00	1.10	1.20	1.30	1.40	1.50	1.60	1.70	1.80
2.2	0.8107	0.7484									
2.4	0.8944	0.8267	0.7659	0.7109							
2.6	0.9816	0.9072	0.8412	0.7820	0.7282						
2.8	1.0731	0.9912	0.9189	0.8545	0.7966	0.7440	0.6956				
3.0	1.1684	1.0790	0.9998	0.9295	0.8666	0.8100	0.7584	0.7110	0.6672		
3.2	1.2648	1.1699	1.0842	1.0075	0.9391	0.8776	0.8222	0.7716	0.7251	0.6820	0.641
3.4	1.3586	1.2612	1.1709	1.0885	1.0143	0.9477	0.8876	0.8332	0.7836	0.7380	0.695
3.6	1.4467	1.3497	1.2574	1.1713	1.0921	1.0202	0.9552	0.8966	0.8433	0.7946	0.749
3.8	1.5274	1.4329	1.3411	1.2534	1.1710	1.0948	1.0252	0.9619	0.9046	0.8524	0.804
4.0	1.5997	1.5092	1.4199	1.3327	1.2492	1.1703	1.0969	1.0294	0.9678	0.9119	0.860
4.2	1.6640	1.5781	1.4923	1.4074	1.3245	1.2446	1.1690	1.0982	1.0329	0.9729	0.918
4.4	1.7208	1.6394	1.5580	1.4763	1.3954	1.3162	1.2399	1.1672	1.0989	1.0355	0.977
4.6			1.6171	1.5390	1.4611	1.3838	1.3081	1.2347	1.1649	1.0990	1.037
4.8			1.6699	1.5958	1.5213	1.4466	1.3725	1.2999	1.2296	1.1622	1.098
5.0				1.6469	1.5757	1.5034	1.4325	1.3614	1.2916	1.2240	1.159
5.2					1.6251	1.5567	1.4879	1.4185	1.3507	1.2836	1.218
5.4						1.6045	1.5386	1.4720	1.4059	1.3400	1.275
5.6							1.5848	1.5211	1.4569	1.3930	1.329
5.8							1.6268	1.5658	1.5043	1.4424	1.380
6.0								1.6068	1.5477	1.4880	1.428
6.2									1.5875	1.5302	1.472
6.4										1.5690	1.513
6.6											1.551

PERCENTAGE POINTS OF PEARSON CURVES

Upper 0.5% points of the standardized deviate $(x_\alpha - \mu)/\sigma$, $(\alpha = .095)$.

β_2 \ β_1	0.00	0.01	0.03	0.05	0.10	0.15	0.20	0.30	0.40	0.50	0.60	0.70
1.2	1.1547	1.2056	1.2326	1.2458								
1.4	1.3229	1.3831	1.4151	1.4308	1.4453	1.4440	1.4348					
1.6	1.5105	1.5820	1.6202	1.6388	1.6561	1.6543	1.6432	1.6063	1.5604			
1.8	1.7147	1.7974	1.8426	1.8655	1.8888	1.8893	1.8778	1.8350	1.7796	1.7200	1.6598	
2.0	1.9175	2.0079	2.0594	2.0870	2.1197	2.1278	2.1219	2.0844	2.0259	1.9569	1.8847	1.8137
2.2	2.1006	2.1946	2.2502	2.2818	2.3242	2.3420	2.3456	2.3250	2.2784	2.2133	2.1366	2.0547
2.4	2.2562	2.3506	2.4084	2.4426	2.4928	2.5192	2.5322	2.5319	2.5052	2.4577	2.3928	2.3145
2.6	2.3846	2.4776	2.5362	2.5719	2.6273	2.6603	2.6810	2.6981	2.6914	2.6652	2.6212	2.5607
2.8	2.4896	2.5805	2.6389	2.6753	2.7340	2.7715	2.7976	2.8280	2.8377	2.8304	2.8072	2.7687
3.0	2.5758	2.6639	2.7217	2.7583	2.8188	2.8594	2.8893	2.9292	2.9509	2.9580	2.9521	2.9333
3.2						2.9294	2.9620	3.0084	3.0386	3.0566	3.0636	3.0601
3.4								3.0712	3.1075	3.1329	3.1492	3.1570
3.6									3.1618	3.1928	3.2157	3.2317
3.8										3.2397	3.2680	3.2898
4.0											3.3092	3.3352
4.2												3.3711

β_2 \ β_1	0.80	0.90	1.00	1.10	1.20	1.30	1.40	1.50	1.60	1.70	1.80
2.0	1.7455										
2.2	1.9733	1.8956	1.8223								
2.4	2.2278	2.1385	2.0515	1.9694	1.8928						
2.6	2.4858	2.3996	2.3069	2.2129	2.1222	2.0373	1.9584				
2.8	2.7151	2.6466	2.5653	2.4741	2.3774	2.2804	2.1875	2.1006			
3.0	2.9010	2.8550	2.7950	2.7213	2.6356	2.5409	2.4415	2.3426	2.2483	2.1602	
3.2	3.0456	3.0197	2.9815	2.9300	2.8655	2.7879	2.6990	2.6019	2.5008	2.4007	2.3056
3.4	3.1561	3.1460	3.1258	3.0949	3.0522	2.9971	2.9289	2.8484	2.7572	2.6583	2.5562
3.6	3.2406	3.2421	3.2360	3.2213	3.1972	3.1624	3.1163	3.0579	2.9869	2.9041	2.8111
3.8	3.3055	3.3158	3.3200	3.3173	3.3072	3.2890	3.2617	3.2239	3.1748	3.1138	3.0406
4.0	3.3565	3.3726	3.3841	3.3902	3.3909	3.3848	3.3717	3.3510	3.3208	3.2805	3.2291
4.2	3.3962	3.4173	3.4339	3.4464	3.4346	3.4579	3.4556	3.4468	3.4318	3.4083	3.3759
4.4	3.4279	3.4513	3.4724	3.4899	3.5038	3.5137	3.5192	3.5202	3.5155	3.5051	3.4875
4.6			3.5031	3.5241	3.5419	3.5564	3.5677	3.5756	3.5793	3.5782	3.5718
4.8			3.5275	3.5510	3.5714	3.5900	3.6053	3.6183	3.6277	3.6335	3.6354
5.0				3.5721	3.5948	3.6156	3.6342	3.6512	3.6647	3.6760	3.6840
5.2					3.6136	3.6364	3.6569	3.6764	3.6937	3.7084	3.7206
5.4						3.6524	3.6755	3.6969	3.7155	3.7328	3.7493
5.6							3.6900	3.7116	3.7333	3.7528	3.7709
5.8							3.7008	3.7245	3.7467	3.7681	3.7873
6.0								3.7341	3.7572	3.7796	3.8010
6.2									3.7645	3.7885	3.8111
6.4										3.7959	3.8183
6.6											3.8247

Miscellaneous Statistical Tables

PERCENTAGE POINTS OF PEARSON CURVES

Lower 0.5% points of the standardized deviate $(x_\alpha - \mu)/\sigma$, $(\alpha = 0.005)$. Note: If $\mu_3 > 0$, the variates in this table are negative.

$\beta_2 \backslash \beta_1$	0.00	0.01	0.03	0.05	0.10	0.15	0.20	0.30	0.40	0.50	0.60	0.70
1.2	1.1547	1.0899	1.0355	0.9954								
1.4	1.3229	1.2468	1.1835	1.1371	1.0477	0.9771	0.9170					
1.6	1.5105	1.4206	1.3459	1.2915	1.1876	1.1064	1.0381	0.9254	0.8331			
1.8	1.7147	1.6113	1.5252	1.4620	1.3411	1.2468	1.1680	1.0396	0.9365	0.8497		
2.0	1.9175	1.8057	1.7120	1.6426	1.5075	1.4001	1.3096	1.1621	1.0451	0.9483	0.8656	0.7934
2.2	1.1006	1.9864	1.8906	1.8190	1.6770	1.5612	1.4613	1.2949	1.1615	1.0521	0.9600	0.8806
2.4	2.2562	2.1437	2.0496	1.9791	1.8375	1.7195	1.6152	1.4356	1.2869	1.1632	1.0596	0.9712
2.6	2.3846	2.2758	2.1854	2.1176	1.9809	1.8654	1.7616	1.5774	1.4185	1.2820	1.1658	1.0668
2.8	2.4896	2.3851	2.2990	2.2348	2.1050	1.9946	1.8943	1.7126	1.5502	1.4056	1.2787	1.1687
3.0	2.5758	2.4758	2.3939	2.3330	2.2105	2.1062	2.0110	1.8362	1.6760	1.5287	1.3952	1.2760
3.2						2.2017	2.1120	1.9460	1.7915	1.6462	1.5107	1.3860
3.4							2.0422	1.8951	1.7547	1.6210	1.4947
3.6								1.9869	1.8528	1.7232	1.5986
3.8									1.9400	1.8161	1.6952
4.0										1.8997	1.7836
4.2											1.8634

$\beta_2 \backslash \beta_1$	0.80	0.90	1.00	1.10	1.20	1.30	1.40	1.50	1.60	1.70	1.80
2.2	0.8107	0.7484									
2.4	0.8944	0.8267	0.7659	0.7109							
2.6	0.9816	0.9072	0.8412	0.7820	0.7282						
2.8	1.0736	0.9912	0.9189	0.8545	0.7966	0.7440	0.6956				
3.0	1.1713	1.0799	1.0000	0.9295	0.8666	0.8100	0.7584	0.7110	0.6672		
3.2	1.2735	1.1735	1.0854	1.0079	0.9391	0.8776	0.8222	0.7716	0.7251	0.6820	0.6419
3.4	1.3776	1.2710	1.1752	1.0901	1.0148	0.9477	0.8876	0.8332	0.7836	0.7380	0.6957
3.6	1.4803	1.3697	1.2682	1.1763	1.0941	1.0208	0.9554	0.8966	0.8433	0.7946	0.7498
3.8	1.5783	1.4667	1.3619	1.2650	1.1768	1.0974	1.0261	0.9623	0.9046	0.8524	0.8046
4.0	1.6700	1.5596	1.4538	1.3542	1.2616	1.1768	1.0999	1.0306	0.9682	0.9119	0.8606
4.2	1.7540	1.6465	1.5420	1.4416	1.3465	1.2578	1.1761	1.1017	1.0343	0.9734	0.9183
4.4	1.8305	1.7263	1.6246	1.5254	1.4296	1.3388	1.2536	1.1749	1.1029	1.0373	0.9779
4.6			1.7014	1.6041	1.5094	1.4180	1.3309	1.2491	1.1732	1.1034	1.0397
4.8			1.7718	1.6774	1.5848	1.4941	1.4064	1.3230	1.2443	1.1710	1.1034
5.0				1.7453	1.6550	1.5661	1.4793	1.3953	1.3150	1.2391	1.1684
5.2					1.7202	1.6336	1.5483	1.4648	1.3842	1.3070	1.2338
5.4						1.6964	1.6132	1.5312	1.4509	1.3731	1.2987
5.6							1.6737	1.5935	1.5144	1.4373	1.3624
5.8							1.7298	1.6519	1.5746	1.4984	1.4238
6.0								1.7061	1.6309	1.5563	1.4827
6.2									1.6836	1.6108	1.5385
6.4										1.6618	1.5912
6.6											1.6408

INTEGRALS

ELEMENTARY FORMS

1. $\int a\,dx = ax$

2. $\int a \cdot f(x)\,dx = a\int f(x)\,dx$

3. $\int \phi(y)\,dx = \int \dfrac{\phi(y)}{y'}\,dy,$ where $y' = \dfrac{dy}{dx}$

4. $\int (u + v)\,dx = \int u\,dx + \int v\,dx,$ where u and v are any functions of x

5. $\int u\,dv = u\int dv - \int v\,du = uv - \int v\,du$

6. $\int u\dfrac{dv}{dx}\,dx = uv - \int v\dfrac{du}{dx}\,dx$

7. $\int x^n\,dx = \dfrac{x^{n+1}}{n+1},$ except $n = -1$

8. $\int \dfrac{f'(x)\,dx}{f(x)} = \log f(x),$ $(df(x) = f'(x)\,dx)$

9. $\int \dfrac{dx}{x} = \log x$

10. $\int \dfrac{f'(x)\,dx}{2\sqrt{f(x)}} = \sqrt{f(x)},$ $(df(x) = f'(x)\,dx)$

11. $\int e^x\,dx = e^x$

12. $\int e^{ax}\,dx = e^{ax}/a$

13. $\int b^{ax}\,dx = \dfrac{b^{ax}}{a\log b},$ $(b > 0)$

14. $\int \log x\,dx = x\log x - x$

15. $\int a^x \log a\,dx = a^x,$ $(a > 0)$

16. $\int \dfrac{dx}{a^2 + x^2} = \dfrac{1}{a}\tan^{-1}\dfrac{x}{a}$

245

INTEGRALS (Continued)

17. $\displaystyle\int \frac{dx}{a^2 - x^2} = \begin{cases} \dfrac{1}{a}\tanh^{-1}\dfrac{x}{a} \\[2mm] \quad\text{or} \\[2mm] \dfrac{1}{2a}\log\dfrac{a+x}{a-x}, \quad (a^2 > x^2) \end{cases}$

18. $\displaystyle\int \frac{dx}{x^2 - a^2} = \begin{cases} -\dfrac{1}{a}\coth^{-1}\dfrac{x}{a} \\[2mm] \quad\text{or} \\[2mm] \dfrac{1}{2a}\log\dfrac{x-a}{x+a}, \quad (x^2 > a^2) \end{cases}$

19. $\displaystyle\int \frac{dx}{\sqrt{a^2 - x^2}} = \begin{cases} \sin^{-1}\dfrac{x}{|a|} \\[2mm] \quad\text{or} \\[2mm] -\cos^{-1}\dfrac{x}{|a|}, \quad (a^2 > x^2) \end{cases}$

20. $\displaystyle\int \frac{dx}{\sqrt{x^2 \pm a^2}} = \log\left(x + \sqrt{x^2 \pm a^2}\right)$

21. $\displaystyle\int \frac{dx}{x\sqrt{x^2 - a^2}} = \frac{1}{|a|}\sec^{-1}\frac{x}{a}$

22. $\displaystyle\int \frac{dx}{x\sqrt{a^2 \pm x^2}} = -\frac{1}{a}\log\left(\frac{a + \sqrt{a^2 \pm x^2}}{x}\right)$

FORMS CONTAINING $(a + bx)$

For forms containing $a + bx$, but not listed in the table, the substitution $u = \dfrac{a + bx}{x}$ may prove helpful.

23. $\displaystyle\int (a + bx)^n\, dx = \frac{(a + bx)^{n+1}}{(n + 1)b}, \quad (n \neq -1)$

24. $\displaystyle\int x(a + bx)^n\, dx$

$$= \frac{1}{b^2(n + 2)}(a + bx)^{n+2} - \frac{a}{b^2(n + 1)}(a + bx)^{n+1}, \quad (n \neq -1, -2)$$

25. $\displaystyle\int x^2(a + bx)^n\, dx = \frac{1}{b^3}\left[\frac{(a + bx)^{n+3}}{n + 3} - 2a\frac{(a + bx)^{n+2}}{n + 2} + a^2\frac{(a + bx)^{n+1}}{n + 1}\right]$

INTEGRALS (Continued)

26. $\int x^m(a + bx)^n\, dx = \begin{cases} \dfrac{x^{m+1}(a + bx)^n}{m + n + 1} + \dfrac{an}{m + n + 1}\displaystyle\int x^m(a + bx)^{n-1}\, dx \\[2mm] \quad\quad\quad\quad\quad\quad \text{or} \\[2mm] \dfrac{1}{a(n + 1)}\left[-x^{m+1}(a + bx)^{n+1} \right. \\[4mm] \quad\quad\quad\quad\quad\quad\quad\quad \left. + (m + n + 2)\displaystyle\int x^m(a + bx)^{n+1}\, dx \right] \\[2mm] \quad\quad\quad\quad\quad\quad \text{or} \\[2mm] \dfrac{1}{b(m + n + 1)}\left[x^m(a + bx)^{n+1} - ma\displaystyle\int x^{m-1}(a + bx)^n\, dx \right] \end{cases}$

27. $\displaystyle\int \frac{dx}{a + bx} = \frac{1}{b}\log(a + bx)$

28. $\displaystyle\int \frac{dx}{(a + bx)^2} = -\frac{1}{b(a + bx)}$

29. $\displaystyle\int \frac{dx}{(a + bx)^3} = -\frac{1}{2b(a + bx)^2}$

30. $\displaystyle\int \frac{x\, dx}{a + bx} = \begin{cases} \dfrac{1}{b^2}[a + bx - a\log(a + bx)] \\[2mm] \quad\quad\quad \text{or} \\[2mm] \dfrac{x}{b} - \dfrac{a}{b^2}\log(a + bx) \end{cases}$

31. $\displaystyle\int \frac{x\, dx}{(a + bx)^2} = \frac{1}{b^2}\left[\log(a + bx) + \frac{a}{a + bx}\right]$

32. $\displaystyle\int \frac{x\, dx}{(a + bx)^n} = \frac{1}{b^2}\left[\frac{-1}{(n - 2)(a + bx)^{n-2}} + \frac{a}{(n - 1)(a + bx)^{n-1}}\right], \quad n \neq 1, 2$

33. $\displaystyle\int \frac{x^2\, dx}{a + bx} = \frac{1}{b^3}\left[\frac{1}{2}(a + bx)^2 - 2a(a + bx) + a^2\log(a + bx)\right]$

34. $\displaystyle\int \frac{x^2\, dx}{(a + bx)^2} = \frac{1}{b^3}\left[a + bx - 2a\log(a + bx) - \frac{a^2}{a + bx}\right]$

35. $\displaystyle\int \frac{x^2\, dx}{(a + bx)^3} = \frac{1}{b^3}\left[\log(a + bx) + \frac{2a}{a + bx} - \frac{a^2}{2(a + bx)^2}\right]$

36. $\displaystyle\int \frac{x^2\, dx}{(a + bx)^n} = \frac{1}{b^3}\left[\frac{-1}{(n - 3)(a + bx)^{n-3}} \right.$
$$\left. + \frac{2a}{(n - 2)(a + bx)^{n-2}} - \frac{a^2}{(n - 1)(a + bx)^{n-1}}\right], \quad n \neq 1, 2, 3$$

INTEGRALS (Continued)

37. $\displaystyle\int \frac{dx}{x(a+bx)} = -\frac{1}{a}\log\frac{a+bx}{x}$

38. $\displaystyle\int \frac{dx}{x(a+bx)^2} = \frac{1}{a(a+bx)} - \frac{1}{a^2}\log\frac{a+bx}{x}$

39. $\displaystyle\int \frac{dx}{x(a+bx)^3} = \frac{1}{a^3}\left[\frac{1}{2}\left(\frac{2a+bx}{a+bx}\right)^2 + \log\frac{x}{a+bx}\right]$

40. $\displaystyle\int \frac{dx}{x^2(a+bx)} = -\frac{1}{ax} + \frac{b}{a^2}\log\frac{a+bx}{x}$

41. $\displaystyle\int \frac{dx}{x^3(a+bx)} = \frac{2bx-a}{2a^2x^2} + \frac{b^2}{a^3}\log\frac{x}{a+bx}$

42. $\displaystyle\int \frac{dx}{x^2(a+bx)^2} = -\frac{a+2bx}{a^2x(a+bx)} + \frac{2b}{a^3}\log\frac{a+bx}{x}$

FORMS CONTAINING $c^2 \pm x^2$, $x^2 - c^2$

43. $\displaystyle\int \frac{dx}{c^2+x^2} = \frac{1}{c}\tan^{-1}\frac{x}{c}$

44. $\displaystyle\int \frac{dx}{c^2-x^2} = \frac{1}{2c}\log\frac{c+x}{c-x}, \qquad (c^2 > x^2)$

45. $\displaystyle\int \frac{dx}{x^2-c^2} = \frac{1}{2c}\log\frac{x-c}{x+c}, \qquad (x^2 > c^2)$

46. $\displaystyle\int \frac{x\,dx}{c^2\pm x^2} = \pm\frac{1}{2}\log(c^2\pm x^2)$

47. $\displaystyle\int \frac{x\,dx}{(c^2\pm x^2)^{n+1}} = \mp\frac{1}{2n(c^2\pm x^2)^n}$

48. $\displaystyle\int \frac{dx}{(c^2\pm x^2)^n} = \frac{1}{2c^2(n-1)}\left[\frac{x}{(c^2\pm x^2)^{n-1}} + (2n-3)\int \frac{dx}{(c^2\pm x^2)^{n-1}}\right]$

49. $\displaystyle\int \frac{dx}{(x^2-c^2)^n} = \frac{1}{2c^2(n-1)}\left[-\frac{x}{(x^2-c^2)^{n-1}} - (2n-3)\int \frac{dx}{(x^2-c^2)^{n-1}}\right]$

50. $\displaystyle\int \frac{x\,dx}{x^2-c^2} = \frac{1}{2}\log(x^2-c^2)$

51. $\displaystyle\int \frac{x\,dx}{(x^2-c^2)^{n+1}} = -\frac{1}{2n(x^2-c^2)^n}$

INTEGRALS (Continued)

FORMS CONTAINING $a + bx$ and $c + dx$

$$u = a + bx, \qquad v = c + dx, \qquad k = ad - bc$$

If $k = 0$, then $v = \dfrac{c}{a}u$

52. $\displaystyle\int \frac{dx}{u \cdot v} = \frac{1}{k} \cdot \log\left(\frac{v}{u}\right)$

53. $\displaystyle\int \frac{x\,dx}{u \cdot v} = \frac{1}{k}\left[\frac{a}{b}\log(u) - \frac{c}{d}\log(v)\right]$

54. $\displaystyle\int \frac{dx}{u^2 \cdot v} = \frac{1}{k}\left(\frac{1}{u} + \frac{d}{k}\log\frac{v}{u}\right)$

55. $\displaystyle\int \frac{x\,dx}{u^2 \cdot v} = \frac{-a}{bku} - \frac{c}{k^2}\log\frac{v}{u}$

56. $\displaystyle\int \frac{x^2\,dx}{u^2 \cdot v} = \frac{a^2}{b^2 ku} + \frac{1}{k^2}\left[\frac{c^2}{d}\log(v) + \frac{a(k - bc)}{b^2}\log(u)\right]$

57. $\displaystyle\int \frac{dx}{u^n \cdot v^m} = \frac{1}{k(m - 1)}\left[\frac{-1}{u^{n-1} \cdot v^{m-1}} - (m + n - 2)b \int \frac{dx}{u^n \cdot v^{m-1}}\right]$

58. $\displaystyle\int \frac{u}{v}\,dx = \frac{bx}{d} + \frac{k}{d^2}\log(v)$

59. $\displaystyle\int \frac{u^m\,dx}{v^n} = \begin{cases} \dfrac{-1}{k(n - 1)}\left[\dfrac{u^{m+1}}{v^{n-1}} + b(n - m - 2)\displaystyle\int \dfrac{u^m}{v^{n-1}}\,dx\right] \\ \qquad\qquad \text{or} \\ \dfrac{-1}{d(n - m - 1)}\left[\dfrac{u^m}{v^{n-1}} + mk\displaystyle\int \dfrac{u^{m-1}}{v^n}\,dx\right] \\ \qquad\qquad \text{or} \\ \dfrac{-1}{d(n - 1)}\left[\dfrac{u^m}{v^{n-1}} - mb\displaystyle\int \dfrac{u^{m-1}}{v^{n-1}}\,dx\right] \end{cases}$

FORMS CONTAINING $(a + bx^n)$

60. $\displaystyle\int \frac{dx}{a + bx^2} = \frac{1}{\sqrt{ab}}\tan^{-1}\frac{x\sqrt{ab}}{a}, \qquad (ab > 0)$

61. $\displaystyle\int \frac{dx}{a + bx^2} = \begin{cases} \dfrac{1}{2\sqrt{-ab}}\log\dfrac{a + x\sqrt{-ab}}{a - x\sqrt{-ab}}, \qquad (ab < 0) \\ \qquad\qquad \text{or} \\ \dfrac{1}{\sqrt{-ab}}\tanh^{-1}\dfrac{x\sqrt{-ab}}{a}, \qquad (ab < 0) \end{cases}$

INTEGRALS (Continued)

62. $\displaystyle\int \frac{dx}{a^2 + b^2 x^2} = \frac{1}{ab} \tan^{-1} \frac{bx}{a}$

63. $\displaystyle\int \frac{x\,dx}{a + bx^2} = \frac{1}{2b} \log(a + bx^2)$

64. $\displaystyle\int \frac{x^2\,dx}{a + bx^2} = \frac{x}{b} - \frac{a}{b} \int \frac{dx}{a + bx^2}$

65. $\displaystyle\int \frac{dx}{(a + bx^2)^2} = \frac{x}{2a(a + bx^2)} + \frac{1}{2a} \int \frac{dx}{a + bx^2}$

66. $\displaystyle\int \frac{dx}{a^2 - b^2 x^2} = \frac{1}{2ab} \log \frac{a + bx}{a - bx}$

67. $\displaystyle\int \frac{dx}{(a + bx^2)^{m+1}} = \begin{cases} \dfrac{1}{2ma} \dfrac{x}{(a + bx^2)^m} + \dfrac{2m - 1}{2ma} \displaystyle\int \dfrac{dx}{(a + bx^2)^m} \\[2mm] \qquad\qquad\text{or} \\[2mm] \dfrac{(2m)!}{(m!)^2} \left[\dfrac{x}{2a} \displaystyle\sum_{r=1}^{m} \dfrac{r!(r - 1)!}{(4a)^{m-r}(2r)!(a + bx^2)^r} + \dfrac{1}{(4a)^m} \displaystyle\int \dfrac{dx}{a + bx^2} \right] \end{cases}$

68. $\displaystyle\int \frac{x\,dx}{(a + bx^2)^{m+1}} = -\frac{1}{2bm(a + bx^2)^m}$

69. $\displaystyle\int \frac{x^2\,dx}{(a + bx^2)^{m+1}} = \frac{-x}{2mb(a + bx^2)^m} + \frac{1}{2mb} \int \frac{dx}{(a + bx^2)^m}$

70. $\displaystyle\int \frac{dx}{x(a + bx^2)} = \frac{1}{2a} \log \frac{x^2}{a + bx^2}$

71. $\displaystyle\int \frac{dx}{x^2(a + bx^2)} = -\frac{1}{ax} - \frac{b}{a} \int \frac{dx}{a + bx^2}$

72. $\displaystyle\int \frac{dx}{x(a + bx^2)^{m+1}} = \begin{cases} \dfrac{1}{2am(a + bx^2)^m} + \dfrac{1}{a} \displaystyle\int \dfrac{dx}{x(a + bx^2)^m} \\[2mm] \qquad\qquad\text{or} \\[2mm] \dfrac{1}{2a^{m+1}} \left[\displaystyle\sum_{r=1}^{m} \dfrac{a^r}{r(a + bx^2)^r} + \log \dfrac{x^2}{a + bx^2} \right] \end{cases}$

73. $\displaystyle\int \frac{dx}{x^2(a + bx^2)^{m+1}} = \frac{1}{a} \int \frac{dx}{x^2(a + bx^2)^m} - \frac{b}{a} \int \frac{dx}{(a + bx^2)^{m+1}}$

74. $\displaystyle\int \frac{dx}{a + bx^3} = \frac{k}{3a} \left[\frac{1}{2} \log \frac{(k + x)^3}{a + bx^3} + \sqrt{3} \tan^{-1} \frac{2x - k}{k\sqrt{3}} \right], \qquad \left(k = \sqrt[3]{\frac{a}{b}} \right)$

75. $\displaystyle\int \frac{x\,dx}{a + bx^3} = \frac{1}{3bk} \left[\frac{1}{2} \log \frac{a + bx^3}{(k + x)^3} + \sqrt{3} \tan^{-1} \frac{2x - k}{k\sqrt{3}} \right], \qquad \left(k = \sqrt[3]{\frac{a}{b}} \right)$

INTEGRALS (Continued)

76. $\displaystyle\int \frac{x^2\,dx}{a + bx^3} = \frac{1}{3b}\log\left(a + bx^3\right)$

77. $\displaystyle\int \frac{dx}{a + bx^4} = \frac{k}{2a}\left[\frac{1}{2}\log\frac{x^2 + 2kx + 2k^2}{x^2 - 2kx + 2k^2} + \tan^{-1}\frac{2kx}{2k^2 - x^2}\right],$

$$\left(ab > 0, k = \sqrt[4]{\frac{a}{4b}}\right)$$

78. $\displaystyle\int \frac{dx}{a + bx^4} = \frac{k}{2a}\left[\frac{1}{2}\log\frac{x + k}{x - k} + \tan^{-1}\frac{x}{k}\right], \qquad \left(ab < 0, k = \sqrt[4]{-\frac{a}{b}}\right)$

79. $\displaystyle\int \frac{x\,dx}{a + bx^4} = \frac{1}{2bk}\tan^{-1}\frac{x^2}{k}, \qquad \left(ab > 0, k = \sqrt{\frac{a}{b}}\right)$

80. $\displaystyle\int \frac{x\,dx}{a + bx^4} = \frac{1}{4bk}\log\frac{x^2 - k}{x^2 + k}, \qquad \left(ab < 0, k = \sqrt{-\frac{a}{b}}\right)$

81. $\displaystyle\int \frac{x^2\,dx}{a + bx^4} = \frac{1}{4bk}\left[\frac{1}{2}\log\frac{x^2 - 2kx + 2k^2}{x^2 + 2kx + 2k^2} + \tan^{-1}\frac{2kx}{2k^2 - x^2}\right],$

$$\left(ab > 0, k = \sqrt[4]{\frac{a}{4b}}\right)$$

82. $\displaystyle\int \frac{x^2\,dx}{a + bx^4} = \frac{1}{4bk}\left[\log\frac{x - k}{x + k} + 2\tan^{-1}\frac{x}{k}\right], \qquad \left(ab < 0, k = \sqrt[4]{-\frac{a}{b}}\right)$

83. $\displaystyle\int \frac{x^3\,dx}{a + bx^4} = \frac{1}{4b}\log\left(a + bx^4\right)$

84. $\displaystyle\int \frac{dx}{x(a + bx^n)} = \frac{1}{an}\log\frac{x^n}{a + bx^n}$

85. $\displaystyle\int \frac{dx}{(a + bx^n)^{m+1}} = \frac{1}{a}\int \frac{dx}{(a + bx^n)^m} - \frac{b}{a}\int \frac{x^n\,dx}{(a + bx^n)^{m+1}}$

86. $\displaystyle\int \frac{x^m\,dx}{(a + bx^n)^{p+1}} = \frac{1}{b}\int \frac{x^{m-n}\,dx}{(a + bx^n)^p} - \frac{a}{b}\int \frac{x^{m-n}\,dx}{(a + bx^n)^{p+1}}$

87. $\displaystyle\int \frac{dx}{x^m(a + bx^n)^{p+1}} = \frac{1}{a}\int \frac{dx}{x^m(a + bx^n)^p} - \frac{b}{a}\int \frac{dx}{x^{m-n}(a + bx^n)^{p+1}}$

INTEGRALS (Continued)

$$88. \int x^m(a + bx^n)^p \, dx = \begin{cases} \dfrac{1}{b(np + m + 1)}\left[x^{m-n+1}(a + bx^n)^{p+1} \right. \\ \qquad\qquad \left. - a(m - n + 1) \int x^{m-n}(a + bx^n)^p \, dx \right] \\ \qquad\text{or} \\ \dfrac{1}{np + m + 1}\left[x^{m+1}(a + bx^n)^p \right. \\ \qquad\qquad \left. + anp \int x^m(a + bx^n)^{p-1} \, dx \right] \\ \qquad\text{or} \\ \dfrac{1}{a(m + 1)}\left[x^{m+1}(a + bx^n)^{p+1} \right. \\ \qquad\qquad \left. - (m + 1 + np + n)b \int x^{m+n}(a + bx^n)^p \, dx \right] \\ \qquad\text{or} \\ \dfrac{1}{an(p + 1)}\left[-x^{m+1}(a + bx^n)^{p+1} \right. \\ \qquad\qquad \left. + (m + 1 + np + n) \int x^m(a + bx^n)^{p+1} \, dx \right] \end{cases}$$

FORMS CONTAINING $c^3 \pm x^3$

$$89. \int \frac{dx}{c^3 \pm x^3} = \pm \frac{1}{6c^2} \log \frac{(c \pm x)^3}{c^3 \pm x^3} + \frac{1}{c^2\sqrt{3}} \tan^{-1} \frac{2x \mp c}{c\sqrt{3}}$$

$$90. \int \frac{dx}{(c^3 \pm x^3)^2} = \frac{x}{3c^3(c^3 \pm x^3)} + \frac{2}{3c^3} \int \frac{dx}{c^3 \pm x^3}$$

$$91. \int \frac{dx}{(c^3 \pm x^3)^{n+1}} = \frac{1}{3nc^3}\left[\frac{x}{(c^3 \pm x^3)^n} + (3n - 1) \int \frac{dx}{(c^3 \pm x^3)^n} \right]$$

$$92. \int \frac{x \, dx}{c^3 \pm x^3} = \frac{1}{6c} \log \frac{c^3 \pm x^3}{(c \pm x)^3} \pm \frac{1}{c\sqrt{3}} \tan^{-1} \frac{2x \mp c}{c\sqrt{3}}$$

$$93. \int \frac{x \, dx}{(c^3 \pm x^3)^2} = \frac{x^2}{3c^3(c^3 \pm x^3)} + \frac{1}{3c^3} \int \frac{x \, dx}{c^3 \pm x^3}$$

$$94. \int \frac{x \, dx}{(c^3 \pm x^3)^{p+1}} = \frac{1}{3nc^3}\left[\frac{x^2}{(c^3 \pm x^3)^n} + (3n - 2) \int \frac{x \, dx}{(c^3 \pm x^3)^n} \right]$$

$$95. \int \frac{x^2 \, dx}{c^3 \pm x^3} = \pm \frac{1}{3} \log (c^3 \pm x^3)$$

INTEGRALS (Continued)

96. $\displaystyle\int \frac{x^2\,dx}{(c^3 \pm x^3)^{n+1}} = \mp\frac{1}{3n(c^3 \pm x^3)^n}$

97. $\displaystyle\int \frac{dx}{x(c^3 \pm x^3)} = \frac{1}{3c^3}\log\frac{x^3}{c^3 \pm x^3}$

98. $\displaystyle\int \frac{dx}{x(c^3 \pm x^3)^2} = \frac{1}{3c^3(c^3 \pm x^3)} + \frac{1}{3c^6}\log\frac{x^3}{c^3 \pm x^3}$

99. $\displaystyle\int \frac{dx}{x(c^3 \pm x^3)^{n+1}} = \frac{1}{3nc^3(c^3 \pm x^3)^n} + \frac{1}{c^3}\int\frac{dx}{x(c^3 \pm x^3)^n}$

100. $\displaystyle\int \frac{dx}{x^2(c^3 \pm x^3)} = -\frac{1}{c^3 x} \mp \frac{1}{c^3}\int\frac{x\,dx}{c^3 \pm x^3}$

101. $\displaystyle\int \frac{dx}{x^2(c^3 \pm x^3)^{n+1}} = \frac{1}{c^3}\int\frac{dx}{x^2(c^3 \pm x^3)^n} \mp \frac{1}{c^3}\int\frac{x\,dx}{(c^3 \pm x^3)^{n+1}}$

FORMS CONTAINING $c^4 \pm x^4$

102. $\displaystyle\int \frac{dx}{c^4 + x^4} = \frac{1}{2c^3\sqrt{2}}\left[\frac{1}{2}\log\frac{x^2 + cx\sqrt{2} + c^2}{x^2 - cx\sqrt{2} + c^2} + \tan^{-1}\frac{cx\sqrt{2}}{c^2 - x^2}\right]$

103. $\displaystyle\int \frac{dx}{c^4 - x^4} = \frac{1}{2c^3}\left[\frac{1}{2}\log\frac{c + x}{c - x} + \tan^{-1}\frac{x}{c}\right]$

104. $\displaystyle\int \frac{x\,dx}{c^4 + x^4} = \frac{1}{2c^2}\tan^{-1}\frac{x^2}{c^2}$

105. $\displaystyle\int \frac{x\,dx}{c^4 - x^4} = \frac{1}{4c^2}\log\frac{c^2 + x^2}{c^2 - x^2}$

106. $\displaystyle\int \frac{x^2\,dx}{c^4 + x^4} = \frac{1}{2c\sqrt{2}}\left[\frac{1}{2}\log\frac{x^2 - cx\sqrt{2} + c^2}{x^2 + cx\sqrt{2} + c^2} + \tan^{-1}\frac{cx\sqrt{2}}{c^2 - x^2}\right]$

107. $\displaystyle\int \frac{x^2\,dx}{c^4 - x^4} = \frac{1}{2c}\left[\frac{1}{2}\log\frac{c + x}{c - x} - \tan^{-1}\frac{x}{c}\right]$

108. $\displaystyle\int \frac{x^3\,dx}{c^4 \pm x^4} = \pm\frac{1}{4}\log(c^4 \pm x^4)$

FORMS CONTAINING $(a + bx + cx^2)$

$$X = a + bx + cx^2 \text{ and } q = 4ac - b^2$$

If $q = 0$, then $X = c\left(x + \dfrac{b}{2c}\right)^2$, and formulas starting with 23 should be used in place of these.

109. $\displaystyle\int \frac{dx}{X} = \frac{2}{\sqrt{q}}\tan^{-1}\frac{2cx + b}{\sqrt{q}}, \qquad (q > 0)$

INTEGRALS (Continued)

110. $\displaystyle\int \frac{dx}{X} = \begin{cases} \dfrac{-2}{\sqrt{-q}} \tanh^{-1} \dfrac{2cx + b}{\sqrt{-q}} \\[2mm] \qquad\qquad \text{or} \\[2mm] \dfrac{1}{\sqrt{-q}} \log \dfrac{2cx + b - \sqrt{-q}}{2cx + b + \sqrt{-q}}, \quad (q < 0) \end{cases}$

111. $\displaystyle\int \frac{dx}{X^2} = \frac{2cx + b}{qX} + \frac{2c}{q} \int \frac{dx}{X}$

112. $\displaystyle\int \frac{dx}{X^3} = \frac{2cx + b}{q} \left(\frac{1}{2X^2} + \frac{3c}{qX} \right) + \frac{6c^2}{q^2} \int \frac{dx}{X}$

113. $\displaystyle\int \frac{dx}{X^{n+1}} = \begin{cases} \dfrac{2cx + b}{nqX^n} + \dfrac{2(2n - 1)c}{qn} \displaystyle\int \dfrac{dx}{X^n} \\[2mm] \qquad\qquad \text{or} \\[2mm] \dfrac{(2n)!}{(n!)^2} \left(\dfrac{c}{q} \right)^n \left[\dfrac{2cx + b}{q} \displaystyle\sum_{r=1}^{n} \left(\dfrac{q}{cX} \right)^r \left(\dfrac{(r-1)!r!}{(2r)!} \right) + \displaystyle\int \dfrac{dx}{X} \right] \end{cases}$

114. $\displaystyle\int \frac{x\,dx}{X} = \frac{1}{2c} \log X - \frac{b}{2c} \int \frac{dx}{X}$

115. $\displaystyle\int \frac{x\,dx}{X^2} = -\frac{bx + 2a}{qX} - \frac{b}{q} \int \frac{dx}{X}$

116. $\displaystyle\int \frac{x\,dx}{X^{n+1}} = -\frac{2a + bx}{nqX^n} - \frac{b(2n - 1)}{nq} \int \frac{dx}{X^n}$

117. $\displaystyle\int \frac{x^2}{X}\,dx = \frac{x}{c} - \frac{b}{2c^2} \log X + \frac{b^2 - 2ac}{2c^2} \int \frac{dx}{X}$

118. $\displaystyle\int \frac{x^2}{X^2}\,dx = \frac{(b^2 - 2ac)x + ab}{cqX} + \frac{2a}{q} \int \frac{dx}{X}$

119. $\displaystyle\int \frac{x^m\,dx}{X^{n+1}} = -\frac{x^{m-1}}{(2n - m + 1)cX^n} - \frac{n - m + 1}{2n - m + 1} \cdot \frac{b}{c} \int \frac{x^{m-1}\,dx}{X^{n+1}}$

$$+ \frac{m - 1}{2n - m + 1} \cdot \frac{a}{c} \int \frac{x^{m-2}\,dx}{X^{n+1}}$$

120. $\displaystyle\int \frac{dx}{xX} = \frac{1}{2a} \log \frac{x^2}{X} - \frac{b}{2a} \int \frac{dx}{X}$

121. $\displaystyle\int \frac{dx}{x^2 X} = \frac{b}{2a^2} \log \frac{X}{x^2} - \frac{1}{ax} + \left(\frac{b^2}{2a^2} - \frac{c}{a} \right) \int \frac{dx}{X}$

122. $\displaystyle\int \frac{dx}{xX^n} = \frac{1}{2a(n - 1)X^{n-1}} - \frac{b}{2a} \int \frac{dx}{X^n} + \frac{1}{a} \int \frac{dx}{xX^{n-1}}$

INTEGRALS (Continued)

123. $\displaystyle\int \frac{dx}{x^m X^{n+1}} = -\frac{1}{(m-1)ax^{m-1}X^n} - \frac{n+m-1}{m-1}\cdot\frac{b}{a}\int \frac{dx}{x^{m-1}X^{n+1}}$

$$-\frac{2n+m-1}{m-1}\cdot\frac{c}{a}\int \frac{dx}{x^{m-2}X^{n+1}}$$

FORMS CONTAINING $\sqrt{a+bx}$

124. $\displaystyle\int \sqrt{a+bx}\,dx = \frac{2}{3b}\sqrt{(a+bx)^3}$

125. $\displaystyle\int x\sqrt{a+bx}\,dx = -\frac{2(2a-3bx)\sqrt{(a+bx)^3}}{15b^2}$

126. $\displaystyle\int x^2\sqrt{a+bx}\,dx \doteq \frac{2(8a^2-12abx+15b^2x^2)\sqrt{(a+bx)^3}}{105b^3}$

127. $\displaystyle\int x^m\sqrt{a+bx}\,dx = \begin{cases} \dfrac{2}{b(2m+3)}\left[x^m\sqrt{(a+bx)^3} - ma\int x^{m-1}\sqrt{a+bx}\,dx\right] \\[4pt] \text{or} \\[4pt] \dfrac{2}{b^{m+1}}\sqrt{a+bx}\displaystyle\sum_{r=0}^{m}\dfrac{m!(-a)^{m-r}}{r!(m-r)!(2r+3)}(a+bx)^{r+1} \end{cases}$

128. $\displaystyle\int \frac{\sqrt{a+bx}}{x}dx = 2\sqrt{a+bx} + a\int \frac{dx}{x\sqrt{a+bx}}$

129. $\displaystyle\int \frac{\sqrt{a+bx}}{x^2}dx = -\frac{\sqrt{a+bx}}{x} + \frac{b}{2}\int \frac{dx}{x\sqrt{a+bx}}$

130. $\displaystyle\int \frac{\sqrt{a+bx}}{x^m}dx = -\frac{1}{(m-1)a}\left[\frac{\sqrt{(a+bx)^3}}{x^{m-1}} + \frac{(2m-5)b}{2}\int \frac{\sqrt{a+bx}}{x^{m-1}}dx\right]$

131. $\displaystyle\int \frac{dx}{\sqrt{a+bx}} = \frac{2\sqrt{a+bx}}{b}$

132. $\displaystyle\int \frac{x\,dx}{\sqrt{a+bx}} = -\frac{2(2a-bx)}{3b^2}\sqrt{a+bx}$

133. $\displaystyle\int \frac{x^2\,dx}{\sqrt{a+bx}} = \frac{2(8a^2-4abx+3b^2x^2)}{15b^3}\sqrt{a+bx}$

INTEGRALS (Continued)

134. $\displaystyle \int \frac{x^m\, dx}{\sqrt{a + bx}} = \begin{cases} \dfrac{2}{(2m + 1)b}\left[x^m\sqrt{a + bx} - ma\displaystyle\int \dfrac{x^{m-1}\, dx}{\sqrt{a + bx}}\right] \\[2mm] \quad\text{or} \\[2mm] \dfrac{2(-a)^m\sqrt{a + bx}}{b^{m+1}}\displaystyle\sum_{r=0}^{m}\dfrac{(-1)^r m!(a + bx)^r}{(2r + 1)r!(m - r)!a^r} \end{cases}$

135. $\displaystyle \int \frac{dx}{x\sqrt{a + bx}} = \frac{1}{\sqrt{a}}\log\left(\frac{\sqrt{a + bx} - \sqrt{a}}{\sqrt{a + bx} + \sqrt{a}}\right), \qquad (a > 0)$

136. $\displaystyle \int \frac{dx}{x\sqrt{a + bx}} = \frac{2}{\sqrt{-a}}\tan^{-1}\sqrt{\frac{a + bx}{a}}, \qquad (a < 0)$

137. $\displaystyle \int \frac{dx}{x^2\sqrt{a + bx}} = -\frac{\sqrt{a + bx}}{ax} - \frac{b}{2a}\int \frac{dx}{x\sqrt{a + bx}}$

138. $\displaystyle \int \frac{dx}{x^n\sqrt{a + bx}} = \begin{cases} -\dfrac{\sqrt{a + bx}}{(n - 1)ax^{n-1}} - \dfrac{(2n - 3)b}{(2n - 2)a}\displaystyle\int \dfrac{dx}{x^{n-1}\sqrt{a + bx}} \\[2mm] \quad\text{or} \\[2mm] \dfrac{(2n - 2)!}{[(n - 1)!]^2}\left[-\dfrac{\sqrt{a + bx}}{a}\displaystyle\sum_{r=1}^{n-1}\dfrac{r!(r - 1)!}{x^r(2r)!}\left(-\dfrac{b}{4a}\right)^{n-r-1}\right. \\[4mm] \qquad\qquad\qquad \left. + \left(-\dfrac{b}{4a}\right)^{n-1}\displaystyle\int \dfrac{dx}{x\sqrt{a + bx}}\right] \end{cases}$

139. $\displaystyle \int (a + bx)^{\pm\frac{n}{2}}\, dx = \frac{2(a + bx)^{\frac{2\pm n}{2}}}{b(2 \pm n)}$

140. $\displaystyle \int x(a + bx)^{\pm\frac{n}{2}}\, dx = \frac{2}{b^2}\left[\frac{(a + bx)^{\frac{4\pm n}{2}}}{4 \pm n} - \frac{a(a + bx)^{\frac{2\pm n}{2}}}{2 \pm n}\right]$

141. $\displaystyle \int \frac{dx}{x(a + bx)^{\frac{m}{2}}} = \frac{1}{a}\int \frac{dx}{x(a + bx)^{\frac{m-2}{2}}} - \frac{b}{a}\int \frac{dx}{(a + bx)^{\frac{m}{2}}}$

142. $\displaystyle \int \frac{(a + bx)^{\frac{n}{2}}\, dx}{x} = b\int (a + bx)^{\frac{n-2}{2}}\, dx + a\int \frac{(a + bx)^{\frac{n-2}{2}}}{x}\, dx$

143. $\displaystyle \int f(x, \sqrt{a + bx})\, dx = \frac{2}{b}\int f\left(\frac{z^2 - a}{b}, z\right)z\, dz, \qquad (z = \sqrt{a + bx})$

FORMS CONTAINING $\sqrt{a + bx}$ and $\sqrt{c + dx}$

$$u = a + bx \qquad v = c + dx \qquad k = ad - bc$$

If $k = 0$, then $v = \dfrac{c}{a}u$, and formulas starting with 124 should be used in place of these.

144. $\displaystyle\int \frac{dx}{\sqrt{uv}} = \begin{cases} \dfrac{2}{\sqrt{bd}} \tanh^{-1} \dfrac{\sqrt{bduv}}{bv} \\[2mm] \text{or} \\[2mm] \dfrac{1}{\sqrt{bd}} \log \dfrac{bv + \sqrt{bduv}}{bv - \sqrt{bduv}} \\[2mm] \text{or} \\[2mm] \dfrac{1}{\sqrt{bd}} \log \dfrac{(bv + \sqrt{bduv})^2}{v}, \qquad (bd > 0) \end{cases}$

145. $\displaystyle\int \frac{dx}{\sqrt{uv}} = \begin{cases} \dfrac{2}{\sqrt{-bd}} \tan^{-1} \dfrac{\sqrt{-bduv}}{bv} \\[2mm] \text{or} \\[2mm] -\dfrac{1}{\sqrt{-bd}} \sin^{-1}\left(\dfrac{2bdx + ad + bc}{|k|}\right), \qquad (bd < 0) \end{cases}$

146. $\displaystyle\int \sqrt{uv}\,dx = \frac{k + 2bv}{4bd} \sqrt{uv} - \frac{k^2}{8bd} \int \frac{dx}{\sqrt{uv}}$

147. $\displaystyle\int \frac{dx}{v\sqrt{u}} = \begin{cases} \dfrac{1}{\sqrt{kd}} \log \dfrac{d\sqrt{u} - \sqrt{kd}}{d\sqrt{u} + \sqrt{kd}} \\[2mm] \text{or} \\[2mm] \dfrac{1}{\sqrt{kd}} \log \dfrac{(d\sqrt{u} - \sqrt{kd})^2}{v}, \qquad (kd > 0) \end{cases}$

148. $\displaystyle\int \frac{dx}{v\sqrt{u}} = \frac{2}{\sqrt{-kd}} \tan^{-1} \frac{d\sqrt{u}}{\sqrt{-kd}}, \qquad (kd < 0)$

149. $\displaystyle\int \frac{x\,dx}{\sqrt{uv}} = \frac{\sqrt{uv}}{bd} - \frac{ad + bc}{2bd} \int \frac{dx}{\sqrt{uv}}$

150. $\displaystyle\int \frac{dx}{v\sqrt{uv}} = \frac{-2\sqrt{uv}}{kv}$

INTEGRALS (Continued)

151. $\displaystyle\int \frac{v\,dx}{\sqrt{uv}} = \frac{\sqrt{uv}}{b} - \frac{k}{2b}\int \frac{dx}{\sqrt{uv}}$

152. $\displaystyle\int \sqrt{\frac{v}{u}}\,dx = \frac{v}{|v|}\int \frac{v\,dx}{\sqrt{uv}}$

153. $\displaystyle\int v^m\sqrt{u}\,dx = \frac{1}{(2m+3)d}\left(2v^{m+1}\sqrt{u} + k\int \frac{v^m\,dx}{\sqrt{u}}\right)$

154. $\displaystyle\int \frac{dx}{v^m\sqrt{u}} = -\frac{1}{(m-1)k}\left(\frac{\sqrt{u}}{v^{m-1}} + \left(m - \frac{3}{2}\right)b\int \frac{dx}{v^{m-1}\sqrt{u}}\right)$

155. $\displaystyle\int \frac{v^m\,dx}{\sqrt{u}} = \begin{cases} \dfrac{2}{b(2m+1)}\left[v^m\sqrt{u} - mk\displaystyle\int \dfrac{v^{m-1}}{\sqrt{u}}\,dx\right] \\ \qquad\text{or} \\ \dfrac{2(m!)^2\sqrt{u}}{b(2m+1)!}\displaystyle\sum_{r=0}^{m}\left(-\dfrac{4k}{b}\right)^{m-r}\dfrac{(2r)!}{(r!)^2}v^r \end{cases}$

FORMS CONTAINING $\sqrt{x^2 \pm a^2}$

156. $\displaystyle\int \sqrt{x^2 \pm a^2}\,dx = \tfrac{1}{2}[x\sqrt{x^2 \pm a^2} \pm a^2\log(x + \sqrt{x^2 \pm a^2})]$

157. $\displaystyle\int \frac{dx}{\sqrt{x^2 \pm a^2}} = \log(x + \sqrt{x^2 \pm a^2})$

158. $\displaystyle\int \frac{dx}{x\sqrt{x^2 - a^2}} = \frac{1}{|a|}\sec^{-1}\frac{x}{a}$

159. $\displaystyle\int \frac{dx}{x\sqrt{x^2 + a^2}} = -\frac{1}{a}\log\left(\frac{a + \sqrt{x^2 + a^2}}{x}\right)$

160. $\displaystyle\int \frac{\sqrt{x^2 + a^2}}{x}\,dx = \sqrt{x^2 + a^2} - a\log\left(\frac{a + \sqrt{x^2 + a^2}}{x}\right)$

161. $\displaystyle\int \frac{\sqrt{x^2 - a^2}}{x}\,dx = \sqrt{x^2 - a^2} - |a|\sec^{-1}\frac{x}{a}$

162. $\displaystyle\int \frac{x\,dx}{\sqrt{x^2 \pm a^2}} = \sqrt{x^2 \pm a^2}$

163. $\displaystyle\int x\sqrt{x^2 \pm a^2}\,dx = \tfrac{1}{3}\sqrt{(x^2 \pm a^2)^3}$

164. $\displaystyle\int \sqrt{(x^2 \pm a^2)^3}\, dx = \frac{1}{4}\left[x\sqrt{(x^2 \pm a^2)^3} \pm \frac{3a^2 x}{2}\sqrt{x^2 \pm a^2}\right.$

$$\left. + \frac{3a^4}{2}\log(x + \sqrt{x^2 \pm a^2})\right]$$

165. $\displaystyle\int \frac{dx}{\sqrt{(x^2 \pm a^2)^3}} = \frac{\pm x}{a^2\sqrt{x^2 \pm a^2}}$

166. $\displaystyle\int \frac{x\, dx}{\sqrt{(x^2 \pm a^2)^3}} = \frac{-1}{\sqrt{x^2 \pm a^2}}$

167. $\displaystyle\int x\sqrt{(x^2 \pm a^2)^3}\, dx = \tfrac{1}{5}\sqrt{(x^2 \pm a^2)^5}$

168. $\displaystyle\int x^2\sqrt{x^2 \pm a^2}\, dx = \frac{x}{4}\sqrt{(x^2 \pm a^2)^3} \mp \frac{a^2}{8}x\sqrt{x^2 \pm a^2} - \frac{a^4}{8}\log(x + \sqrt{x^2 \pm a^2})$

169. $\displaystyle\int x^3\sqrt{x^2 + a^2}\, dx = (\tfrac{1}{5}x^2 - \tfrac{2}{15}a^2)\sqrt{(a^2 + x^2)^3}$

170. $\displaystyle\int x^3\sqrt{x^2 - a^2}\, dx = \frac{1}{5}\sqrt{(x^2 - a^2)^5} + \frac{a^2}{3}\sqrt{(x^2 - a^2)^3}$

171. $\displaystyle\int \frac{x^2\, dx}{\sqrt{x^2 \pm a^2}} = \frac{x}{2}\sqrt{x^2 \pm a^2} \mp \frac{a^2}{2}\log(x + \sqrt{x^2 \pm a^2})$

172. $\displaystyle\int \frac{x^3\, dx}{\sqrt{x^2 \pm a^2}} = \frac{1}{3}\sqrt{(x^2 \pm a^2)^3} \mp a^2\sqrt{x^2 \pm a^2}$

173. $\displaystyle\int \frac{dx}{x^2\sqrt{x^2 \pm a^2}} = \mp \frac{\sqrt{x^2 \pm a^2}}{a^2 x}$

174. $\displaystyle\int \frac{dx}{x^3\sqrt{x^2 + a^2}} = -\frac{\sqrt{x^2 + a^2}}{2a^2 x^2} + \frac{1}{2a^3}\log\frac{a + \sqrt{x^2 + a^2}}{x}$

175. $\displaystyle\int \frac{dx}{x^3\sqrt{x^2 - a^2}} = \frac{\sqrt{x^2 - a^2}}{2a^2 x^2} + \frac{1}{2|a^3|}\sec^{-1}\frac{x}{a}$

176. $\displaystyle\int x^2\sqrt{(x^2 \pm a^2)^3}\, dx = \frac{x}{6}\sqrt{(x^2 \pm a^2)^5} \mp \frac{a^2 x}{24}\sqrt{(x^2 \pm a^2)^3} - \frac{a^4 x}{16}\sqrt{x^2 \pm a^2}$

$$\mp \frac{a^6}{16}\log(x + \sqrt{x^2 \pm a^2})$$

177. $\displaystyle\int x^3\sqrt{(x^2 \pm a^2)^3}\, dx = \frac{1}{7}\sqrt{(x^2 \pm a^2)^7} \mp \frac{a^2}{5}\sqrt{(x^2 \pm a^2)^5}$

INTEGRALS (Continued)

178. $\displaystyle\int \frac{\sqrt{x^2 \pm a^2}\,dx}{x^2} = -\frac{\sqrt{x^2 \pm a^2}}{x} + \log(x + \sqrt{x^2 \pm a^2})$

179. $\displaystyle\int \frac{\sqrt{x^2 + a^2}}{x^3}\,dx = -\frac{\sqrt{x^2 + a^2}}{2x^2} - \frac{1}{2a}\log\frac{a + \sqrt{x^2 + a^2}}{x}$

180. $\displaystyle\int \frac{\sqrt{x^2 - a^2}}{x^3}\,dx = -\frac{\sqrt{x^2 - a^2}}{2x^2} + \frac{1}{2|a|}\sec^{-1}\frac{x}{a}$

181. $\displaystyle\int \frac{\sqrt{x^2 \pm a^2}}{x^4}\,dx = \mp\frac{\sqrt{(x^2 \pm a^2)^3}}{3a^2 x^3}$

182. $\displaystyle\int \frac{x^2\,dx}{\sqrt{(x^2 \pm a^2)^3}} = \frac{-x}{\sqrt{x^2 \pm a^2}} + \log(x + \sqrt{x^2 \pm a^2})$

183. $\displaystyle\int \frac{x^3\,dx}{\sqrt{(x^2 \pm a^2)^3}} = \sqrt{x^2 \pm a^2} \pm \frac{a^2}{\sqrt{x^2 \pm a^2}}$

184. $\displaystyle\int \frac{dx}{x\sqrt{(x^2 + a^2)^3}} = \frac{1}{a^2\sqrt{x^2 + a^2}} - \frac{1}{a^3}\log\frac{a + \sqrt{x^2 + a^2}}{x}$

185. $\displaystyle\int \frac{dx}{x\sqrt{(x^2 - a^2)^3}} = -\frac{1}{a^2\sqrt{x^2 - a^2}} - \frac{1}{|a^3|}\sec^{-1}\frac{x}{a}$

186. $\displaystyle\int \frac{dx}{x^2\sqrt{(x^2 \pm a^2)^3}} = -\frac{1}{a^4}\left[\frac{\sqrt{x^2 \pm a^2}}{x} + \frac{x}{\sqrt{x^2 \pm a^2}}\right]$

187. $\displaystyle\int \frac{dx}{x^3\sqrt{(x^2 + a^2)^3}} = -\frac{1}{2a^2 x^2\sqrt{x^2 + a^2}} - \frac{3}{2a^4\sqrt{x^2 + a^2}}$

$$+ \frac{3}{2a^5}\log\frac{a + \sqrt{x^2 + a^2}}{x}$$

188. $\displaystyle\int \frac{dx}{x^3\sqrt{(x^2 - a^2)^3}} = \frac{1}{2a^2 x^2\sqrt{x^2 - a^2}} - \frac{3}{2a^4\sqrt{x^2 - a^2}} - \frac{3}{2|a^5|}\sec^{-1}\frac{x}{a}$

189. $\displaystyle\int \frac{x^m}{\sqrt{x^2 \pm a^2}}\,dx = \frac{1}{m}x^{m-1}\sqrt{x^2 \pm a^2} \mp \frac{m-1}{m}a^2\int \frac{x^{m-2}}{\sqrt{x^2 \pm a^2}}\,dx$

190. $\displaystyle\int \frac{x^{2m}}{\sqrt{x^2 \pm a^2}}\,dx = \frac{(2m)!}{2^{2m}(m!)^2}\left[\sqrt{x^2 \pm a^2}\sum_{r=1}^{m}\frac{r!(r-1)!}{(2r)!}(\mp a^2)^{m-r}(2x)^{2r-1}\right.$

$$\left. + (\mp a^2)^m \log(x + \sqrt{x^2 \pm a^2})\right]$$

191. $\displaystyle\int \frac{x^{2m+1}}{\sqrt{x^2 \pm a^2}}\,dx = \sqrt{x^2 \pm a^2}\sum_{r=0}^{m}\frac{(2r)!(m!)^2}{(2m+1)!(r!)^2}(\mp 4a^2)^{m-r}x^{2r}$

INTEGRALS (Continued)

192. $\displaystyle \int \frac{dx}{x^m\sqrt{x^2 \pm a^2}} = \mp \frac{\sqrt{x^2 \pm a^2}}{(m-1)a^2 x^{m-1}} \mp \frac{(m-2)}{(m-1)a^2} \int \frac{dx}{x^{m-2}\sqrt{x^2 \pm a^2}}$

193. $\displaystyle \int \frac{dx}{x^{2m}\sqrt{x^2 \pm a^2}} = \sqrt{x^2 \pm a^2} \sum_{r=0}^{m-1} \frac{(m-1)!\,m!\,(2r)!\,2^{2m-2r-1}}{(r!)^2 (2m)!\,(\mp a^2)^{m-r} x^{2r+1}}$

194. $\displaystyle \int \frac{dx}{x^{2m+1}\sqrt{x^2 + a^2}} = \frac{(2m)!}{(m!)^2}\left[\frac{\sqrt{x^2+a^2}}{a^2} \sum_{r=1}^{m} (-1)^{m-r+1} \frac{r!(r-1)!}{2(2r)!(4a^2)^{m-r} x^{2r}}\right.$

$$\left. + \frac{(-1)^{m+1}}{2^{2m}a^{2m+1}}\log \frac{\sqrt{x^2+a^2}+a}{x}\right]$$

195. $\displaystyle \int \frac{dx}{x^{2m+1}\sqrt{x^2 - a^2}} = \frac{(2m)!}{(m!)^2}\left[\frac{\sqrt{x^2-a^2}}{a^2} \sum_{r=1}^{m} \frac{r!(r-1)!}{2(2r)!(4a^2)^{m-r} x^{2r}}\right.$

$$\left. + \frac{1}{2^{2m}|a|^{2m+1}}\sec^{-1}\frac{x}{a}\right]$$

196. $\displaystyle \int \frac{dx}{(x-a)\sqrt{x^2-a^2}} = -\frac{\sqrt{x^2-a^2}}{a(x-a)}$

197. $\displaystyle \int \frac{dx}{(x+a)\sqrt{x^2-a^2}} = \frac{\sqrt{x^2-a^2}}{a(x+a)}$

198. $\displaystyle \int f(x,\sqrt{x^2+a^2})\,dx = a \int f(a\tan u, a\sec u)\sec^2 u\,du, \qquad \left(u = \tan^{-1}\frac{x}{a}, a > 0\right)$

199. $\displaystyle \int f(x,\sqrt{x^2-a^2})\,dx = a \int f(a\sec u, a\tan u)\sec u \tan u\,du, \qquad \left(u = \sec^{-1}\frac{x}{a},\right.$

$$\left. a > 0\right)$$

FORMS CONTAINING $\sqrt{a^2 - x^2}$

200. $\displaystyle \int \sqrt{a^2 - x^2}\,dx = \frac{1}{2}\left[x\sqrt{a^2-x^2} + a^2\sin^{-1}\frac{x}{|a|}\right]$

201. $\displaystyle \int \frac{dx}{\sqrt{a^2 - x^2}} = \begin{cases} \sin^{-1}\dfrac{x}{|a|} \\ \text{or} \\ -\cos^{-1}\dfrac{x}{|a|} \end{cases}$

202. $\displaystyle \int \frac{dx}{x\sqrt{a^2 - x^2}} = -\frac{1}{a}\log\left(\frac{a+\sqrt{a^2-x^2}}{x}\right)$

INTEGRALS (Continued)

203. $\displaystyle\int \frac{\sqrt{a^2 - x^2}}{x}\,dx = \sqrt{a^2 - x^2} - a\log\left(\frac{a + \sqrt{a^2 - x^2}}{x}\right)$

204. $\displaystyle\int \frac{x\,dx}{\sqrt{a^2 - x^2}} = -\sqrt{a^2 - x^2}$

205. $\displaystyle\int x\sqrt{a^2 - x^2}\,dx = -\tfrac{1}{3}\sqrt{(a^2 - x^2)^3}$

206. $\displaystyle\int \sqrt{(a^2 - x^2)^3}\,dx = \frac{1}{4}\left[x\sqrt{(a^2 - x^2)^3} + \frac{3a^2 x}{2}\sqrt{a^2 - x^2} + \frac{3a^4}{2}\sin^{-1}\frac{x}{|a|}\right]$

207. $\displaystyle\int \frac{dx}{\sqrt{(a^2 - x^2)^3}} = \frac{x}{a^2\sqrt{a^2 - x^2}}$

208. $\displaystyle\int \frac{x\,dx}{\sqrt{(a^2 - x^2)^3}} = \frac{1}{\sqrt{a^2 - x^2}}$

209. $\displaystyle\int x\sqrt{(a^2 - x^2)^3}\,dx = -\tfrac{1}{5}\sqrt{(a^2 - x^2)^5}$

210. $\displaystyle\int x^2\sqrt{a^2 - x^2}\,dx = -\frac{x}{4}\sqrt{(a^2 - x^2)^3} + \frac{a^2}{8}\left(x\sqrt{a^2 - x^2} + a^2\sin^{-1}\frac{x}{|a|}\right)$

211. $\displaystyle\int x^3\sqrt{a^2 - x^2}\,dx = \left(-\tfrac{1}{5}x^2 - \tfrac{2}{15}a^2\right)\sqrt{(a^2 - x^2)^3}$

212. $\displaystyle\int x^2\sqrt{(a^2 - x^2)^3}\,dx = -\frac{1}{6}x\sqrt{(a^2 - x^2)^5} + \frac{a^2 x}{24}\sqrt{(a^2 - x^2)^3}$

$$+ \frac{a^4 x}{16}\sqrt{a^2 - x^2} + \frac{a^6}{16}\sin^{-1}\frac{x}{|a|}$$

213. $\displaystyle\int x^3\sqrt{(a^2 - x^2)^3}\,dx = \frac{1}{7}\sqrt{(a^2 - x^2)^7} - \frac{a^2}{5}\sqrt{(a^2 - x^2)^5}$

214. $\displaystyle\int \frac{x^2\,dx}{\sqrt{a^2 - x^2}} = -\frac{x}{2}\sqrt{a^2 - x^2} + \frac{a^2}{2}\sin^{-1}\frac{x}{|a|}$

215. $\displaystyle\int \frac{dx}{x^2\sqrt{a^2 - x^2}} = -\frac{\sqrt{a^2 - x^2}}{a^2 x}$

216. $\displaystyle\int \frac{\sqrt{a^2 - x^2}}{x^2}\,dx = -\frac{\sqrt{a^2 - x^2}}{x} - \sin^{-1}\frac{x}{|a|}$

217. $\displaystyle\int \frac{\sqrt{a^2 - x^2}}{x^3}\,dx = -\frac{\sqrt{a^2 - x^2}}{2x^2} + \frac{1}{2a}\log\frac{a + \sqrt{a^2 - x^2}}{x}$

218. $\displaystyle\int \frac{\sqrt{a^2 - x^2}}{x^4}\,dx = -\frac{\sqrt{(a^2 - x^2)^3}}{3a^2 x^3}$

INTEGRALS (Continued)

219. $\displaystyle\int \frac{x^2\,dx}{\sqrt{(a^2-x^2)^3}} = \frac{x}{\sqrt{a^2-x^2}} - \sin^{-1}\frac{x}{|a|}$

220. $\displaystyle\int \frac{x^3\,dx}{\sqrt{a^2-x^2}} = -\frac{2}{3}(a^2-x^2)^{\frac{3}{2}} - x^2(a^2-x^2)^{\frac{1}{2}} - \frac{1}{3}\sqrt{a^2-x^2}(x^2+2a^2)$

221. $\displaystyle\int \frac{x^3\,dx}{\sqrt{(a^2-x^2)^3}} = 2(a^2-x^2)^{\frac{1}{2}} + \frac{x^2}{(a^2-x^2)^{\frac{1}{2}}} = \frac{a^2}{\sqrt{a^2-x^2}} + \sqrt{a^2-x^2}$

222. $\displaystyle\int \frac{dx}{x^3\sqrt{a^2-x^2}} = -\frac{\sqrt{a^2-x^2}}{2a^2x^2} - \frac{1}{2a^3}\log\frac{a+\sqrt{a^2-x^2}}{x}$

223. $\displaystyle\int \frac{dx}{x\sqrt{(a^2-x^2)^3}} = \frac{1}{a^2\sqrt{a^2-x^2}} - \frac{1}{a^3}\log\frac{a+\sqrt{a^2-x^2}}{x}$

224. $\displaystyle\int \frac{dx}{x^2\sqrt{(a^2-x^2)^3}} = \frac{1}{a^4}\left[-\frac{\sqrt{a^2-x^2}}{x} + \frac{x}{\sqrt{a^2-x^2}}\right]$

225. $\displaystyle\int \frac{dx}{x^3\sqrt{(a^2-x^2)^3}} = -\frac{1}{2a^2x^2\sqrt{a^2-x^2}} + \frac{3}{2a^4\sqrt{a^2-x^2}}$

$$-\frac{3}{2a^5}\log\frac{a+\sqrt{a^2-x^2}}{x}$$

226. $\displaystyle\int \frac{x^m}{\sqrt{a^2-x^2}}\,dx = -\frac{x^{m-1}\sqrt{a^2-x^2}}{m} + \frac{(m-1)a^2}{m}\int\frac{x^{m-2}}{\sqrt{a^2-x^2}}\,dx$

227. $\displaystyle\int \frac{x^{2m}}{\sqrt{a^2-x^2}}\,dx = \frac{(2m)!}{(m!)^2}\left[-\sqrt{a^2-x^2}\sum_{r=1}^{m}\frac{r!(r-1)!}{2^{2m-2r+1}(2r)!}a^{2m-2r}x^{2r-1}\right.$

$$\left.+\frac{a^{2m}}{2^{2m}}\sin^{-1}\frac{x}{|a|}\right]$$

228. $\displaystyle\int \frac{x^{2m+1}}{\sqrt{a^2-x^2}}\,dx = -\sqrt{a^2-x^2}\sum_{r=0}^{m}\frac{(2r)!(m!)^2}{(2m+1)!(r!)^2}(4a^2)^{m-r}x^{2r}$

229. $\displaystyle\int \frac{dx}{x^m\sqrt{a^2-x^2}} = -\frac{\sqrt{a^2-x^2}}{(m-1)a^2x^{m-1}} + \frac{m-2}{(m-1)a^2}\int\frac{dx}{x^{m-2}\sqrt{a^2-x^2}}$

230. $\displaystyle\int \frac{dx}{x^{2m}\sqrt{a^2-x^2}} = -\sqrt{a^2-x^2}\sum_{r=0}^{m-1}\frac{(m-1)!m!(2r)!2^{2m-2r-1}}{(r!)^2(2m)!a^{2m-2r}x^{2r+1}}$

231. $\displaystyle\int \frac{dx}{x^{2m+1}\sqrt{a^2-x^2}} = \frac{(2m)!}{(m!)^2}\left[-\frac{\sqrt{a^2-x^2}}{a^2}\sum_{r=1}^{m}\frac{r!(r-1)!}{2(2r)!(4a^2)^{m-r}x^{2r}}\right.$

$$\left.+\frac{1}{2^{2m}a^{2m+1}}\log\frac{a-\sqrt{a^2-x^2}}{x}\right]$$

INTEGRALS (Continued)

232. $\int \dfrac{dx}{(b^2 - x^2)\sqrt{a^2 - x^2}} = \dfrac{1}{2b\sqrt{a^2 - b^2}} \log \dfrac{(b\sqrt{a^2 - x^2} + x\sqrt{a^2 - b^2})^2}{b^2 - x^2},$

$$(a^2 > b^2)$$

233. $\int \dfrac{dx}{(b^2 - x^2)\sqrt{a^2 - x^2}} = \dfrac{1}{b\sqrt{b^2 - a^2}} \tan^{-1} \dfrac{x\sqrt{b^2 - a^2}}{b\sqrt{a^2 - x^2}}, \qquad (b^2 > a^2)$

234. $\int \dfrac{dx}{(b^2 + x^2)\sqrt{a^2 - x^2}} = \dfrac{1}{b\sqrt{a^2 + b^2}} \tan^{-1} \dfrac{x\sqrt{a^2 + b^2}}{b\sqrt{a^2 - x^2}}$

235. $\int \dfrac{\sqrt{a^2 - x^2}}{b^2 + x^2} \, dx = \dfrac{\sqrt{a^2 + b^2}}{|b|} \sin^{-1} \dfrac{x\sqrt{a^2 + b^2}}{|a|\sqrt{x^2 + b^2}} - \sin^{-1} \dfrac{x}{|a|}$

236. $\int f(x, \sqrt{a^2 - x^2}) \, dx = a \int f(a \sin u, a \cos u) \cos u \, du, \qquad \left(u = \sin^{-1} \dfrac{x}{a}, a > 0 \right)$

FORMS CONTAINING $\sqrt{a + bx + cx^2}$

$$X = a + bx + cx^2, q = 4ac - b^2, \text{ and } k = \dfrac{4c}{q}$$

If $q = 0$, then $\sqrt{X} = \sqrt{c} \left| x + \dfrac{b}{2c} \right|$

237. $\int \dfrac{dx}{\sqrt{X}} = \begin{cases} \dfrac{1}{\sqrt{c}} \log (2\sqrt{cX} + 2cx + b) \\ \text{or} \\ \dfrac{1}{\sqrt{c}} \sinh^{-1} \dfrac{2cx + b}{\sqrt{q}}, \qquad (c > 0) \end{cases}$

238. $\int \dfrac{dx}{\sqrt{X}} = -\dfrac{1}{\sqrt{-c}} \sin^{-1} \dfrac{2cx + b}{\sqrt{-q}}, \qquad (c < 0)$

239. $\int \dfrac{dx}{X\sqrt{X}} = \dfrac{2(2cx + b)}{q\sqrt{X}}$

240. $\int \dfrac{dx}{X^2\sqrt{X}} = \dfrac{2(2cx + b)}{3q\sqrt{X}} \left(\dfrac{1}{X} + 2k \right)$

241. $\int \dfrac{dx}{X^n\sqrt{X}} = \begin{cases} \dfrac{2(2cx + b)\sqrt{X}}{(2n - 1)qX^n} + \dfrac{2k(n - 1)}{2n - 1} \displaystyle\int \dfrac{dx}{X^{n-1}\sqrt{X}} \\ \text{or} \\ \dfrac{(2cx + b)(n!)(n - 1)!4^n k^{n-1}}{q[(2n)!]\sqrt{X}} \displaystyle\sum_{r=0}^{n-1} \dfrac{(2r)!}{(4kX)^r(r!)^2} \end{cases}$

INTEGRALS (Continued)

242. $\displaystyle \int \sqrt{X}\, dx = \frac{(2cx + b)\sqrt{X}}{4c} + \frac{1}{2k} \int \frac{dx}{\sqrt{X}}$

243. $\displaystyle \int X\sqrt{X}\, dx = \frac{(2cx + b)\sqrt{X}}{8c}\left(X + \frac{3}{2k}\right) + \frac{3}{8k^2}\int \frac{dx}{\sqrt{X}}$

244. $\displaystyle \int X^2\sqrt{X}\, dx = \frac{(2cx + b)\sqrt{X}}{12c}\left(X^2 + \frac{5X}{4k} + \frac{15}{8k^2}\right) + \frac{5}{16k^3}\int \frac{dx}{\sqrt{X}}$

245. $\displaystyle \int X^n\sqrt{X}\, dx = \begin{cases} \dfrac{(2cx + b)X^n\sqrt{X}}{4(n + 1)c} + \dfrac{2n + 1}{2(n + 1)k}\displaystyle\int X^{n-1}\sqrt{X}\, dx \\[2mm] \text{or} \\[2mm] \dfrac{(2n + 2)!}{[(n + 1)!]^2(4k)^{n+1}}\left[\dfrac{k(2cx + b)\sqrt{X}}{c}\displaystyle\sum_{r=0}^{n} \dfrac{r!(r + 1)!(4kX)^r}{(2r + 2)!} \right. \\[4mm] \qquad\qquad\qquad\qquad\qquad\qquad\qquad\qquad \left. + \displaystyle\int \dfrac{dx}{\sqrt{X}}\right] \end{cases}$

246. $\displaystyle \int \frac{x\, dx}{\sqrt{X}} = \frac{\sqrt{X}}{c} - \frac{b}{2c}\int \frac{dx}{\sqrt{X}}$

247. $\displaystyle \int \frac{x\, dx}{X\sqrt{X}} = -\frac{2(bx + 2a)}{q\sqrt{X}}$

248. $\displaystyle \int \frac{x\, dx}{X^n\sqrt{X}} = -\frac{\sqrt{X}}{(2n - 1)cX^n} - \frac{b}{2c}\int \frac{dx}{X^n\sqrt{X}}$

249. $\displaystyle \int \frac{x^2\, dx}{\sqrt{X}} = \left(\frac{x}{2c} - \frac{3b}{4c^2}\right)\sqrt{X} + \frac{3b^2 - 4ac}{8c^2}\int \frac{dx}{\sqrt{X}}$

250. $\displaystyle \int \frac{x^2\, dx}{X\sqrt{X}} = \frac{(2b^2 - 4ac)x + 2ab}{cq\sqrt{X}} + \frac{1}{c}\int \frac{dx}{\sqrt{X}}$

251. $\displaystyle \int \frac{x^2\, dx}{X^n\sqrt{X}} = \frac{(2b^2 - 4ac)x + 2ab}{(2n - 1)cqX^{n-1}\sqrt{X}} + \frac{4ac + (2n - 3)b^2}{(2n - 1)cq}\int \frac{dx}{X^{n-1}\sqrt{X}}$

252. $\displaystyle \int \frac{x^3\, dx}{\sqrt{X}} = \left(\frac{x^2}{3c} - \frac{5bx}{12c^2} + \frac{5b^2}{8c^3} - \frac{2a}{3c^2}\right)\sqrt{X} + \left(\frac{3ab}{4c^2} - \frac{5b^3}{16c^3}\right)\int \frac{dx}{\sqrt{X}}$

253. $\displaystyle \int \frac{x^n\, dx}{\sqrt{X}} = \frac{1}{nc}x^{n-1}\sqrt{X} - \frac{(2n - 1)b}{2nc}\int \frac{x^{n-1}\, dx}{\sqrt{X}} - \frac{(n - 1)a}{nc}\int \frac{x^{n-2}\, dx}{\sqrt{X}}$

254. $\displaystyle \int x\sqrt{X}\,dx = \frac{X\sqrt{X}}{3c} - \frac{b(2cx+b)}{8c^2}\sqrt{X} - \frac{b}{4ck}\int \frac{dx}{\sqrt{X}}$

255. $\displaystyle \int xX\sqrt{X}\,dx = \frac{X^2\sqrt{X}}{5c} - \frac{b}{2c}\int X\sqrt{X}\,dx$

256. $\displaystyle \int xX^n\sqrt{X}\,dx = \frac{X^{n+1}\sqrt{X}}{(2n+3)c} - \frac{b}{2c}\int X^n\sqrt{X}\,dx$

257. $\displaystyle \int x^2\sqrt{X}\,dx = \left(x - \frac{5b}{6c}\right)\frac{X\sqrt{X}}{4c} + \frac{5b^2-4ac}{16c^2}\int \sqrt{X}\,dx$

258. $\displaystyle \int \frac{dx}{x\sqrt{X}} = -\frac{1}{\sqrt{a}}\log\frac{2\sqrt{aX}+bx+2a}{x}, \qquad (a>0)$

259. $\displaystyle \int \frac{dx}{x\sqrt{X}} = \frac{1}{\sqrt{-a}}\sin^{-1}\left(\frac{bx+2a}{|x|\sqrt{-q}}\right), \qquad (a<0)$

260. $\displaystyle \int \frac{dx}{x\sqrt{X}} = -\frac{2\sqrt{X}}{bx}, \qquad (a=0)$

261. $\displaystyle \int \frac{dx}{x^2\sqrt{X}} = -\frac{\sqrt{X}}{ax} - \frac{b}{2a}\int \frac{dx}{x\sqrt{X}}$

262. $\displaystyle \int \frac{\sqrt{X}\,dx}{x} = \sqrt{X} + \frac{b}{2}\int \frac{dx}{\sqrt{X}} + a\int \frac{dx}{x\sqrt{X}}$

263. $\displaystyle \int \frac{\sqrt{X}\,dx}{x^2} = -\frac{\sqrt{X}}{x} + \frac{b}{2}\int \frac{dx}{x\sqrt{X}} + c\int \frac{dx}{\sqrt{X}}$

FORMS INVOLVING $\sqrt{2ax-x^2}$

264. $\displaystyle \int \sqrt{2ax-x^2}\,dx = \frac{1}{2}\left[(x-a)\sqrt{2ax-x^2} + a^2\sin^{-1}\frac{x-a}{|a|}\right]$

265. $\displaystyle \int \frac{dx}{\sqrt{2ax-x^2}} = \begin{cases} \cos^{-1}\dfrac{a-x}{|a|} \\ \quad\text{or} \\ \sin^{-1}\dfrac{x-a}{|a|} \end{cases}$

INTEGRALS (Continued)

266. $\displaystyle\int x^n \sqrt{2ax - x^2}\, dx = \begin{cases} -\dfrac{x^{n-1}(2ax - x^2)^{\frac{3}{2}}}{n + 2} + \dfrac{(2n + 1)a}{n + 2}\displaystyle\int x^{n-1}\sqrt{2ax - x^2}\, dx \\[2mm] \quad\text{or} \\[2mm] \sqrt{2ax - x^2}\left[\dfrac{x^{n+1}}{n + 2} - \displaystyle\sum_{r=0}^{n}\dfrac{(2n + 1)!(r!)^2 a^{n-r+1}}{2^{n-r}(2r + 1)!(n + 2)!n!}\, x^r\right] \\[2mm] \quad\quad\quad\quad + \dfrac{(2n + 1)!a^{n+2}}{2^n n!(n + 2)!}\sin^{-1}\dfrac{x - a}{|a|} \end{cases}$

267. $\displaystyle\int \dfrac{\sqrt{2ax - x^2}}{x^n}\, dx = \dfrac{(2ax - x^2)^{\frac{3}{2}}}{(3 - 2n)ax^n} + \dfrac{n - 3}{(2n - 3)a}\int \dfrac{\sqrt{2ax - x^2}}{x^{n-1}}\, dx$

268. $\displaystyle\int \dfrac{x^n\, dx}{\sqrt{2ax - x^2}} = \begin{cases} \dfrac{-x^{n-1}\sqrt{2ax - x^2}}{n} + \dfrac{a(2n - 1)}{n}\displaystyle\int \dfrac{x^{n-1}}{\sqrt{2ax - x^2}}\, dx \\[2mm] \quad\text{or} \\[2mm] -\sqrt{2ax - x^2}\displaystyle\sum_{r=1}^{n}\dfrac{(2n)!r!(r - 1)!a^{n-r}}{2^{n-r}(2r)!(n!)^2}\, x^{r-1} \\[2mm] \quad\quad\quad\quad + \dfrac{(2n)!a^n}{2^n(n!)^2}\sin^{-1}\dfrac{x - a}{|a|} \end{cases}$

269. $\displaystyle\int \dfrac{dx}{x^n\sqrt{2ax - x^2}} = \begin{cases} \dfrac{\sqrt{2ax - x^2}}{a(1 - 2n)x^n} + \dfrac{n - 1}{(2n - 1)a}\displaystyle\int \dfrac{dx}{x^{n-1}\sqrt{2ax - x^2}} \\[2mm] \quad\text{or} \\[2mm] -\sqrt{2ax - x^2}\displaystyle\sum_{r=0}^{n-1}\dfrac{2^{n-r}(n - 1)!n!(2r)!}{(2n)!(r!)^2 a^{n-r}x^{r+1}} \end{cases}$

270. $\displaystyle\int \dfrac{dx}{(2ax - x^2)^{\frac{3}{2}}} = \dfrac{x - a}{a^2\sqrt{2ax - x^2}}$

271. $\displaystyle\int \dfrac{x\, dx}{(2ax - x^2)^{\frac{3}{2}}} = \dfrac{x}{a\sqrt{2ax - x^2}}$

MISCELLANEOUS ALGEBRAIC FORMS

272. $\displaystyle\int \dfrac{dx}{\sqrt{2ax + x^2}} = \log\left(x + a + \sqrt{2ax + x^2}\right)$

273. $\displaystyle\int \sqrt{ax^2 + c}\, dx = \dfrac{x}{2}\sqrt{ax^2 + c} + \dfrac{c}{2\sqrt{a}}\log\left(x\sqrt{a} + \sqrt{ax^2 + c}\right), \quad (a > 0)$

274. $\displaystyle\int \sqrt{ax^2 + c}\, dx = \dfrac{x}{2}\sqrt{ax^2 + c} + \dfrac{c}{2\sqrt{-a}}\sin^{-1}\left(x\sqrt{-\dfrac{a}{c}}\right), \quad (a < 0)$

INTEGRALS (Continued)

275. $\displaystyle\int \sqrt{\frac{1+x}{1-x}}\,dx = \sin^{-1} x - \sqrt{1-x^2}$

276. $\displaystyle\int \frac{dx}{x\sqrt{ax^n + c}} = \begin{cases} \dfrac{1}{n\sqrt{c}} \log \dfrac{\sqrt{ax^n + c} - \sqrt{c}}{\sqrt{ax^n + c} + \sqrt{c}} \\[2mm] \qquad\qquad \text{or} \\[2mm] \dfrac{2}{n\sqrt{c}} \log \dfrac{\sqrt{ax^n + c} - \sqrt{c}}{\sqrt{x^n}}, \qquad (c > 0) \end{cases}$

277. $\displaystyle\int \frac{dx}{x\sqrt{ax^n + c}} = \frac{2}{n\sqrt{-c}} \sec^{-1}\sqrt{-\frac{ax^n}{c}}, \qquad (c < 0)$

278. $\displaystyle\int \frac{dx}{\sqrt{ax^2 + c}} = \frac{1}{\sqrt{a}} \log\left(x\sqrt{a} + \sqrt{ax^2 + c}\right), \qquad (a > 0)$

279. $\displaystyle\int \frac{dx}{\sqrt{ax^2 + c}} = \frac{1}{\sqrt{-a}} \sin^{-1}\left(x\sqrt{-\frac{a}{c}}\right), \qquad (a < 0)$

280. $\displaystyle\int (ax^2 + c)^{m+\frac{1}{2}}\,dx = \begin{cases} \dfrac{x(ax^2 + c)^{m+\frac{1}{2}}}{2(m+1)} + \dfrac{(2m+1)c}{2(m+1)} \displaystyle\int (ax^2 + c)^{m-\frac{1}{2}}\,dx \\[2mm] \qquad\qquad \text{or} \\[2mm] x\sqrt{ax^2 + c}\displaystyle\sum_{r=0}^{m} \dfrac{(2m+1)!(r!)^2 c^{m-r}}{2^{2m-2r+1}m!(m+1)!(2r+1)!}(ax^2 + c)^r \\[2mm] \qquad + \dfrac{(2m+1)!c^{m+1}}{2^{2m+1}m!(m+1)!}\displaystyle\int \dfrac{dx}{\sqrt{ax^2 + c}} \end{cases}$

281. $\displaystyle\int x(ax^2 + c)^{m+\frac{1}{2}}\,dx = \frac{(ax^2 + c)^{m+\frac{3}{2}}}{(2m+3)a}$

282. $\displaystyle\int \frac{(ax^2 + c)^{m+\frac{1}{2}}}{x}\,dx = \begin{cases} \dfrac{(ax^2 + c)^{m+\frac{1}{2}}}{2m+1} + c\displaystyle\int \dfrac{(ax^2 + c)^{m-\frac{1}{2}}}{x}\,dx \\[2mm] \qquad\qquad \text{or} \\[2mm] \sqrt{ax^2 + c}\displaystyle\sum_{r=0}^{m} \dfrac{c^{m-r}(ax^2 + c)^r}{2r+1} + c^{m+1}\displaystyle\int \dfrac{dx}{x\sqrt{ax^2 + c}} \end{cases}$

283. $\displaystyle\int \frac{dx}{(ax^2 + c)^{m+\frac{1}{2}}} = \begin{cases} \dfrac{x}{(2m-1)c(ax^2 + c)^{m-\frac{1}{2}}} + \dfrac{2m-2}{(2m-1)c}\displaystyle\int \dfrac{dx}{(ax^2 + c)^{m-\frac{1}{2}}} \\[2mm] \qquad\qquad \text{or} \\[2mm] \dfrac{x}{\sqrt{ax^2 + c}}\displaystyle\sum_{r=0}^{m-1} \dfrac{2^{2m-2r-1}(m-1)!m!(2r)!}{(2m)!(r!)^2 c^{m-r}(ax^2 + c)^r} \end{cases}$

INTEGRALS (Continued)

284. $\displaystyle\int \frac{dx}{x^m\sqrt{ax^2+c}} = -\frac{\sqrt{ax^2+c}}{(m-1)cx^{m-1}} - \frac{(m-2)a}{(m-1)c}\int \frac{dx}{x^{m-2}\sqrt{ax^2+c}}$

285. $\displaystyle\int \frac{1+x^2}{(1-x^2)\sqrt{1+x^4}}\,dx = \frac{1}{\sqrt{2}}\log\frac{x\sqrt{2}+\sqrt{1+x^4}}{1-x^2}$

286. $\displaystyle\int \frac{1-x^2}{(1+x^2)\sqrt{1+x^4}}\,dx = \frac{1}{\sqrt{2}}\tan^{-1}\frac{x\sqrt{2}}{\sqrt{1+x^4}}$

287. $\displaystyle\int \frac{dx}{x\sqrt{x^n+a^2}} = -\frac{2}{na}\log\frac{a+\sqrt{x^n+a^2}}{\sqrt{x^n}}$

288. $\displaystyle\int \frac{dx}{x\sqrt{x^n-a^2}} = -\frac{2}{na}\sin^{-1}\frac{a}{\sqrt{x^n}}$

289. $\displaystyle\int \sqrt{\frac{x}{a^3-x^3}}\,dx = \frac{2}{3}\sin^{-1}\left(\frac{x}{a}\right)^{\frac{3}{2}}$

FORMS INVOLVING TRIGONOMETRIC FUNCTIONS

290. $\displaystyle\int (\sin ax)\,dx = -\frac{1}{a}\cos ax$

291. $\displaystyle\int (\cos ax)\,dx = \frac{1}{a}\sin ax$

292. $\displaystyle\int (\tan ax)\,dx = -\frac{1}{a}\log\cos ax = \frac{1}{a}\log\sec ax$

293. $\displaystyle\int (\cot ax)\,dx = \frac{1}{a}\log\sin ax = -\frac{1}{a}\log\csc ax$

294. $\displaystyle\int (\sec ax)\,dx = \frac{1}{a}\log(\sec ax + \tan ax) = \frac{1}{a}\log\tan\left(\frac{\pi}{4}+\frac{ax}{2}\right)$

295. $\displaystyle\int (\csc ax)\,dx = \frac{1}{a}\log(\csc ax - \cot ax) = \frac{1}{a}\log\tan\frac{ax}{2}$

296. $\displaystyle\int (\sin^2 ax)\,dx = -\frac{1}{2a}\cos ax\sin ax + \frac{1}{2}x = \frac{1}{2}x - \frac{1}{4a}\sin 2ax$

297. $\displaystyle\int (\sin^3 ax)\,dx = -\frac{1}{3a}(\cos ax)(\sin^2 ax + 2)$

298. $\displaystyle\int (\sin^4 ax)\,dx = \frac{3x}{8} - \frac{\sin 2ax}{4a} + \frac{\sin 4ax}{32a}$

299. $\displaystyle\int (\sin^n ax)\,dx = -\frac{\sin^{n-1} ax\cos ax}{na} + \frac{n-1}{n}\int(\sin^{n-2} ax)\,dx$

INTEGRALS (Continued)

300. $\displaystyle \int (\sin^{2m} ax)\, dx = -\frac{\cos ax}{a} \sum_{r=0}^{m-1} \frac{(2m)!(r!)^2}{2^{2m-2r}(2r+1)!(m!)^2} \sin^{2r+1} ax + \frac{(2m)!}{2^{2m}(m!)^2} x$

301. $\displaystyle \int (\sin^{2m+1} ax)\, dx = -\frac{\cos ax}{a} \sum_{r=0}^{m} \frac{2^{2m-2r}(m!)^2(2r)!}{(2m+1)!(r!)^2} \sin^{2r} ax$

302. $\displaystyle \int (\cos^2 ax)\, dx = \frac{1}{2a}\sin ax \cos ax + \frac{1}{2}x = \frac{1}{2}x + \frac{1}{4a}\sin 2ax$

303. $\displaystyle \int (\cos^3 ax)\, dx = \frac{1}{3a}(\sin ax)(\cos^2 ax + 2)$

304. $\displaystyle \int (\cos^4 ax)\, dx = \frac{3x}{8} + \frac{\sin 2ax}{4a} + \frac{\sin 4ax}{32a}$

305. $\displaystyle \int (\cos^n ax)\, dx = \frac{1}{na}\cos^{n-1} ax \sin ax + \frac{n-1}{n}\int (\cos^{n-2} ax)\, dx$

306. $\displaystyle \int (\cos^{2m} ax)\, dx = \frac{\sin ax}{a} \sum_{r=0}^{m-1} \frac{(2m)!(r!)^2}{2^{2m-2r}(2r+1)!(m!)^2} \cos^{2r+1} ax + \frac{(2m)!}{2^{2m}(m!)^2} x$

307. $\displaystyle \int (\cos^{2m+1} ax)\, dx = \frac{\sin ax}{a} \sum_{r=0}^{m} \frac{2^{2m-2r}(m!)^2(2r)!}{(2m+1)!(r!)^2} \cos^{2r} ax$

308. $\displaystyle \int \frac{dx}{\sin^2 ax} = \int (\csc^2 ax)\, dx = -\frac{1}{a}\cot ax$

309. $\displaystyle \int \frac{dx}{\sin^m ax} = \int (\csc^m ax)\, dx = -\frac{1}{(m-1)a}\cdot\frac{\cos ax}{\sin^{m-1} ax} + \frac{m-2}{m-1}\int \frac{dx}{\sin^{m-2} ax}$

310. $\displaystyle \int \frac{dx}{\sin^{2m} ax} = \int (\csc^{2m} ax)\, dx = -\frac{1}{a}\cos ax \sum_{r=0}^{m-1} \frac{2^{2m-2r-1}(m-1)!m!(2r)!}{(2m)!(r!)^2 \sin^{2r+1} ax}$

311. $\displaystyle \int \frac{dx}{\sin^{2m+1} ax} = \int (\csc^{2m+1} ax)\, dx =$

$$-\frac{1}{a}\cos ax \sum_{r=0}^{m-1} \frac{(2m)!(r!)^2}{2^{2m-2r}(m!)^2(2r+1)!\sin^{2r+2} ax} + \frac{1}{a}\cdot\frac{(2m)!}{2^{2m}(m!)^2}\log\tan\frac{ax}{2}$$

312. $\displaystyle \int \frac{dx}{\cos^2 ax} = \int (\sec^2 ax)\, dx = \frac{1}{a}\tan ax$

313. $\displaystyle \int \frac{dx}{\cos^n ax} = \int (\sec^n ax)\, dx = \frac{1}{(n-1)a}\cdot\frac{\sin ax}{\cos^{n-1} ax} + \frac{n-2}{n-1}\int \frac{dx}{\cos^{n-2} ax}$

314. $\displaystyle \int \frac{dx}{\cos^{2m} ax} = \int (\sec^{2m} ax)\, dx = \frac{1}{a}\sin ax \sum_{r=0}^{m-1} \frac{2^{2m-2r-1}(m-1)!m!(2r)!}{(2m)!(r!)^2 \cos^{2r+1} ax}$

INTEGRALS (Continued)

315. $\displaystyle\int \frac{dx}{\cos^{2m+1} ax} = \int (\sec^{2m+1} ax)\, dx =$

$$\frac{1}{a}\sin ax \sum_{r=0}^{m-1} \frac{(2m)!(r!)^2}{2^{2m-2r}(m!)^2(2r+1)!\cos^{2r+2} ax}$$

$$+ \frac{1}{a}\cdot\frac{(2m)!}{2^{2m}(m!)^2}\log(\sec ax + \tan ax)$$

316. $\displaystyle\int (\sin mx)(\sin nx)\, dx = \frac{\sin(m-n)x}{2(m-n)} - \frac{\sin(m+n)x}{2(m+n)}, \qquad (m^2 \neq n^2)$

317. $\displaystyle\int (\cos mx)(\cos nx)\, dx = \frac{\sin(m-n)x}{2(m-n)} + \frac{\sin(m+n)x}{2(m+n)}, \qquad (m^2 \neq n^2)$

318. $\displaystyle\int (\sin ax)(\cos ax)\, dx = \frac{1}{2a}\sin^2 ax$

319. $\displaystyle\int (\sin mx)(\cos nx)\, dx = -\frac{\cos(m-n)x}{2(m-n)} - \frac{\cos(m+n)x}{2(m+n)}, \qquad (m^2 \neq n^2)$

320. $\displaystyle\int (\sin^2 ax)(\cos^2 ax)\, dx = -\frac{1}{32a}\sin 4ax + \frac{x}{8}$

321. $\displaystyle\int (\sin ax)(\cos^m ax)\, dx = -\frac{\cos^{m+1} ax}{(m+1)a}$

322. $\displaystyle\int (\sin^m ax)(\cos ax)\, dx = \frac{\sin^{m+1} ax}{(m+1)a}$

323. $\displaystyle\int (\cos^m ax)(\sin^n ax)\, dx = \begin{cases} \dfrac{\cos^{m-1} ax \sin^{n+1} ax}{(m+n)a} \\[2ex] \qquad + \dfrac{m-1}{m+n}\displaystyle\int (\cos^{m-2} ax)(\sin^n ax)\, dx \\[2ex] \text{or} \\[2ex] -\dfrac{\sin^{n-1} ax \cos^{m+1} ax}{(m+n)a} \\[2ex] \qquad + \dfrac{n-1}{m+n}\displaystyle\int (\cos^m ax)(\sin^{n-2} ax)\, dx \end{cases}$

324. $\displaystyle\int \frac{\cos^m ax}{\sin^n ax}\, dx = \begin{cases} -\dfrac{\cos^{m+1} ax}{(n-1)a\sin^{n-1} ax} - \dfrac{m-n+2}{n-1}\displaystyle\int \dfrac{\cos^m ax}{\sin^{n-2} ax}\, dx \\[2ex] \text{or} \\[2ex] \dfrac{\cos^{m-1} ax}{a(m-n)\sin^{n-1} ax} + \dfrac{m-1}{m-n}\displaystyle\int \dfrac{\cos^{m-2} ax}{\sin^n ax}\, dx \end{cases}$

INTEGRALS (Continued)

325. $\displaystyle\int \frac{\sin^m ax}{\cos^n ax}\,dx = \begin{cases} \dfrac{\sin^{m+1} ax}{a(n-1)\cos^{n-1} ax} + \dfrac{m-n+2}{n-1}\displaystyle\int \dfrac{\sin^m ax}{\cos^{n-2} ax}\,dx \\[2ex] \text{or} \\[1ex] -\dfrac{\sin^{m-1} ax}{a(m-n)\cos^{n-1} ax} - \dfrac{m-1}{m-n}\displaystyle\int \dfrac{\sin^{m-2} ax}{\cos^n ax}\,dx \end{cases}$

326. $\displaystyle\int \frac{\sin ax}{\cos^2 ax}\,dx = \frac{1}{a\cos ax} = \frac{\sec ax}{a}$

327. $\displaystyle\int \frac{\sin^2 ax}{\cos ax}\,dx = -\frac{1}{a}\sin ax + \frac{1}{a}\log \tan\left(\frac{\pi}{4} + \frac{ax}{2}\right)$

328. $\displaystyle\int \frac{\cos ax}{\sin^2 ax}\,dx = -\frac{1}{a\sin ax} = -\frac{\csc ax}{a}$

329. $\displaystyle\int \frac{dx}{(\sin ax)(\cos ax)} = \frac{1}{a}\log \tan ax$

330. $\displaystyle\int \frac{dx}{(\sin ax)(\cos^2 ax)} = \frac{1}{a}\left(\sec ax + \log \tan \frac{ax}{2}\right)$

331. $\displaystyle\int \frac{dx}{(\sin ax)(\cos^n ax)} = \frac{1}{a(n-1)\cos^{n-1} ax} + \int \frac{dx}{(\sin ax)(\cos^{n-2} ax)}$

332. $\displaystyle\int \frac{dx}{(\sin^2 ax)(\cos ax)} = -\frac{1}{a}\csc ax + \frac{1}{a}\log \tan\left(\frac{\pi}{4} + \frac{ax}{2}\right)$

333. $\displaystyle\int \frac{dx}{(\sin^2 ax)(\cos^2 ax)} = -\frac{2}{a}\cot 2ax$

334. $\displaystyle\int \frac{dx}{\sin^m ax \cos^n ax} = \begin{cases} -\dfrac{1}{a(m-1)(\sin^{m-1} ax)(\cos^{n-1} ax)} \\[2ex] \qquad + \dfrac{m+n-2}{m-1}\displaystyle\int \dfrac{dx}{(\sin^{m-2} ax)(\cos^n ax)} \\[2ex] \text{or} \\[1ex] \dfrac{1}{a(n-1)\sin^{m-1} ax \cos^{n-1} ax} \\[2ex] \qquad - \dfrac{m+n-2}{n-1}\displaystyle\int \dfrac{dx}{\sin^m ax \cos^{n-2} ax} \end{cases}$

335. $\displaystyle\int \sin(a+bx)\,dx = -\frac{1}{b}\cos(a+bx)$

336. $\displaystyle\int \cos(a+bx)\,dx = \frac{1}{b}\sin(a+bx)$

337. $\displaystyle\int \frac{dx}{1 \pm \sin ax} = \mp\frac{1}{a}\tan\left(\frac{\pi}{4} \mp \frac{ax}{2}\right)$

INTEGRALS (Continued)

338. $\displaystyle\int \frac{dx}{1 + \cos ax} = \frac{1}{a} \tan \frac{ax}{2}$

339. $\displaystyle\int \frac{dx}{1 - \cos ax} = -\frac{1}{a} \cot \frac{ax}{2}$

*340. $\displaystyle\int \frac{dx}{a + b \sin x} = \begin{cases} \dfrac{2}{\sqrt{a^2 - b^2}} \tan^{-1} \dfrac{a \tan \dfrac{x}{2} + b}{\sqrt{a^2 - b^2}} \\[3ex] \text{or} \\[2ex] \dfrac{1}{\sqrt{b^2 - a^2}} \log \dfrac{a \tan \dfrac{x}{2} + b - \sqrt{b^2 - a^2}}{a \tan \dfrac{x}{2} + b + \sqrt{b^2 - a^2}} \end{cases}$

*341. $\displaystyle\int \frac{dx}{a + b \cos x} = \begin{cases} \dfrac{2}{\sqrt{a^2 - b^2}} \tan^{-1} \dfrac{\sqrt{a^2 - b^2} \tan \dfrac{x}{2}}{a + b} \\[3ex] \text{or} \\[2ex] \dfrac{1}{\sqrt{b^2 - a^2}} \log \left(\dfrac{\sqrt{b^2 - a^2} \tan \dfrac{x}{2} + a + b}{\sqrt{b^2 - a^2} \tan \dfrac{x}{2} - a - b} \right) \end{cases}$

*342. $\displaystyle\int \frac{dx}{a + b \sin x + c \cos x}$

$$= \begin{cases} \dfrac{1}{\sqrt{b^2 + c^2 - a^2}} \log \dfrac{b - \sqrt{b^2 + c^2 - a^2} + (a - c) \tan \dfrac{x}{2}}{b + \sqrt{b^2 + c^2 - a^2} + (a - c) \tan \dfrac{x}{2}}, & \text{if } a^2 < b^2 + c^2, a \neq c \\[3ex] \text{or} \\[2ex] \dfrac{2}{\sqrt{a^2 - b^2 - c^2}} \tan^{-1} \dfrac{b + (a - c) \tan \dfrac{x}{2}}{\sqrt{a^2 - b^2 - c^2}}, & \text{if } a^2 > b^2 + c^2 \\[3ex] \text{or} \\[2ex] \dfrac{1}{a} \left[\dfrac{a - (b + c) \cos x - (b - c) \sin x}{a - (b - c) \cos x + (b + c) \sin x} \right], & \text{if } a^2 = b^2 + c^2, a \neq c. \end{cases}$$

*343. $\displaystyle\int \frac{\sin^2 x \, dx}{a + b \cos^2 x} = \frac{1}{b} \sqrt{\frac{a + b}{a}} \tan^{-1} \left(\sqrt{\frac{a}{a + b}} \tan x \right) - \frac{x}{b}, \qquad (ab > 0, \text{ or } |a| > |b|)$

* See note p. 403

274　　　　　　　　　　　　　　*Calculus*

INTEGRALS (Continued)

*344. $\displaystyle\int \frac{dx}{a^2 \cos^2 x + b^2 \sin^2 x} = \frac{1}{ab}\tan^{-1}\left(\frac{b\tan x}{a}\right)$

*345. $\displaystyle\int \frac{\cos^2 cx}{a^2 + b^2 \sin^2 cx}\,dx = \frac{\sqrt{a^2 + b^2}}{ab^2 c}\tan^{-1}\frac{\sqrt{a^2 + b^2}\,\tan cx}{a} - \frac{x}{b^2}$

346. $\displaystyle\int \frac{\sin cx \cos cx}{a\cos^2 cx + b\sin^2 cx}\,dx = \frac{1}{2c(b - a)}\log(a\cos^2 cx + b\sin^2 cx)$

347. $\displaystyle\int \frac{\cos cx}{a\cos cx + b\sin cx}\,dx = \int \frac{dx}{a + b\tan cx} =$

$$\frac{1}{c(a^2 + b^2)}[acx + b\log(a\cos cx + b\sin cx)]$$

348. $\displaystyle\int \frac{\sin cx}{a\sin cx + b\cos cx}\,dx = \int \frac{dx}{a + b\cot cx} =$

$$\frac{1}{c(a^2 + b^2)}[acx - b\log(a\sin cx + b\cos cx)]$$

*349. $\displaystyle\int \frac{dx}{a\cos^2 x + 2b\cos x \sin x + c\sin^2 x} =$
$$\begin{cases} \dfrac{1}{2\sqrt{b^2 - ac}}\log\dfrac{c\tan x + b - \sqrt{b^2 - ac}}{c\tan x + b + \sqrt{b^2 - ac}}, \\ \qquad\qquad\qquad\qquad (b^2 > ac) \\[2pt] \text{or} \\ \dfrac{1}{\sqrt{ac - b^2}}\tan^{-1}\dfrac{c\tan x + b}{\sqrt{ac - b^2}}, \quad (b^2 < ac) \\[2pt] \text{or} \\ -\dfrac{1}{c\tan x + b}, \qquad (b^2 = ac) \end{cases}$$

350. $\displaystyle\int \frac{\sin ax}{1 \pm \sin ax}\,dx = \pm x + \frac{1}{a}\tan\left(\frac{\pi}{4} \mp \frac{ax}{2}\right)$

351. $\displaystyle\int \frac{dx}{(\sin ax)(1 \pm \sin ax)} = \frac{1}{a}\tan\left(\frac{\pi}{4} \mp \frac{ax}{2}\right) + \frac{1}{a}\log\tan\frac{ax}{2}$

352. $\displaystyle\int \frac{dx}{(1 + \sin ax)^2} = -\frac{1}{2a}\tan\left(\frac{\pi}{4} - \frac{ax}{2}\right) - \frac{1}{6a}\tan^3\left(\frac{\pi}{4} - \frac{ax}{2}\right)$

353. $\displaystyle\int \frac{dx}{(1 - \sin ax)^2} = \frac{1}{2a}\cot\left(\frac{\pi}{4} - \frac{ax}{2}\right) + \frac{1}{6a}\cot^3\left(\frac{\pi}{4} - \frac{ax}{2}\right)$

354. $\displaystyle\int \frac{\sin ax}{(1 + \sin ax)^2}\,dx = -\frac{1}{2a}\tan\left(\frac{\pi}{4} - \frac{ax}{2}\right) + \frac{1}{6a}\tan^3\left(\frac{\pi}{4} - \frac{ax}{2}\right)$

* See note p. 403.

INTEGRALS (Continued)

355. $\displaystyle\int \frac{\sin ax}{(1 - \sin ax)^2}\,dx = -\frac{1}{2a}\cot\left(\frac{\pi}{4} - \frac{ax}{2}\right) + \frac{1}{6a}\cot^3\left(\frac{\pi}{4} - \frac{ax}{2}\right)$

356. $\displaystyle\int \frac{\sin x\,dx}{a + b\sin x} = \frac{x}{b} - \frac{a}{b}\int \frac{dx}{a + b\sin x}$

357. $\displaystyle\int \frac{dx}{(\sin x)(a + b\sin x)} = \frac{1}{a}\log\tan\frac{x}{2} - \frac{b}{a}\int \frac{dx}{a + b\sin x}$

358. $\displaystyle\int \frac{dx}{(a + b\sin x)^2} = \frac{b\cos x}{(a^2 - b^2)(a + b\sin x)} + \frac{a}{a^2 - b^2}\int \frac{dx}{a + b\sin x}$

359. $\displaystyle\int \frac{\sin x\,dx}{(a + b\sin x)^2} = \frac{a\cos x}{(b^2 - a^2)(a + b\sin x)} + \frac{b}{b^2 - a^2}\int \frac{dx}{a + b\sin x}$

*360. $\displaystyle\int \frac{dx}{a^2 + b^2\sin^2 cx} = \frac{1}{ac\sqrt{a^2 + b^2}}\tan^{-1}\frac{\sqrt{a^2 + b^2}\,\tan cx}{a}$

*361. $\displaystyle\int \frac{dx}{a^2 - b^2\sin^2 cx} = \begin{cases} \dfrac{1}{ac\sqrt{a^2 - b^2}}\tan^{-1}\dfrac{\sqrt{a^2 - b^2}\,\tan cx}{a}, & (a^2 > b^2) \\[2mm] \text{or} \\[2mm] \dfrac{1}{2ac\sqrt{b^2 - a^2}}\log\dfrac{\sqrt{b^2 - a^2}\,\tan cx + a}{\sqrt{b^2 - a^2}\,\tan cx - a}, & (a^2 < b^2) \end{cases}$

362. $\displaystyle\int \frac{\cos ax}{1 + \cos ax}\,dx = x - \frac{1}{a}\tan\frac{ax}{2}$

363. $\displaystyle\int \frac{\cos ax}{1 - \cos ax}\,dx = -x - \frac{1}{a}\cot\frac{ax}{2}$

364. $\displaystyle\int \frac{dx}{(\cos ax)(1 + \cos ax)} = \frac{1}{a}\log\tan\left(\frac{\pi}{4} + \frac{ax}{2}\right) - \frac{1}{a}\tan\frac{ax}{2}$

365. $\displaystyle\int \frac{dx}{(\cos ax)(1 - \cos ax)} = \frac{1}{a}\log\tan\left(\frac{\pi}{4} + \frac{ax}{2}\right) - \frac{1}{a}\cot\frac{ax}{2}$

366. $\displaystyle\int \frac{dx}{(1 + \cos ax)^2} = \frac{1}{2a}\tan\frac{ax}{2} + \frac{1}{6a}\tan^3\frac{ax}{2}$

367. $\displaystyle\int \frac{dx}{(1 - \cos ax)^2} = -\frac{1}{2a}\cot\frac{ax}{2} - \frac{1}{6a}\cot^3\frac{ax}{2}$

368. $\displaystyle\int \frac{\cos ax}{(1 + \cos ax)^2}\,dx = \frac{1}{2a}\tan\frac{ax}{2} - \frac{1}{6a}\tan^3\frac{ax}{2}$

369. $\displaystyle\int \frac{\cos ax}{(1 - \cos ax)^2}\,dx = \frac{1}{2a}\cot\frac{ax}{2} - \frac{1}{6a}\cot^3\frac{ax}{2}$

* See note p. 403.

INTEGRALS (Continued)

370. $\displaystyle\int \frac{\cos x\,dx}{a + b\cos x} = \frac{x}{b} - \frac{a}{b}\int \frac{dx}{a + b\cos x}$

371. $\displaystyle\int \frac{dx}{(\cos x)(a + b\cos x)} = \frac{1}{a}\log\tan\left(\frac{x}{2} + \frac{\pi}{4}\right) - \frac{b}{a}\int \frac{dx}{a + b\cos x}$

372. $\displaystyle\int \frac{dx}{(a + b\cos x)^2} = \frac{b\sin x}{(b^2 - a^2)(a + b\cos x)} - \frac{a}{b^2 - a^2}\int \frac{dx}{a + b\cos x}$

373. $\displaystyle\int \frac{\cos x}{(a + b\cos x)^2}\,dx = \frac{a\sin x}{(a^2 - b^2)(a + b\cos x)} - \frac{b}{a^2 - b^2}\int \frac{dx}{a + b\cos x}$

*374. $\displaystyle\int \frac{dx}{a^2 + b^2 - 2ab\cos cx} = \frac{2}{c(a^2 - b^2)}\tan^{-1}\left(\frac{a + b}{a - b}\tan\frac{cx}{2}\right)$

*375. $\displaystyle\int \frac{dx}{a^2 + b^2\cos^2 cx} = \frac{1}{ac\sqrt{a^2 + b^2}}\tan^{-1}\frac{a\tan cx}{\sqrt{a^2 + b^2}}$

*376. $\displaystyle\int \frac{dx}{a^2 - b^2\cos^2 cx} = \begin{cases} \dfrac{1}{ac\sqrt{a^2 - b^2}}\tan^{-1}\dfrac{a\tan cx}{\sqrt{a^2 - b^2}}, & (a^2 > b^2) \\[2ex] \text{or} \\[2ex] \dfrac{1}{2ac\sqrt{b^2 - a^2}}\log\dfrac{a\tan cx - \sqrt{b^2 - a^2}}{a\tan cx + \sqrt{b^2 - a^2}}, & (b^2 > a^2) \end{cases}$

377. $\displaystyle\int \frac{\sin ax}{1 \pm \cos ax}\,dx = \mp\frac{1}{a}\log(1 \pm \cos ax)$

378. $\displaystyle\int \frac{\cos ax}{1 \pm \sin ax}\,dx = \pm\frac{1}{a}\log(1 \pm \sin ax)$

379. $\displaystyle\int \frac{dx}{(\sin ax)(1 \pm \cos ax)} = \pm\frac{1}{2a(1 \pm \cos ax)} + \frac{1}{2a}\log\tan\frac{ax}{2}$

380. $\displaystyle\int \frac{dx}{(\cos ax)(1 \pm \sin ax)} = \mp\frac{1}{2a(1 \pm \sin ax)} + \frac{1}{2a}\log\tan\left(\frac{\pi}{4} + \frac{ax}{2}\right)$

381. $\displaystyle\int \frac{\sin ax}{(\cos ax)(1 \pm \cos ax)}\,dx = \frac{1}{a}\log(\sec ax \pm 1)$

382. $\displaystyle\int \frac{\cos ax}{(\sin ax)(1 \pm \sin ax)}\,dx = -\frac{1}{a}\log(\csc ax \pm 1)$

383. $\displaystyle\int \frac{\sin ax}{(\cos ax)(1 \pm \sin ax)}\,dx = \frac{1}{2a(1 \pm \sin ax)} \pm \frac{1}{2a}\log\tan\left(\frac{\pi}{4} + \frac{ax}{2}\right)$

384. $\displaystyle\int \frac{\cos ax}{(\sin ax)(1 \pm \cos ax)}\,dx = -\frac{1}{2a(1 \pm \cos ax)} \pm \frac{1}{2a}\log\tan\frac{ax}{2}$

* See note p. 403.

INTEGRALS (Continued)

385. $\displaystyle\int \frac{dx}{\sin ax \pm \cos ax} = \frac{1}{a\sqrt{2}} \log \tan\left(\frac{ax}{2} \pm \frac{\pi}{8}\right)$

386. $\displaystyle\int \frac{dx}{(\sin ax \pm \cos ax)^2} = \frac{1}{2a} \tan\left(ax \mp \frac{\pi}{4}\right)$

387. $\displaystyle\int \frac{dx}{1 + \cos ax \pm \sin ax} = \pm\frac{1}{a} \log\left(1 \pm \tan\frac{ax}{2}\right)$

388. $\displaystyle\int \frac{dx}{a^2 \cos^2 cx - b^2 \sin^2 cx} = \frac{1}{2abc} \log \frac{b \tan cx + a}{b \tan cx - a}$

389. $\displaystyle\int x(\sin ax)\,dx = \frac{1}{a^2} \sin ax - \frac{x}{a} \cos ax$

390. $\displaystyle\int x^2(\sin ax)\,dx = \frac{2x}{a^2} \sin ax - \frac{a^2 x^2 - 2}{a^3} \cos ax$

391. $\displaystyle\int x^3(\sin ax)\,dx = \frac{3a^2 x^2 - 6}{a^4} \sin ax - \frac{a^2 x^3 - 6x}{a^3} \cos ax$

392. $\displaystyle\int x^m \sin ax\,dx = \begin{cases} -\dfrac{1}{a}x^m \cos ax + \dfrac{m}{a}\displaystyle\int x^{m-1} \cos ax\,dx \\[2mm] \qquad\qquad \text{or} \\[2mm] \cos ax \displaystyle\sum_{r=0}^{\left[\frac{m}{2}\right]} (-1)^{r+1} \dfrac{m!}{(m-2r)!} \cdot \dfrac{x^{m-2r}}{a^{2r+1}} \\[4mm] + \sin ax \displaystyle\sum_{r=0}^{\left[\frac{m-1}{2}\right]} (-1)^{r} \dfrac{m!}{(m-2r-1)!} \cdot \dfrac{x^{m-2r-1}}{a^{2r+2}} \end{cases}$

Note: $[s]$ means greatest integer $\le s$; $[3\frac{1}{2}] = 3$, $[\frac{1}{2}] = 0$, etc.

393. $\displaystyle\int x(\cos ax)\,dx = \frac{1}{a^2} \cos ax + \frac{x}{a} \sin ax$

394. $\displaystyle\int x^2(\cos ax)\,dx = \frac{2x \cos ax}{a^2} + \frac{a^2 x^2 - 2}{a^3} \sin ax$

395. $\displaystyle\int x^3(\cos ax)\,dx = \frac{3a^2 x^2 - 6}{a^4} \cos ax + \frac{a^2 x^3 - 6x}{a^3} \sin ax$

396. $\displaystyle\int x^m(\cos ax)\,dx = \begin{cases} \dfrac{x^m \sin ax}{a} - \dfrac{m}{a}\displaystyle\int x^{m-1} \sin ax\,dx \\[2mm] \qquad\qquad \text{or} \\[2mm] \sin ax \displaystyle\sum_{r=0}^{\left[\frac{m}{2}\right]} (-1)^{r} \dfrac{m!}{(m-2r)!} \cdot \dfrac{x^{m-2r}}{a^{2r+1}} \\[4mm] + \cos ax \displaystyle\sum_{r=0}^{\left[\frac{m-1}{2}\right]} (-1)^{r} \dfrac{m!}{(m-2r-1)!} \cdot \dfrac{x^{m-2r-1}}{a^{2r+2}} \end{cases}$

See note integral 392.

INTEGRALS (Continued)

397. $\displaystyle\int \frac{\sin ax}{x}\,dx = \sum_{n=0}^{\infty} (-1)^n \frac{(ax)^{2n+1}}{(2n+1)(2n+1)!}$

398. $\displaystyle\int \frac{\cos ax}{x}\,dx = \log x + \sum_{n=1}^{\infty} (-1)^n \frac{(ax)^{2n}}{2n(2n)!}$

399. $\displaystyle\int x(\sin^2 ax)\,dx = \frac{x^2}{4} - \frac{x \sin 2ax}{4a} - \frac{\cos 2ax}{8a^2}$

400. $\displaystyle\int x^2(\sin^2 ax)\,dx = \frac{x^3}{6} - \left(\frac{x^2}{4a} - \frac{1}{8a^3}\right) \sin 2ax - \frac{x \cos 2ax}{4a^2}$

401. $\displaystyle\int x(\sin^3 ax)\,dx = \frac{x \cos 3ax}{12a} - \frac{\sin 3ax}{36a^2} - \frac{3x \cos ax}{4a} + \frac{3 \sin ax}{4a^2}$

402. $\displaystyle\int x(\cos^2 ax)\,dx = \frac{x^2}{4} + \frac{x \sin 2ax}{4a} + \frac{\cos 2ax}{8a^2}$

403. $\displaystyle\int x^2(\cos^2 ax)\,dx = \frac{x^3}{6} + \left(\frac{x^2}{4a} - \frac{1}{8a^3}\right) \sin 2ax + \frac{x \cos 2ax}{4a^2}$

404. $\displaystyle\int x(\cos^3 ax)\,dx = \frac{x \sin 3ax}{12a} + \frac{\cos 3ax}{36a^2} + \frac{3x \sin ax}{4a} + \frac{3 \cos ax}{4a^2}$

405. $\displaystyle\int \frac{\sin ax}{x^m}\,dx = -\frac{\sin ax}{(m-1)x^{m-1}} + \frac{a}{m-1} \int \frac{\cos ax}{x^{m-1}}\,dx$

406. $\displaystyle\int \frac{\cos ax}{x^m}\,dx = -\frac{\cos ax}{(m-1)x^{m-1}} - \frac{a}{m-1} \int \frac{\sin ax}{x^{m-1}}\,dx$

407. $\displaystyle\int \frac{x}{1 \pm \sin ax}\,dx = \mp \frac{x \cos ax}{a(1 \pm \sin ax)} + \frac{1}{a^2} \log (1 \pm \sin ax)$

408. $\displaystyle\int \frac{x}{1 + \cos ax}\,dx = \frac{x}{a} \tan \frac{ax}{2} + \frac{2}{a^2} \log \cos \frac{ax}{2}$

409. $\displaystyle\int \frac{x}{1 - \cos ax}\,dx = -\frac{x}{a} \cot \frac{ax}{2} + \frac{2}{a^2} \log \sin \frac{ax}{2}$

410. $\displaystyle\int \frac{x + \sin x}{1 + \cos x}\,dx = x \tan \frac{x}{2}$

411. $\displaystyle\int \frac{x - \sin x}{1 - \cos x}\,dx = -x \cot \frac{x}{2}$

412. $\displaystyle\int \sqrt{1 - \cos ax}\,dx = -\frac{2 \sin ax}{a\sqrt{1 - \cos ax}}$

413. $\displaystyle\int \sqrt{1 + \cos ax}\,dx = \frac{2 \sin ax}{a\sqrt{1 + \cos ax}}$

INTEGRALS (Continued)

414. $\int \sqrt{1 + \sin x}\, dx = \pm 2 \left(\sin\dfrac{x}{2} - \cos\dfrac{x}{2} \right),$

[use + if $(8k - 1)\dfrac{\pi}{2} < x \leq (8k + 3)\dfrac{\pi}{2}$, otherwise −; k an integer]

415. $\int \sqrt{1 - \sin x}\, dx = \pm 2 \left(\sin\dfrac{x}{2} + \cos\dfrac{x}{2} \right),$

[use + if $(8k - 3)\dfrac{\pi}{2} < x \leq (8k + 1)\dfrac{\pi}{2}$, otherwise −; k an integer]

416. $\int \dfrac{dx}{\sqrt{1 - \cos x}} = \pm \sqrt{2} \log \tan \dfrac{x}{4},$

[use + if $4k\pi < x < (4k + 2)\pi$, otherwise −; k an integer]

417. $\int \dfrac{dx}{\sqrt{1 + \cos x}} = \pm \sqrt{2} \log \tan \left(\dfrac{x + \pi}{4} \right),$

[use + if $(4k - 1)\pi < x < (4k + 1)\pi$, otherwise −; k an integer]

418. $\int \dfrac{dx}{\sqrt{1 - \sin x}} = \pm \sqrt{2} \log \tan \left(\dfrac{x}{4} - \dfrac{\pi}{8} \right),$

[use + if $(8k + 1)\dfrac{\pi}{2} < x < (8k + 5)\dfrac{\pi}{2}$, otherwise −; k an integer]

419. $\int \dfrac{dx}{\sqrt{1 + \sin x}} = \pm \sqrt{2} \log \tan \left(\dfrac{x}{4} + \dfrac{\pi}{8} \right),$

[use + if $(8k - 1)\dfrac{\pi}{2} < x < (8k + 3)\dfrac{\pi}{2}$, otherwise −; k an integer]

420. $\int (\tan^2 ax)\, dx = \dfrac{1}{a} \tan ax - x$

421. $\int (\tan^3 ax)\, dx = \dfrac{1}{2a} \tan^2 ax + \dfrac{1}{a} \log \cos ax$

422. $\int (\tan^4 ax)\, dx = \dfrac{\tan^3 ax}{3a} - \dfrac{1}{a} \tan x + x$

423. $\int (\tan^n ax)\, dx = \dfrac{\tan^{n-1} ax}{a(n - 1)} - \int (\tan^{n-2} ax)\, dx$

424. $\int (\cot^2 ax)\, dx = -\dfrac{1}{a} \cot ax - x$

425. $\int (\cot^3 ax)\, dx = -\dfrac{1}{2a} \cot^2 ax - \dfrac{1}{a} \log \sin ax$

426. $\int (\cot^4 ax)\, dx = -\dfrac{1}{3a} \cot^3 ax + \dfrac{1}{a} \cot ax + x$

INTEGRALS (Continued)

427. $\displaystyle\int (\cot^n ax)\, dx = -\frac{\cot^{n-1} ax}{a(n-1)} - \int (\cot^{n-2} ax)\, dx$

428. $\displaystyle\int \frac{x}{\sin^2 ax}\, dx = \int x(\csc^2 ax)\, dx = -\frac{x \cot ax}{a} + \frac{1}{a^2}\log \sin ax$

429. $\displaystyle\int \frac{x}{\sin^n ax}\, dx = \int x(\csc^n ax)\, dx = -\frac{x \cos ax}{a(n-1)\sin^{n-1} ax}$

$$-\frac{1}{a^2(n-1)(n-2)\sin^{n-2} ax} + \frac{(n-2)}{(n-1)}\int \frac{x}{\sin^{n-2} ax}\, d.$$

430. $\displaystyle\int \frac{x}{\cos^2 ax}\, dx = \int x(\sec^2 ax)\, dx = \frac{1}{a} x \tan ax + \frac{1}{a^2}\log \cos ax$

431. $\displaystyle\int \frac{x}{\cos^n ax}\, dx = \int x(\sec^n ax)\, dx = \frac{x \sin ax}{a(n-1)\cos^{n-1} ax}$

$$-\frac{1}{a^2(n-1)(n-2)\cos^{n-2} ax} + \frac{n-2}{n-1}\int \frac{x}{\cos^{n-2} ax}\, d$$

432. $\displaystyle\int \frac{\sin ax}{\sqrt{1 + b^2 \sin^2 ax}}\, dx = -\frac{1}{ab}\sin^{-1}\frac{b \cos ax}{\sqrt{1 + b^2}}$

433. $\displaystyle\int \frac{\sin ax}{\sqrt{1 - b^2 \sin^2 ax}}\, dx = -\frac{1}{ab}\log (b \cos ax + \sqrt{1 - b^2 \sin^2 ax})$

434. $\displaystyle\int (\sin ax)\sqrt{1 + b^2 \sin^2 ax}\, dx = -\frac{\cos ax}{2a}\sqrt{1 + b^2 \sin^2 ax}$

$$-\frac{1 + b^2}{2ab}\sin^{-1}\frac{b \cos a.}{\sqrt{1 + }}$$

435. $\displaystyle\int (\sin ax)\sqrt{1 - b^2 \sin^2 ax}\, dx = -\frac{\cos ax}{2a}\sqrt{1 - b^2 \sin^2 ax}$

$$-\frac{1 - b^2}{2ab}\log (b \cos ax + \sqrt{1 - b^2 \sin^2 a.}$$

436. $\displaystyle\int \frac{\cos ax}{\sqrt{1 + b^2 \sin^2 ax}}\, dx = \frac{1}{ab}\log (b \sin ax + \sqrt{1 + b^2 \sin^2 ax})$

437. $\displaystyle\int \frac{\cos ax}{\sqrt{1 - b^2 \sin^2 ax}}\, dx = \frac{1}{ab}\sin^{-1}(b \sin ax)$

438. $\displaystyle\int (\cos ax)\sqrt{1 + b^2 \sin^2 ax}\, dx = \frac{\sin ax}{2a}\sqrt{1 + b^2 \sin^2 ax}$

$$+\frac{1}{2ab}\log (b \sin ax + \sqrt{1 + b^2 \sin^2 a}$$

INTEGRALS (Continued)

439. $\int (\cos ax) \sqrt{1 - b^2 \sin^2 ax}\, dx = \dfrac{\sin ax}{2a} \sqrt{1 - b^2 \sin^2 ax} + \dfrac{1}{2ab} \sin^{-1}(b \sin ax)$

440. $\int \dfrac{dx}{\sqrt{a + b \tan^2 cx}} = \dfrac{\pm 1}{c\sqrt{a - b}} \sin^{-1}\left(\sqrt{\dfrac{a - b}{a}}\, \sin cx\right), \qquad (a > |b|)$

[use $+$ if $(2k - 1)\dfrac{\pi}{2} < x \le (2k + 1)\dfrac{\pi}{2}$, otherwise $-$; k an integer]

FORMS INVOLVING INVERSE TRIGONOMETRIC FUNCTIONS

441. $\int (\sin^{-1} ax)\, dx = x \sin^{-1} ax + \dfrac{\sqrt{1 - a^2 x^2}}{a}$

442. $\int (\cos^{-1} ax)\, dx = x \cos^{-1} ax - \dfrac{\sqrt{1 - a^2 x^2}}{a}$

443. $\int (\tan^{-1} ax)\, dx = x \tan^{-1} ax - \dfrac{1}{2a} \log(1 + a^2 x^2)$

444. $\int (\cot^{-1} ax)\, dx = x \cot^{-1} ax + \dfrac{1}{2a} \log(1 + a^2 x^2)$

445. $\int (\sec^{-1} ax)\, dx = x \sec^{-1} ax - \dfrac{1}{a} \log(ax + \sqrt{a^2 x^2 - 1})$

446. $\int (\csc^{-1} ax)\, dx = x \csc^{-1} ax + \dfrac{1}{a} \log(ax + \sqrt{a^2 x^2 - 1})$

447. $\int \left(\sin^{-1} \dfrac{x}{a}\right) dx = x \sin^{-1} \dfrac{x}{a} + \sqrt{a^2 - x^2}, \qquad (a > 0)$

448. $\int \left(\cos^{-1} \dfrac{x}{a}\right) dx = x \cos^{-1} \dfrac{x}{a} - \sqrt{a^2 - x^2}, \qquad (a > 0)$

449. $\int \left(\tan^{-1} \dfrac{x}{a}\right) dx = x \tan^{-1} \dfrac{x}{a} - \dfrac{a}{2} \log(a^2 + x^2)$

450. $\int \left(\cot^{-1} \dfrac{x}{a}\right) dx = x \cot^{-1} \dfrac{x}{a} + \dfrac{a}{2} \log(a^2 + x^2)$

451. $\int x[\sin^{-1}(ax)]\, dx = \dfrac{1}{4a^2}[(2a^2 x^2 - 1)\sin^{-1}(ax) + ax\sqrt{1 - a^2 x^2}]$

452. $\int x[\cos^{-1}(ax)]\, dx = \dfrac{1}{4a^2}[(2a^2 x^2 - 1)\cos^{-1}(ax) - ax\sqrt{1 - a^2 x^2}]$

INTEGRALS (Continued)

453. $\displaystyle \int x^n[\sin^{-1}(ax)]\,dx = \frac{x^{n+1}}{n+1}\sin^{-1}(ax) - \frac{a}{n+1}\int \frac{x^{n+1}\,dx}{\sqrt{1-a^2x^2}}, \qquad (n \neq -1)$

454. $\displaystyle \int x^n[\cos^{-1}(ax)]\,dx = \frac{x^{n+1}}{n+1}\cos^{-1}(ax) + \frac{a}{n+1}\int \frac{x^{n+1}\,dx}{\sqrt{1-a^2x^2}}, \qquad (n \neq -1)$

455. $\displaystyle \int x(\tan^{-1} ax)\,dx = \frac{1+a^2x^2}{2a^2}\tan^{-1} ax - \frac{x}{2a}$

456. $\displaystyle \int x^n(\tan^{-1} ax)\,dx = \frac{x^{n+1}}{n+1}\tan^{-1} ax - \frac{a}{n+1}\int \frac{x^{n+1}}{1+a^2x^2}\,dx$

457. $\displaystyle \int x(\cot^{-1} ax)\,dx = \frac{1+a^2x^2}{2a^2}\cot^{-1} ax + \frac{x}{2a}$

458. $\displaystyle \int x^n(\cot^{-1} ax)\,dx = \frac{x^{n+1}}{n+1}\cot^{-1} ax + \frac{a}{n+1}\int \frac{x^{n+1}}{1+a^2x^2}\,dx$

459. $\displaystyle \int \frac{\sin^{-1}(ax)}{x^2}\,dx = a\log\left(\frac{1-\sqrt{1-a^2x^2}}{x}\right) - \frac{\sin^{-1}(ax)}{x}$

460. $\displaystyle \int \frac{\cos^{-1}(ax)\,dx}{x^2} = -\frac{1}{x}\cos^{-1}(ax) + a\log\frac{1+\sqrt{1-a^2x^2}}{x}$

461. $\displaystyle \int \frac{\tan^{-1}(ax)\,dx}{x^2} = -\frac{1}{x}\tan^{-1}(ax) - \frac{a}{2}\log\frac{1+a^2x^2}{x^2}$

462. $\displaystyle \int \frac{\cot^{-1} ax}{x^2}\,dx = -\frac{1}{x}\cot^{-1} ax - \frac{a}{2}\log\frac{x^2}{a^2x^2+1}$

463. $\displaystyle \int (\sin^{-1} ax)^2\,dx = x(\sin^{-1} ax)^2 - 2x + \frac{2\sqrt{1-a^2x^2}}{a}\sin^{-1} ax$

464. $\displaystyle \int (\cos^{-1} ax)^2\,dx = x(\cos^{-1} ax)^2 - 2x - \frac{2\sqrt{1-a^2x^2}}{a}\cos^{-1} ax$

465. $\displaystyle \int (\sin^{-1} ax)^n\,dx = \begin{cases} x(\sin^{-1} ax)^n + \dfrac{n\sqrt{1-a^2x^2}}{a}(\sin^{-1} ax)^{n-1} \\ \qquad\qquad -n(n-1)\displaystyle\int (\sin^{-1} ax)^{n-2}\,dx \\[2mm] \text{or} \\ \displaystyle\sum_{r=0}^{\left[\frac{n}{2}\right]} (-1)^r \frac{n!}{(n-2r)!} x(\sin^{-1} ax)^{n-2r} \\[2mm] \qquad + \displaystyle\sum_{r=0}^{\left[\frac{n-1}{2}\right]} (-1)^r \frac{n!\sqrt{1-a^2x^2}}{(n-2r-1)!a}(\sin^{-1} ax)^{n-2r-1} \end{cases}$

Note: [s] means greatest integer $\leq s$. Thus [3.5] means 3; [5] = 5, $[\frac{1}{2}] = 0$.

466. $\displaystyle \int (\cos^{-1} ax)^n \, dx = \begin{cases} x(\cos^{-1} ax)^n - \dfrac{n\sqrt{1 - a^2x^2}}{a}(\cos^{-1} ax)^{n-1} \\ \qquad\qquad\qquad\qquad -n(n-1)\displaystyle\int (\cos^{-1} ax)^{n-2} \, dx \\ \qquad\qquad\text{or} \\ \displaystyle\sum_{r=0}^{\left[\frac{n}{2}\right]} (-1)^r \dfrac{n!}{(n-2r)!} x(\cos^{-1} ax)^{n-2r} \\ \qquad -\displaystyle\sum_{r=0}^{\left[\frac{n-1}{2}\right]} (-1)^r \dfrac{n!\sqrt{1 - a^2x^2}}{(n-2r-1)!a}(\cos^{-1} ax)^{n-2r-1} \end{cases}$

467. $\displaystyle \int \frac{1}{\sqrt{1 - a^2x^2}}(\sin^{-1} ax) \, dx = \frac{1}{2a}(\sin^{-1} ax)^2$

468. $\displaystyle \int \frac{x^n}{\sqrt{1 - a^2x^2}}(\sin^{-1} ax) \, dx = -\frac{x^{n-1}}{na^2}\sqrt{1 - a^2x^2}\,\sin^{-1} ax + \frac{x^n}{n^2a}$
$$+\frac{n-1}{na^2}\int \frac{x^{n-2}}{\sqrt{1 - a^2x^2}}\sin^{-1} ax \, dx$$

469. $\displaystyle \int \frac{1}{\sqrt{1 - a^2x^2}}(\cos^{-1} ax) \, dx = -\frac{1}{2a}(\cos^{-1} ax)^2$

470. $\displaystyle \int \frac{x^n}{\sqrt{1 - a^2x^2}}(\cos^{-1} ax) \, dx = -\frac{x^{n-1}}{na^2}\sqrt{1 - a^2x^2}\,\cos^{-1} ax - \frac{x^n}{n^2a}$
$$+\frac{n-1}{na^2}\int \frac{x^{n-2}}{\sqrt{1 - a^2x^2}}\cos^{-1} ax \, dx$$

471. $\displaystyle \int \frac{\tan^{-1} ax}{a^2x^2 + 1} \, dx = \frac{1}{2a}(\tan^{-1} ax)^2$

472. $\displaystyle \int \frac{\cot^{-1} ax}{a^2x^2 + 1} \, dx = -\frac{1}{2a}(\cot^{-1} ax)^2$

473. $\displaystyle \int x\sec^{-1} ax \, dx = \frac{x^2}{2}\sec^{-1} ax - \frac{1}{2a^2}\sqrt{a^2x^2 - 1}$

474. $\displaystyle \int x^n\sec^{-1} ax \, dx = \frac{x^{n+1}}{n+1}\sec^{-1} ax - \frac{1}{n+1}\int \frac{x^n \, dx}{\sqrt{a^2x^2 - 1}}$

475. $\displaystyle \int \frac{\sec^{-1} ax}{x^2} \, dx = -\frac{\sec^{-1} ax}{x} + \frac{\sqrt{a^2x^2 - 1}}{x}$

476. $\displaystyle \int x\csc^{-1} ax \, dx = \frac{x^2}{2}\csc^{-1} ax + \frac{1}{2a^2}\sqrt{a^2x^2 - 1}$

477. $\displaystyle \int x^n\csc^{-1} ax \, dx = \frac{x^{n+1}}{n+1}\csc^{-1} ax + \frac{1}{n+1}\int \frac{x^n \, dx}{\sqrt{a^2x^2 - 1}}$

INTEGRALS (Continued)

478. $\displaystyle\int \frac{\csc^{-1} ax}{x^2}\,dx = -\frac{\csc^{-1} ax}{x} - \frac{\sqrt{a^2 x^2 - 1}}{x}$

FORMS INVOLVING TRIGONOMETRIC SUBSTITUTIONS

479. $\displaystyle\int f(\sin x)\,dx = 2\int f\left(\frac{2z}{1 + z^2}\right)\frac{dz}{1 + z^2}, \qquad \left(z = \tan\frac{x}{2}\right)$

480. $\displaystyle\int f(\cos x)\,dx = 2\int f\left(\frac{1 - z^2}{1 + z^2}\right)\frac{dz}{1 + z^2}, \qquad \left(z = \tan\frac{x}{2}\right)$

*481. $\displaystyle\int f(\sin x)\,dx = \int f(u)\frac{du}{\sqrt{1 - u^2}}, \qquad (u = \sin x)$

*482. $\displaystyle\int f(\cos x)\,dx = -\int f(u)\frac{du}{\sqrt{1 - u^2}}, \qquad (u = \cos x)$

*483. $\displaystyle\int f(\sin x, \cos x)\,dx = \int f(u, \sqrt{1 - u^2})\frac{du}{\sqrt{1 - u^2}}, \qquad (u = \sin x)$

484. $\displaystyle\int f(\sin x, \cos x)\,dx = 2\int f\left(\frac{2z}{1 + z^2}, \frac{1 - z^2}{1 + z^2}\right)\frac{dz}{1 + z^2}, \qquad \left(= \tan\frac{x}{2}\right)$

LOGARITHMIC FORMS

485. $\displaystyle\int (\log x)\,dx = x\log x - x$

486. $\displaystyle\int x(\log x)\,dx = \frac{x^2}{2}\log x - \frac{x^2}{4}$

487. $\displaystyle\int x^2(\log x)\,dx = \frac{x^3}{3}\log x - \frac{x^3}{9}$

488. $\displaystyle\int x^n(\log ax)\,dx = \frac{x^{n+1}}{n + 1}\log ax - \frac{x^{n+1}}{(n + 1)^2}$

489. $\displaystyle\int (\log x)^2\,dx = x(\log x)^2 - 2x\log x + 2x$

490. $\displaystyle\int (\log x)^n\,dx = \begin{cases} x(\log x)^n - n\displaystyle\int (\log x)^{n-1}\,dx, & (n \neq -1) \\ \text{or} \\ (-1)^n n!\, x\displaystyle\sum_{r=0}^{n}\frac{(-\log x)^r}{r!} \end{cases}$

---

* The square roots appearing in these formulas may be plus or minus, depending on the quadrant of x. Care must be used to give them the proper sign.

INTEGRALS (Continued)

91. $\displaystyle \int \frac{(\log x)^n}{x}\, dx = \frac{1}{n+1}(\log x)^{n+1}$

92. $\displaystyle \int \frac{dx}{\log x} = \log(\log x) + \log x + \frac{(\log x)^2}{2\cdot 2!} + \frac{(\log x)^3}{3\cdot 3!} + \cdots$

93. $\displaystyle \int \frac{dx}{x\log x} = \log(\log x)$

94. $\displaystyle \int \frac{dx}{x(\log x)^n} = -\frac{1}{(n-1)(\log x)^{n-1}}$

95. $\displaystyle \int \frac{x^m\, dx}{(\log x)^n} = -\frac{x^{m+1}}{(n-1)(\log x)^{n-1}} + \frac{m+1}{n-1}\int \frac{x^m\, dx}{(\log x)^{n-1}}$

96. $\displaystyle \int x^m(\log x)^n\, dx = \begin{cases} \dfrac{x^{m+1}(\log x)^n}{m+1} - \dfrac{n}{m+1}\displaystyle\int x^m(\log x)^{n-1}\, dx \\[2mm] \text{or} \\[2mm] (-1)^n \dfrac{n!}{m+1}x^{m+1}\displaystyle\sum_{r=0}^{n} \frac{(-\log x)^r}{r!(m+1)^{n-r}} \end{cases}$

97. $\displaystyle \int \sin(\log x)\, dx = \tfrac{1}{2}x\sin(\log x) - \tfrac{1}{2}x\cos(\log x)$

98. $\displaystyle \int \cos(\log x)\, dx = \tfrac{1}{2}x\sin(\log x) + \tfrac{1}{2}x\cos(\log x)$

99. $\displaystyle \int [\log(ax+b)]\, dx = \frac{ax+b}{a}\log(ax+b) - x$

500. $\displaystyle \int \frac{\log(ax+b)}{x^2}\, dx = \frac{a}{b}\log x - \frac{ax+b}{bx}\log(ax+b)$

501. $\displaystyle \int x^m[\log(ax+b)]\, dx = \frac{1}{m+1}\left[x^{m+1} - \left(-\frac{b}{a}\right)^{m+1} \right]\log(ax+b)$
$$- \frac{1}{m+1}\left(-\frac{b}{a}\right)^{m+1}\sum_{r=1}^{m+1}\frac{1}{r}\left(-\frac{ax}{b}\right)^r$$

502. $\displaystyle \int \frac{\log(ax+b)}{x^m}\, dx = -\frac{1}{m-1}\frac{\log(ax+b)}{x^{m-1}} + \frac{1}{m-1}\left(-\frac{a}{b}\right)^{m-1}\log\frac{ax+b}{x}$
$$+ \frac{1}{m-1}\left(-\frac{a}{b}\right)^{m-1}\sum_{r=1}^{m-2}\frac{1}{r}\left(-\frac{b}{ax}\right)^r, (m>2)$$

503. $\displaystyle \int \left[\log\frac{x+a}{x-a}\right] dx = (x+a)\log(x+a) - (x-a)\log(x-a)$

504. $\displaystyle \int x^m\left[\log\frac{x+a}{x-a}\right] dx = \frac{x^{m+1}-(-a)^{m+1}}{m+1}\log(x+a) - \frac{x^{m+1}-a^{m+1}}{m+1}\log(x-a)$
$$+ \frac{2a^{m+1}}{m+1}\sum_{r=1}^{\left[\frac{m+1}{2}\right]}\frac{1}{m-2r+2}\left(\frac{x}{a}\right)^{m-2r+2}$$

INTEGRALS (Continued)

505. $\displaystyle\int \frac{1}{x^2}\left[\log\frac{x+a}{x-a}\right]dx = \frac{1}{x}\log\frac{x-a}{x+a} - \frac{1}{a}\log\frac{x^2-a^2}{x^2}$

506. $\displaystyle\int (\log X)\,dx = \begin{cases} \left(x+\dfrac{b}{2c}\right)\log X - 2x + \dfrac{\sqrt{4ac-b^2}}{c}\tan^{-1}\dfrac{2cx+b}{\sqrt{4ac-b^2}}, \\ \hspace{7cm}(b^2-4ac<0) \\[2mm] \text{or} \\[2mm] \left(x+\dfrac{b}{2c}\right)\log X - 2x + \dfrac{\sqrt{b^2-4ac}}{c}\tanh^{-1}\dfrac{2cx+b}{\sqrt{b^2-4ac}}, \\ \hspace{7cm}(b^2-4ac>0) \\[2mm] \text{where} \\[2mm] X = a + bx + cx^2 \end{cases}$

507. $\displaystyle\int x^n(\log X)\,dx = \frac{x^{n+1}}{n+1}\log X - \frac{2c}{n+1}\int\frac{x^{n+2}}{X}\,dx - \frac{b}{n+1}\int\frac{x^{n+1}}{X}\,dx$

where $X = a + bx + cx^2$

508. $\displaystyle\int [\log(x^2+a^2)]\,dx = x\log(x^2+a^2) - 2x + 2a\tan^{-1}\frac{x}{a}$

509. $\displaystyle\int [\log(x^2-a^2)]\,dx = x\log(x^2-a^2) - 2x + a\log\frac{x+a}{x-a}$

510. $\displaystyle\int x[\log(x^2\pm a^2)]\,dx = \tfrac{1}{2}(x^2\pm a^2)\log(x^2\pm a^2) - \tfrac{1}{2}x^2$

511. $\displaystyle\int [\log(x+\sqrt{x^2\pm a^2})]\,dx = x\log(x+\sqrt{x^2\pm a^2}) - \sqrt{x^2\pm a^2}$

512. $\displaystyle\int x[\log(x+\sqrt{x^2\pm a^2})]\,dx = \left(\frac{x^2}{2}\pm\frac{a^2}{4}\right)\log(x+\sqrt{x^2\pm a^2}) - \frac{x\sqrt{x^2\pm a^2}}{4}$

513. $\displaystyle\int x^m[\log(x+\sqrt{x^2\pm a^2})]\,dx = \frac{x^{m+1}}{m+1}\log(x+\sqrt{x^2\pm a^2})$

$$-\frac{1}{m+1}\int\frac{x^{m+1}}{\sqrt{x^2\pm a^2}}\,dx$$

514. $\displaystyle\int\frac{\log(x+\sqrt{x^2+a^2})}{x^2}\,dx = -\frac{\log(x+\sqrt{x^2+a^2})}{x} - \frac{1}{a}\log\frac{a+\sqrt{x^2+a^2}}{x}$

515. $\displaystyle\int\frac{\log(x+\sqrt{x^2-a^2})}{x^2}\,dx = -\frac{\log(x+\sqrt{x^2-a^2})}{x} + \frac{1}{|a|}\sec^{-1}\frac{x}{a}$

INTEGRALS (Continued)

516. $\displaystyle\int x^n \log (x^2 - a^2)\, dx = \frac{1}{n+1}\Bigg[x^{n+1} \log (x^2 - a^2) - a^{n+1} \log (x - a)$

$$-(-a)^{n+1} \log (x + a) - 2 \sum_{r=0}^{\left[\frac{n}{2}\right]} \frac{a^{2r} x^{n-2r+1}}{n - 2r + 1} \Bigg]$$

EXPONENTIAL FORMS

517. $\displaystyle\int e^x\, dx = e^x$

518. $\displaystyle\int e^{-x}\, dx = -e^{-x}$

519. $\displaystyle\int e^{ax}\, dx = \frac{e^{ax}}{a}$

520. $\displaystyle\int x\, e^{ax}\, dx = \frac{e^{ax}}{a^2}(ax - 1)$

521. $\displaystyle\int x^m\, e^{ax}\, dx = \begin{cases} \dfrac{x^m e^{ax}}{a} - \dfrac{m}{a}\displaystyle\int x^{m-1}\, e^{ax}\, dx \\[2mm] \qquad\qquad \text{or} \\[2mm] e^{ax}\displaystyle\sum_{r=0}^{m}(-1)^r \dfrac{m!\, x^{m-r}}{(m-r)!\, a^{r+1}} \end{cases}$

522. $\displaystyle\int \frac{e^{ax}\, dx}{x} = \log x + \frac{ax}{1!} + \frac{a^2 x^2}{2 \cdot 2!} + \frac{a^3 x^3}{3 \cdot 3!} + \cdots$

523. $\displaystyle\int \frac{e^{ax}}{x^m}\, dx = -\frac{1}{m-1}\frac{e^{ax}}{x^{m-1}} + \frac{a}{m-1}\int \frac{e^{ax}}{x^{m-1}}\, dx$

524. $\displaystyle\int e^{ax} \log x\, dx = \frac{e^{ax} \log x}{a} - \frac{1}{a}\int \frac{e^{ax}}{x}\, dx$

525. $\displaystyle\int \frac{dx}{1 + e^x} = x - \log (1 + e^x) = \log \frac{e^x}{1 + e^x}$

526. $\displaystyle\int \frac{dx}{a + be^{px}} = \frac{x}{a} - \frac{1}{ap}\log (a + be^{px})$

527. $\displaystyle\int \frac{dx}{ae^{mx} + be^{-mx}} = \frac{1}{m\sqrt{ab}}\tan^{-1}\left(e^{mx}\sqrt{\frac{a}{b}}\right), \qquad (a > 0, b > 0)$

528. $\displaystyle\int \frac{dx}{ae^{mx} - be^{-mx}} = \begin{cases} \dfrac{1}{2m\sqrt{ab}}\log \dfrac{\sqrt{a}\, e^{mx} - \sqrt{b}}{\sqrt{a}\, e^{mx} + \sqrt{b}} \\[3mm] \qquad\qquad \text{or} \\[3mm] \dfrac{-1}{m\sqrt{ab}}\tanh^{-1}\left(\sqrt{\dfrac{a}{b}}\, e^{mx}\right), \qquad (a > 0, b > 0) \end{cases}$

INTEGRALS (Continued)

529. $\displaystyle\int (a^x - a^{-x})\,dx = \frac{a^x + a^{-x}}{\log a}$

530. $\displaystyle\int \frac{e^{ax}}{b + ce^{ax}}\,dx = \frac{1}{ac}\log\,(b + ce^{ax})$

531. $\displaystyle\int \frac{x\,e^{ax}}{(1 + ax)^2}\,dx = \frac{e^{ax}}{a^2(1 + ax)}$

532. $\displaystyle\int x\,e^{-x^2}\,dx = -\tfrac{1}{2}e^{-x^2}$

533. $\displaystyle\int e^{ax}\,[\sin\,(bx)]\,dx = \frac{e^{ax}[a\sin\,(bx) - b\cos\,(bx)]}{a^2 + b^2}$

534. $\displaystyle\int e^{ax}\,[\sin\,(bx)]\,[\sin\,(cx)]\,dx = \frac{e^{ax}[(b - c)\sin\,(b - c)x + a\cos\,(b - c)x]}{2[a^2 + (b - c)^2]}$

$$-\frac{e^{ax}[(b + c)\sin\,(b + c)x + a\cos\,(b + c)x]}{2[a^2 + (b + c)^2]}$$

535. $\displaystyle\int e^{ax}[\sin\,(bx)]\,[\cos\,(cx)]\,dx = \begin{cases} \dfrac{e^{ax}[a\sin\,(b - c)x - (b - c)\cos\,(b - c)x]}{2[a^2 + (b - c)^2]} \\[2mm] \quad + \dfrac{e^{ax}[a\sin\,(b + c)x - (b + c)\cos\,(b + c)x]}{2[a^2 + (b + c)^2]} \\[2mm] \qquad\qquad\qquad \text{or} \\[2mm] \dfrac{e^{ax}}{\rho}[(a\sin bx - b\cos bx)[\cos\,(cx - \alpha)] \\[2mm] \qquad\qquad\qquad\qquad -c(\sin bx)\sin\,(cx - \alpha)] \\[2mm] \text{where} \\[2mm] \rho = \sqrt{(a^2 + b^2 - c^2)^2 + 4a^2c^2}, \\[2mm] \quad \rho\cos\alpha = a^2 + b^2 - c^2, \qquad \rho\sin\alpha = 2ac \end{cases}$

536. $\displaystyle\int e^{ax}[\sin\,(bx)]\,[\sin\,(bx + c)]\,dx$

$$= \frac{e^{ax}\cos c}{2a} - \frac{e^{ax}[a\cos\,(2bx + c) + 2b\sin\,(2bx + c)]}{2(a^2 + 4b^2)}$$

537. $\displaystyle\int e^{ax}[\sin\,(bx)]\,[\cos\,(bx + c)]\,dx$

$$= \frac{-e^{ax}\sin c}{2a} + \frac{e^{ax}[a\sin\,(2bx + c) - 2b\cos\,(2bx + c)]}{2(a^2 + 4b^2)}$$

538. $\displaystyle\int e^{ax}[\cos\,(bx)]\,dx = \frac{e^{ax}}{a^2 + b^2}[a\cos\,(bx) + b\sin\,(bx)]$

INTEGRALS (Continued)

539. $\displaystyle\int e^{ax}[\cos{(bx)}][\cos{(cx)}]\,dx = \frac{e^{ax}[(b-c)\sin{(b-c)x} + a\cos{(b-c)x}]}{2[a^2 + (b-c)^2]}$

$$+ \frac{e^{ax}[(b+c)\sin{(b+c)x} + a\cos{(b+c)x}]}{2[a^2 + (b+c)^2]}$$

540. $\displaystyle\int e^{ax}[\cos{(bx)}][\cos{(bx+c)}]\,dx$

$$= \frac{e^{ax}\cos c}{2a} + \frac{e^{ax}[a\cos{(2bx+c)} + 2b\sin{(2bx+c)}]}{2(a^2 + 4b^2)}$$

541. $\displaystyle\int e^{ax}[\cos{(bx)}][\sin{(bx+c)}]\,dx$

$$= \frac{e^{ax}\sin c}{2a} + \frac{e^{ax}[a\sin{(2bx+c)} - 2b\cos{(2bx+c)}]}{2(a^2 + 4b^2)}$$

542. $\displaystyle\int e^{ax}[\sin^n bx]\,dx = \frac{1}{a^2 + n^2 b^2}\left[(a\sin bx - nb\cos bx)\,e^{ax}\sin^{n-1} bx\right.$

$$\left. + n(n-1)b^2 \int e^{ax}[\sin^{n-2} bx]\,dx\right]$$

543. $\displaystyle\int e^{ax}[\cos^n bx]\,dx = \frac{1}{a^2 + n^2 b^2}\left[(a\cos bx + nb\sin bx)\,e^{ax}\cos^{n-1} bx\right.$

$$\left. + n(n-1)b^2 \int e^{ax}[\cos^{n-2} bx]\,dx\right]$$

544. $\displaystyle\int x^m e^x \sin x\,dx = \frac{1}{2}x^m e^x(\sin x - \cos x) - \frac{m}{2}\int x^{m-1} e^x \sin x\,dx$

$$+ \frac{m}{2}\int x^{m-1} e^x \cos x\,dx$$

545. $\displaystyle\int x^m e^{ax}[\sin bx]\,dx = \begin{cases} x^m e^{ax}\dfrac{a\sin bx - b\cos bx}{a^2 + b^2} \\ \qquad - \dfrac{m}{a^2 + b^2}\displaystyle\int x^{m-1} e^{ax}(a\sin bx - b\cos bx)\,dx \\ \qquad\qquad \text{or} \\ e^{ax}\displaystyle\sum_{r=0}^{m} \dfrac{(-1)^r m!\,x^{m-r}}{\rho^{r+1}(m-r)!}\sin{[bx - (r+1)\alpha]} \\ \qquad\qquad \text{where} \\ \rho = \sqrt{a^2 + b^2}, \qquad \rho\cos\alpha = a, \qquad \rho\sin\alpha = b \end{cases}$

546. $\displaystyle\int x^m e^x \cos x\,dx = \frac{1}{2}x^m e^x(\sin x + \cos x)$

$$- \frac{m}{2}\int x^{m-1} e^x \sin x\,dx - \frac{m}{2}\int x^{m-1} e^x \cos x\,dx$$

INTEGRALS (Continued)

547. $\displaystyle\int x^m e^{ax} \cos bx \, dx =$

$$\begin{cases} x^m e^{ax} \dfrac{a \cos bx + b \sin bx}{a^2 + b^2} \\[2ex] \qquad - \dfrac{m}{a^2 + b^2} \displaystyle\int x^{m-1} e^{ax}(a \cos bx + b \sin bx)\,dx \\[2ex] \text{or} \\[1ex] e^{ax} \displaystyle\sum_{r=0}^{m} \dfrac{(-1)^r m!\, x^{m-r}}{\rho^{r+1}(m-r)!} \cos\left[bx - (r+1)\alpha\right] \\[2ex] \text{where} \\[1ex] \rho = \sqrt{a^2 + b^2}, \qquad \rho \cos\alpha = a, \qquad \rho \sin\alpha = b \end{cases}$$

548. $\displaystyle\int e^{ax}(\cos^m x)(\sin^n x)\,dx =$

$$\begin{cases}
\dfrac{e^{ax} \cos^{m-1} x \sin^n x \left[a \cos x + (m+n)\sin x\right]}{(m+n)^2 + a^2} \\[2ex]
\quad - \dfrac{na}{(m+n)^2 + a^2} \displaystyle\int e^{ax}(\cos^{m-1} x)(\sin^{n-1} x)\,dx \\[2ex]
\quad + \dfrac{(m-1)(m+n)}{(m+n)^2 + a^2} \displaystyle\int e^{ax}(\cos^{m-2} x)(\sin^n x)\,dx \\[2ex]
\text{or} \\[1ex]
\dfrac{e^{ax} \cos^m x \sin^{n-1} x \left[a \sin x - (m+n)\cos x\right]}{(m+n)^2 + a^2} \\[2ex]
\quad + \dfrac{ma}{(m+n)^2 + a^2} \displaystyle\int e^{ax}(\cos^{m-1} x)(\sin^{n-1} x)\,dx \\[2ex]
\quad + \dfrac{(n-1)(m+n)}{(m+n)^2 + a^2} \displaystyle\int e^{ax}(\cos^m x)(\sin^{n-2} x)\,dx \\[2ex]
\text{or} \\[1ex]
\dfrac{e^{ax}(\cos^{m-1} x)(\sin^{n-1} x)(a \sin x \cos x + m \sin^2 x - n \cos^2 x)}{(m+n)^2 + a^2} \\[2ex]
\quad + \dfrac{m(m-1)}{(m+n)^2 + a^2} \displaystyle\int e^{ax}(\cos^{m-2} x)(\sin^n x)\,dx \\[2ex]
\quad + \dfrac{n(n-1)}{(m+n)^2 + a^2} \displaystyle\int e^{ax}(\cos^m x)(\sin^{n-2} x)\,dx \\[2ex]
\text{or} \\[1ex]
\dfrac{e^{ax}(\cos^{m-1} x)(\sin^{n-1} x)(a \cos x \sin x + m \sin^2 x - n \cos^2 x)}{(m+n)^2 + a^2} \\[2ex]
\quad + \dfrac{m(m-1)}{(m+n)^2 + a^2} \displaystyle\int e^{ax}(\cos^{m-2} x)(\sin^{n-2} x)\,dx \\[2ex]
\quad + \dfrac{(n-m)(n+m-1)}{(m+n)^2 + a^2} \displaystyle\int e^{ax}(\cos^m x)(\sin^{n-2} x)\,dx
\end{cases}$$

INTEGRALS (Continued)

549. $\int x\,e^{ax}(\sin bx)\,dx = \dfrac{x\,e^{ax}}{a^2 + b^2}(a\sin bx - b\cos bx)$

$$-\dfrac{e^{ax}}{(a^2 + b^2)^2}[(a^2 - b^2)\sin bx - 2ab\cos bx]$$

550. $\int x\,e^{ax}(\cos bx)\,dx = \dfrac{x\,e^{ax}}{a^2 + b^2}(a\cos bx + b\sin bx)$

$$-\dfrac{e^{ax}}{(a^2 + b^2)^2}[(a^2 - b^2)\cos bx + 2ab\sin bx]$$

551. $\int \dfrac{e^{ax}}{\sin^n x}\,dx = -\dfrac{e^{ax}[a\sin x + (n-2)\cos x]}{(n-1)(n-2)\sin^{n-1} x} + \dfrac{a^2 + (n-2)^2}{(n-1)(n-2)}\int \dfrac{e^{ax}}{\sin^{n-2} x}\,dx$

552. $\int \dfrac{e^{ax}}{\cos^n x}\,dx = -\dfrac{e^{ax}[a\cos x - (n-2)\sin x]}{(n-1)(n-2)\cos^{n-1} x} + \dfrac{a^2 + (n-2)^2}{(n-1)(n-2)}\int \dfrac{e^{ax}}{\cos^{n-2} x}\,dx$

553. $\int e^{ax}\tan^n x\,dx = e^{ax}\dfrac{\tan^{n-1} x}{n-1} - \dfrac{a}{n-1}\int e^{ax}\tan^{n-1} x\,dx - \int e^{ax}\tan^{n-2} x\,dx$

HYPERBOLIC FORMS

554. $\int (\sinh x)\,dx = \cosh x$

555. $\int (\cosh x)\,dx = \sinh x$

556. $\int (\tanh x)\,dx = \log \cosh x$

557. $\int (\coth x)\,dx = \log \sinh x$

558. $\int (\text{sech}\, x)\,dx = \tan^{-1}(\sinh x)$

559. $\int \text{csch}\, x\,dx = \log \tanh\left(\dfrac{x}{2}\right)$

560. $\int x(\sinh x)\,dx = x\cosh x - \sinh x$

561. $\int x^n(\sinh x)\,dx = x^n \cosh x - n\int x^{n-1}(\cosh x)\,dx$

562. $\int x(\cosh x)\,dx = x\sinh x - \cosh x$

563. $\int x^n(\cosh x)\,dx = x^n \sinh x - n\int x^{n-1}(\sinh x)\,dx$

INTEGRALS (Continued)

564. $\int (\text{sech } x)(\tanh x)\, dx = -\text{sech } x$

565. $\int (\text{csch } x)(\coth x)\, dx = -\text{csch } x$

566. $\int (\sinh^2 x)\, dx = \dfrac{\sinh 2x}{4} - \dfrac{x}{2}$

567. $\int (\sinh^m x)(\cosh^n x)\, dx = \begin{cases} \dfrac{1}{m+n}(\sinh^{m+1} x)(\cosh^{n-1} x) \\[2mm] \qquad + \dfrac{n-1}{m+n}\displaystyle\int (\sinh^m x)(\cosh^{n-2} x)\, dx \\[2mm] \qquad\text{or} \\[2mm] \dfrac{1}{m+n}\sinh^{m-1} x \cosh^{n+1} x \\[2mm] \qquad - \dfrac{m-1}{m+n}\displaystyle\int (\sinh^{m-2} x)(\cosh^n x)\, dx, \quad (m+n \neq 0) \end{cases}$

568. $\int \dfrac{dx}{(\sinh^m x)(\cosh^n x)} = \begin{cases} -\dfrac{1}{(m-1)(\sinh^{m-1} x)(\cosh^{n-1} x)} \\[2mm] \qquad - \dfrac{m+n-2}{m-1}\displaystyle\int \dfrac{dx}{(\sinh^{m-2} x)(\cosh^n x)}, \quad (m \neq 1) \\[2mm] \qquad\text{or} \\[2mm] \dfrac{1}{(n-1)\sinh^{m-1} x \cosh^{n-1} x} \\[2mm] \qquad + \dfrac{m+n-2}{n-1}\displaystyle\int \dfrac{dx}{(\sinh^m x)(\cosh^{n-2} x)}, \quad (n \neq 1) \end{cases}$

569. $\int (\tanh^2 x)\, dx = x - \tanh x$

570. $\int (\tanh^n x)\, dx = -\dfrac{\tanh^{n-1} x}{n-1} + \int (\tanh^{n-2} x)\, dx, \quad (n \neq 1)$

571. $\int (\text{sech}^2 x)\, dx = \tanh x$

572. $\int (\cosh^2 x)\, dx = \dfrac{\sinh 2x}{4} + \dfrac{x}{2}$

573. $\int (\coth^2 x)\, dx = x - \coth x$

574. $\int (\coth^n x)\, dx = -\dfrac{\coth^{n-1} x}{n-1} + \int \coth^{n-2} x\, dx, \quad (n \neq 1)$

INTEGRALS (Continued)

575. $\int (\operatorname{csch}^2 x)\, dx = -\operatorname{ctnh} x$

576. $\int (\sinh mx)(\sinh nx)\, dx = \dfrac{\sinh (m + n)x}{2(m + n)} - \dfrac{\sinh (m - n)x}{2(m - n)}, \qquad (m^2 \neq n^2)$

577. $\int (\cosh mx)(\cosh nx)\, dx = \dfrac{\sinh (m + n)x}{2(m + n)} + \dfrac{\sinh (m - n)x}{2(m - n)}, \qquad (m^2 \neq n^2)$

578. $\int (\sinh mx)(\cosh nx)\, dx = \dfrac{\cosh (m + n)x}{2(m + n)} + \dfrac{\cosh (m - n)x}{2(m - n)}, \qquad (m^2 \neq n^2)$

579. $\int \left(\sinh^{-1} \dfrac{x}{a}\right) dx = x \sinh^{-1} \dfrac{x}{a} - \sqrt{x^2 + a^2}, \qquad (a > 0)$

580. $\int x \left(\sinh^{-1} \dfrac{x}{a}\right) dx = \left(\dfrac{x^2}{2} + \dfrac{a^2}{4}\right) \sinh^{-1} \dfrac{x}{a} - \dfrac{x}{4}\sqrt{x^2 + a^2}, \qquad (a > 0)$

581. $\int x^n (\sinh^{-1} x)\, dx = \dfrac{x^{n+1}}{n + 1} \sinh^{-1} x - \dfrac{1}{n + 1} \int \dfrac{x^{n+1}}{(1 + x^2)^{\frac{1}{2}}}\, dx, \qquad (n \neq -1)$

582. $\int \left(\cosh^{-1} \dfrac{x}{a}\right) dx = \begin{cases} x \cosh^{-1} \dfrac{x}{a} - \sqrt{x^2 - a^2}, & \left(\cosh^{-1} \dfrac{x}{a} > 0\right) \\[2mm] \text{or} \\[2mm] x \cosh^{-1} \dfrac{x}{a} + \sqrt{x^2 - a^2}, & \left(\cosh^{-1} \dfrac{x}{a} < 0\right), \qquad (a > 0) \end{cases}$

583. $\int x \left(\cosh^{-1} \dfrac{x}{a}\right) dx = \dfrac{2x^2 - a^2}{4} \cosh^{-1} \dfrac{x}{a} - \dfrac{x}{4}(x^2 - a^2)^{\frac{1}{2}}$

584. $\int x^n (\cosh^{-1} x)\, dx = \dfrac{x^{n+1}}{n + 1} \cosh^{-1} x - \dfrac{1}{n + 1} \int \dfrac{x^{n+1}}{(x^2 - 1)^{\frac{1}{2}}}\, dx, \qquad (n \neq -1)$

585. $\int \left(\tanh^{-1} \dfrac{x}{a}\right) dx = x \tanh^{-1} \dfrac{x}{a} + \dfrac{a}{2} \log (a^2 - x^2), \qquad \left(\left|\dfrac{x}{a}\right| < 1\right)$

586. $\int \left(\coth^{-1} \dfrac{x}{a}\right) dx = x \coth^{-1} \dfrac{x}{a} + \dfrac{a}{2} \log (x^2 - a^2), \qquad \left(\left|\dfrac{x}{a}\right| > 1\right)$

587. $\int x \left(\tanh^{-1} \dfrac{x}{a}\right) dx = \dfrac{x^2 - a^2}{2} \tanh^{-1} \dfrac{x}{a} + \dfrac{ax}{2}, \qquad \left(\left|\dfrac{x}{a}\right| < 1\right)$

588. $\int x^n \left(\tanh^{-1} x\right) dx = \dfrac{x^{n+1}}{n + 1} \tanh^{-1} x - \dfrac{1}{n + 1} \int \dfrac{x^{n+1}}{1 - x^2}\, dx, \qquad (n \neq -1)$

589. $\int x \left(\coth^{-1} \dfrac{x}{a}\right) dx = \dfrac{x^2 - a^2}{2} \coth^{-1} \dfrac{x}{a} + \dfrac{ax}{2}, \qquad \left(\left|\dfrac{x}{a}\right| > 1\right)$

590. $\int x^n (\coth^{-1} x)\, dx = \dfrac{x^{n+1}}{n + 1} \coth^{-1} x + \dfrac{1}{n + 1} \int \dfrac{x^{n+1}}{x^2 - 1}\, dx, \qquad (n \neq -1)$

DEFINITE INTEGRALS

591. $\displaystyle\int (\text{sech}^{-1} x)\,dx = x\,\text{sech}^{-1} x + \sin^{-1} x$

592. $\displaystyle\int x\,\text{sech}^{-1} x\,dx = \frac{x^2}{2}\,\text{sech}^{-1} x - \frac{1}{2}(1 - x^2)$

593. $\displaystyle\int x^n\,\text{sech}^{-1} x\,dx = \frac{x^{n+1}}{n+1}\,\text{sech}^{-1} x + \frac{1}{n+1}\int \frac{x^n}{(1 - x^2)^{\frac{1}{2}}}\,dx, \qquad (n \neq -1)$

594. $\displaystyle\int \text{csch}^{-1} x\,dx = x\,\text{csch}^{-1} x + \frac{x}{|x|}\sinh^{-1} x$

595. $\displaystyle\int x\,\text{csch}^{-1} x\,dx = \frac{x^2}{2}\,\text{csch}^{-1} x + \frac{1}{2}\frac{x}{|x|}\sqrt{1 + x^2}$

596. $\displaystyle\int x^n\,\text{csch}^{-1} x\,dx = \frac{x^{n+1}}{n+1}\,\text{csch}^{-1} x + \frac{1}{n+1}\frac{x}{|x|}\int \frac{x^n}{(x^2 + 1)^{\frac{1}{2}}}\,dx, \qquad (n \neq -1)$

DEFINITE INTEGRALS

597. $\displaystyle\int_0^\infty x^{n-1} e^{-x}\,dx = \int_0^1 \left(\log\frac{1}{x}\right)^{n-1}\,dx = \frac{1}{n}\prod_{m=1}^\infty \frac{\left(1 + \dfrac{1}{m}\right)^n}{1 + \dfrac{n}{m}}$

$$= \Gamma(n), n \neq 0, -1, -2, -3, \ldots \qquad \text{(Gamma Function)}$$

598. $\displaystyle\int_0^\infty t^n p^{-t}\,dt = \frac{n!}{(\log p)^{n+1}}, \qquad (n = 0, 1, 2, 3, \ldots \text{ and } p > 0)$

599. $\displaystyle\int_0^\infty t^{n-1} e^{-(a+1)t}\,dt = \frac{\Gamma(n)}{(a+1)^n}, \qquad (n > 0, a > -1)$

600. $\displaystyle\int_0^1 x^m \left(\log\frac{1}{x}\right)^n\,dx = \frac{\Gamma(n+1)}{(m+1)^{n+1}}, \qquad (m > -1, n > -1)$

601. $\Gamma(n)$ is finite if $n > 0$, $\Gamma(n + 1) = n\Gamma(n)$

602. $\displaystyle\Gamma(n)\cdot\Gamma(1 - n) = \frac{\pi}{\sin n\pi}$

603. $\Gamma(n) = (n - 1)!$ if $n = \text{integer} > 0$

604. $\displaystyle\Gamma(\tfrac{1}{2}) = 2\int_0^\infty e^{-t^2}\,dt = \sqrt{\pi} = 1.7724538509\cdots = (-\tfrac{1}{2})!$

605. $\displaystyle\Gamma(n + \tfrac{1}{2}) = \frac{1 \cdot 3 \cdot 5 \ldots (2n - 1)}{2^n}\sqrt{\pi} \qquad n = 1, 2, 3, \ldots$

606. $\displaystyle\Gamma(-n + \tfrac{1}{2}) = \frac{(-1)^n 2^n \sqrt{\pi}}{1 \cdot 3 \cdot 5 \ldots (2n - 1)} \qquad n = 1, 2, 3, \ldots$

DEFINITE INTEGRALS (Continued)

607. $\displaystyle\int_0^1 x^{m-1}(1-x)^{n-1}\,dx = \int_0^\infty \frac{x^{m-1}}{(1+x)^{m+n}}\,dx = \frac{\Gamma(m)\Gamma(n)}{\Gamma(m+n)} = B(m,n)$

(Beta function)

608. $B(m,n) = B(n,m) = \dfrac{\Gamma(m)\Gamma(n)}{\Gamma(m+n)}$, where m and n are any positive real numbers.

609. $\displaystyle\int_a^b (x-a)^m(b-x)^n\,dx = (b-a)^{m+n+1}\frac{\Gamma(m+1)\cdot\Gamma(n+1)}{\Gamma(m+n+2)},$

$(m > -1, n > -1, b > a)$

610. $\displaystyle\int_1^\infty \frac{dx}{x^m} = \frac{1}{m-1}, \qquad [m > 1]$

611. $\displaystyle\int_0^\infty \frac{dx}{(1+x)x^p} = \pi \csc p\pi, \qquad [p < 1]$

612. $\displaystyle\int_0^\infty \frac{dx}{(1-x)x^p} = -\pi \cot p\pi, \qquad [p < 1]$

613. $\displaystyle\int_0^\infty \frac{x^{p-1}\,dx}{1+x} = \frac{\pi}{\sin p\pi}$

$\qquad\qquad = B(p, 1-p) = \Gamma(p)\Gamma(1-p), \qquad [0 < p < 1]$

614. $\displaystyle\int_0^\infty \frac{x^{m-1}\,dx}{1+x^n} = \frac{\pi}{n\sin\dfrac{m\pi}{n}}, \qquad [0 < m < n]$

615. $\displaystyle\int_0^\infty \frac{x^a\,dx}{(m+x^b)^c} = \frac{m^{\frac{a+1-bc}{b}}}{b}\cdot\frac{\Gamma\left(\dfrac{a+1}{b}\right)\Gamma\left(c-\dfrac{a+1}{b}\right)}{\Gamma(c)},$

$\left(a > -1, b > 0, m > 0, c > \dfrac{a+1}{b}\right)$

616. $\displaystyle\int_0^\infty \frac{dx}{(1+x)\sqrt{x}} = \pi$

617. $\displaystyle\int_0^\infty \frac{a\,dx}{a^2+x^2} = \frac{\pi}{2}, \text{ if } a > 0; 0, \text{ if } a = 0; -\frac{\pi}{2}, \text{ if } a < 0$

618. $\displaystyle\int_0^a (a^2-x^2)^{\frac{n}{2}}\,dx = \frac{1}{2}\int_{-a}^a (a^2-x^2)^{\frac{n}{2}}\,dx = \frac{1\cdot3\cdot5\ldots n}{2\cdot4\cdot6\ldots(n+1)}\cdot\frac{\pi}{2}\cdot a^{n+1}$ (n odd)

619. $\displaystyle\int_0^a x^m(a^2-x^2)^{\frac{n}{2}}\,dx = \begin{cases} \dfrac{1}{2}a^{m+n+1}B\left(\dfrac{m+1}{2}, \dfrac{n+2}{2}\right) \\[2mm] \qquad\qquad \text{or} \\[2mm] \dfrac{1}{2}a^{m+n+1}\dfrac{\Gamma\left(\dfrac{m+1}{2}\right)\Gamma\left(\dfrac{n+2}{2}\right)}{\Gamma\left(\dfrac{m+n+3}{2}\right)} \end{cases}$

DEFINITE INTEGRALS (Continued)

620. $\displaystyle\int_0^{\pi/2} (\sin^n x)\, dx = \begin{cases} \displaystyle\int_0^{\pi/2} (\cos^n x)\, dx \\[2mm] \text{or} \\[2mm] \dfrac{1\cdot 3\cdot 5\cdot 7\ldots(n-1)}{2\cdot 4\cdot 6\cdot 8\ldots(n)}\,\dfrac{\pi}{2}, \quad (n \text{ an even integer, } n \neq 0) \\[3mm] \text{or} \\[2mm] \dfrac{2\cdot 4\cdot 6\cdot 8\ldots(n-1)}{1\cdot 3\cdot 5\cdot 7\ldots(n)}, \quad (n \text{ an odd integer, } n \neq 1) \\[3mm] \text{or} \\[2mm] \dfrac{\sqrt{\pi}}{2}\dfrac{\Gamma\left(\dfrac{n+1}{2}\right)}{\Gamma\left(\dfrac{n}{2}+1\right)}, \quad (n > -1) \end{cases}$

621. $\displaystyle\int_0^\infty \frac{\sin mx\, dx}{x} = \frac{\pi}{2}$, if $m > 0$; 0, if $m = 0$; $-\frac{\pi}{2}$, if $m < 0$

622. $\displaystyle\int_0^\infty \frac{\cos x\, dx}{x} = \infty$

623. $\displaystyle\int_0^\infty \frac{\tan x\, dx}{x} = \frac{\pi}{2}$

624. $\displaystyle\int_0^\pi \sin ax \cdot \sin bx\, dx = \int_0^\pi \cos ax \cdot \cos bx\, dx = 0$, $\quad (a \neq b; a, b \text{ integers})$

625. $\displaystyle\int_0^{\pi/a} [\sin(ax)][\cos(ax)]\, dx = \int_0^\pi [\sin(ax)][\cos(ax)]\, dx = 0$

626. $\displaystyle\int_0^\pi [\sin(ax)][\cos(bx)]\, dx = \frac{2a}{a^2 - b^2}$, if $a - b$ is odd, or 0 if $a - b$ is even

627. $\displaystyle\int_0^\infty \frac{\sin x \cos mx\, dx}{x}$

$\qquad = 0$, if $m < -1$ or $m > 1$; $\dfrac{\pi}{4}$, if $m = \pm 1$; $\dfrac{\pi}{2}$, if $m^2 < 1$

628. $\displaystyle\int_0^\infty \frac{\sin ax \sin bx}{x^2}\, dx = \frac{\pi a}{2}$, $\quad (a \leq b)$

629. $\displaystyle\int_0^\pi \sin^2 mx\, dx = \int_0^\pi \cos^2 mx\, dx = \frac{\pi}{2}$

630. $\displaystyle\int_0^\infty \frac{\sin^2(px)}{x^2}\, dx = \frac{\pi p}{2}$

DEFINITE INTEGRALS (Continued)

631. $\displaystyle\int_0^\infty \frac{\sin x}{x^p}\,dx = \frac{\pi}{2\Gamma(p)\sin(p\pi/2)}, \qquad 0 < p < 1$

632. $\displaystyle\int_0^\infty \frac{\cos x}{x^p}\,dx = \frac{\pi}{2\Gamma(p)\cos(p\pi/2)}, \qquad 0 < p < 1$

633. $\displaystyle\int_0^\infty \frac{1 - \cos px}{x^2}\,dx = \frac{\pi p}{2}$

634. $\displaystyle\int_0^\infty \frac{\sin px \cos qx}{x}\,dx = \left\{0, \quad q > p > 0; \ \frac{\pi}{2}, \ p > q > 0; \ \frac{\pi}{4}, \ p = q > 0\right\}$

635. $\displaystyle\int_0^\infty \frac{\cos(mx)}{x^2 + a^2}\,dx = \frac{\pi}{2a}e^{-|m|a}, \qquad (a > 0)$

636. $\displaystyle\int_0^\infty \cos(x^2)\,dx = \int_0^\infty \sin(x^2)\,dx = \frac{1}{2}\sqrt{\frac{\pi}{2}}$

637. $\displaystyle\int_0^\infty \sin ax^n\,dx = \frac{1}{na^{1/n}}\Gamma(1/n)\sin\frac{\pi}{2n}, \qquad n > 1$

638. $\displaystyle\int_0^\infty \cos ax^n\,dx = \frac{1}{na^{1/n}}\Gamma(1/n)\cos\frac{\pi}{2n}, \qquad n > 1$

639. $\displaystyle\int_0^\infty \frac{\sin x}{\sqrt{x}}\,dx = \int_0^\infty \frac{\cos x}{\sqrt{x}}\,dx = \sqrt{\frac{\pi}{2}}$

640. $\displaystyle\int_0^\infty \frac{\sin^3 x}{x^2}\,dx = \frac{3}{4}\log 3$

641. $\displaystyle\int_0^\infty \frac{\sin^3 x}{x^3}\,dx = \frac{3\pi}{8}$

642. $\displaystyle\int_0^\infty \frac{\sin^4 x}{x^4}\,dx = \frac{\pi}{3}$

643. $\displaystyle\int_0^{\pi/2} \frac{dx}{1 + a\cos x} = \frac{\cos^{-1} a}{\sqrt{1 - a^2}}, \qquad (a < 1)$

644. $\displaystyle\int_0^\pi \frac{dx}{a + b\cos x} = \frac{\pi}{\sqrt{a^2 - b^2}}, \qquad (a > b \geq 0)$

645. $\displaystyle\int_0^{2\pi} \frac{dx}{1 + a\cos x} = \frac{2\pi}{\sqrt{1 - a^2}}, \qquad (a^2 < 1)$

646. $\displaystyle\int_0^\infty \frac{\cos ax - \cos bx}{x}\,dx = \log\frac{b}{a}$

647. $\displaystyle\int_0^{\pi/2} \frac{dx}{a^2\sin^2 x + b^2\cos^2 x} = \frac{\pi}{2ab}$

DEFINITE INTEGRALS (Continued)

648. $\displaystyle\int_0^{\pi/2} \frac{dx}{(a^2 \sin^2 x + b^2 \cos^2 x)^2} = \frac{\pi(a^2 + b^2)}{4a^3 b^3}$, $(a, b > 0)$

649. $\displaystyle\int_0^{\pi/2} \sin^{n-1} x \cos^{m-1} x \, dx = \frac{1}{2} B\left(\frac{n}{2}, \frac{m}{2}\right)$, m and n positive integers

650. $\displaystyle\int_0^{\pi/2} (\sin^{2n+1} \theta) \, d\theta = \frac{2 \cdot 4 \cdot 6 \ldots (2n)}{1 \cdot 3 \cdot 5 \ldots (2n + 1)}$, $(n = 1, 2, 3 \ldots)$

651. $\displaystyle\int_0^{\pi/2} (\sin^{2n} \theta) \, d\theta = \frac{1 \cdot 3 \cdot 5 \ldots (2n - 1)}{2 \cdot 4 \ldots (2n)}\left(\frac{\pi}{2}\right)$, $(n = 1, 2, 3 \ldots)$

652. $\displaystyle\int_0^{\pi/2} \frac{x}{\sin x} \, dx = 2\left\{\frac{1}{1^2} - \frac{1}{3^2} + \frac{1}{5^2} - \frac{1}{7^2} + \cdots\right\}$

653. $\displaystyle\int_0^{\pi/2} \frac{dx}{1 + \tan^m x} = \frac{\pi}{4}$

654. $\displaystyle\int_0^{\pi/2} \sqrt{\cos \theta} \, d\theta = \frac{(2\pi)^{\frac{3}{2}}}{[\Gamma(\frac{1}{4})]^2}$

655. $\displaystyle\int_0^{\pi/2} (\tan^h \theta) \, d\theta = \frac{\pi}{2 \cos\left(\dfrac{h\pi}{2}\right)}$, $(0 < h < 1)$

656. $\displaystyle\int_0^{\infty} \frac{\tan^{-1}(ax) - \tan^{-1}(bx)}{x} \, dx = \frac{\pi}{2} \log \frac{a}{b}$, $(a, b > 0)$

657. The area enclosed by a curve defined through the equation $x^{\frac{b}{c}} + y^{\frac{b}{c}} = a^{\frac{b}{c}}$ where $a > 0$, c a positive odd integer and b a positive even integer is given by

$$\frac{\left[\Gamma\left(\dfrac{c}{b}\right)\right]^2}{\Gamma\left(\dfrac{2c}{b}\right)} \left(\frac{2ca^2}{b}\right)$$

658. $\displaystyle I = \iiint_R x^{h-1} y^{m-1} z^{n-1} \, dv$, where R denotes the region of space bounded by

the co-ordinate planes and that portion of the surface $\left(\dfrac{x}{a}\right)^p + \left(\dfrac{y}{b}\right)^q + \left(\dfrac{z}{c}\right)^k = 1$,

which lies in the first octant and where $h, m, n, p, q, k, a, b, c$, denote positive real numbers is given by

$$\int_0^a x^{h-1} \, dx \int_0^{b\left[1 - \left(\frac{x}{a}\right)^p\right]^{\frac{1}{q}}} y^m \, dy \int_0^{c\left[1 - \left(\frac{x}{a}\right)^p - \left(\frac{y}{b}\right)^q\right]^{\frac{1}{k}}} z^{n-1} \, dz$$

$$= \frac{a^h b^m c^n}{pqk} \frac{\Gamma\left(\dfrac{h}{p}\right)\Gamma\left(\dfrac{m}{q}\right)\Gamma\left(\dfrac{n}{k}\right)}{\Gamma\left(\dfrac{h}{p} + \dfrac{m}{q} + \dfrac{n}{k} + 1\right)}$$

DEFINITE INTEGRALS (Continued)

659. $\displaystyle\int_0^\infty e^{-ax}\,dx = \frac{1}{a}, \qquad (a > 0)$

660. $\displaystyle\int_0^\infty \frac{e^{-ax} - e^{-bx}}{x}\,dx = \log\frac{b}{a}, \qquad (a, b > 0)$

661. $\displaystyle\int_0^\infty x^n e^{-ax}\,dx = \begin{cases} \dfrac{\Gamma(n+1)}{a^{n+1}}, & (n > -1, a > 0) \\[2mm] \qquad\text{or} \\[2mm] \dfrac{n!}{a^{r+1}}, & (a > 0, n\text{ positive integer}) \end{cases}$

662. $\displaystyle\int_0^\infty x^n \exp(-ax^p)\,dx = \frac{\Gamma(k)}{pa^k}, \qquad \left(n > -1, p > 0, a > 0, k = \frac{n+1}{p}\right)$

663. $\displaystyle\int_0^\infty e^{-a^2x^2}\,dx = \frac{1}{2a}\sqrt{\pi} = \frac{1}{2a}\Gamma\left(\frac{1}{2}\right), \qquad (a > 0)$

664. $\displaystyle\int_0^\infty x e^{-x^2}\,dx = \tfrac{1}{2}$

665. $\displaystyle\int_0^\infty x^2 e^{-x^2}\,dx = \frac{\sqrt{\pi}}{4}$

666. $\displaystyle\int_0^\infty x^{2n} e^{-ax^2}\,dx = \frac{1 \cdot 3 \cdot 5 \ldots (2n-1)}{2^{n+1}a^n}\sqrt{\frac{\pi}{a}}$

667. $\displaystyle\int_0^\infty x^{2n+1} e^{-ax^2}\,dx = \frac{n!}{2a^{n+1}}, \qquad (a > 0)$

668. $\displaystyle\int_0^1 x^m e^{-ax}\,dx = \frac{m!}{a^{m+1}}\left[1 - e^{-a}\sum_{r=0}^m \frac{a^r}{r!}\right]$

669. $\displaystyle\int_0^\infty e^{\left(-x^2 - \frac{a^2}{x^2}\right)}\,dx = \frac{e^{-2a}\sqrt{\pi}}{2}, \qquad (a \geq 0)$

670. $\displaystyle\int_0^\infty e^{-nx}\sqrt{x}\,dx = \frac{1}{2n}\sqrt{\frac{\pi}{n}}$

671. $\displaystyle\int_0^\infty \frac{e^{-nx}}{\sqrt{x}}\,dx = \sqrt{\frac{\pi}{n}}$

672. $\displaystyle\int_0^\infty e^{-ax}(\cos mx)\,dx = \frac{a}{a^2 + m^2}, \qquad (a > 0)$

673. $\displaystyle\int_0^\infty e^{-ax}(\sin mx)\,dx = \frac{m}{a^2 + m^2}, \qquad (a > 0)$

DEFINITE INTEGRALS (Continued)

674. $$\int_0^\infty x\,e^{-ax}[\sin(bx)]\,dx = \frac{2ab}{(a^2+b^2)^2}, \qquad (a>0)$$

675. $$\int_0^\infty x\,e^{-ax}[\cos(bx)]\,dx = \frac{a^2-b^2}{(a^2+b^2)^2}, \qquad (a>0)$$

676. $$\int_0^\infty x^n e^{-ax}[\sin(bx)]\,dx = \frac{n![(a-ib)^{n+1}-(a+ib)^{n+1}]}{2(a^2+b^2)^{n+1}}, \qquad (i^2=-1, a>0)$$

677. $$\int_0^\infty x^n e^{-ax}[\cos(bx)]\,dx = \frac{n![(a-ib)^{n+1}+(a+ib)^{n+1}]}{2(a^2+b^2)^{n+1}}, \qquad (i^2=-1, a>0)$$

678. $$\int_0^\infty \frac{e^{-ax}\sin x}{x}\,dx = \cot^{-1} a, \qquad (a>0)$$

679. $$\int_0^\infty e^{-a^2x^2}\cos bx\,dx = \frac{\sqrt{\pi}}{2a}\exp\left(-\frac{b^2}{4a^2}\right), \qquad (ab\neq 0)$$

680. $$\int_0^\infty e^{-t\cos\phi}\,t^{b-1}\sin(t\sin\phi)\,dt = [\Gamma(b)]\sin(b\phi), \qquad \left(b>0, -\frac{\pi}{2}<\phi<\frac{\pi}{2}\right)$$

681. $$\int_0^\infty e^{-t\cos\phi}\,t^{b-1}[\cos(t\sin\phi)]\,dt = [\Gamma(b)]\cos(b\phi), \qquad \left(b>0, -\frac{\pi}{2}<\phi<\frac{\pi}{2}\right)$$

682. $$\int_0^\infty t^{b-1}\cos t\,dt = [\Gamma(b)]\cos\left(\frac{b\pi}{2}\right), \qquad (0<b<1)$$

683. $$\int_0^\infty t^{b-1}(\sin t)\,dt = [\Gamma(b)]\sin\left(\frac{b\pi}{2}\right), \qquad (0<b<1)$$

684. $$\int_0^1 (\log x)^n\,dx = (-1)^n\cdot n!$$

685. $$\int_0^1 \left(\log\frac{1}{x}\right)^{\frac{1}{2}} dx = \frac{\sqrt{\pi}}{2}$$

686. $$\int_0^1 \left(\log\frac{1}{x}\right)^{-\frac{1}{2}} dx = \sqrt{\pi}$$

687. $$\int_0^1 \left(\log\frac{1}{x}\right)^{n} dx = n!$$

688. $$\int_0^1 x\log(1-x)\,dx = -\tfrac{3}{4}$$

689. $$\int_0^1 x\log(1+x)\,dx = \tfrac{1}{4}$$

690. $$\int_0^1 x^m(\log x)^n\,dx = \frac{(-1)^n n!}{(m+1)^{n+1}}, \qquad m>-1, n=0,1,2,\ldots$$

If $n\neq 0,1,2,\ldots$ replace $n!$ by $\Gamma(n+1)$.

DEFINITE INTEGRALS (Continued)

691. $\displaystyle\int_0^1 \frac{\log x}{1+x}\,dx = -\frac{\pi^2}{12}$

692. $\displaystyle\int_0^1 \frac{\log x}{1-x}\,dx = -\frac{\pi^2}{6}$

693. $\displaystyle\int_0^1 \frac{\log(1+x)}{x}\,dx = \frac{\pi^2}{12}$

694. $\displaystyle\int_0^1 \frac{\log(1-x)}{x}\,dx = -\frac{\pi^2}{6}$

695. $\displaystyle\int_0^1 (\log x)[\log(1+x)]\,dx = 2 - 2\log 2 - \frac{\pi^2}{12}$

696. $\displaystyle\int_0^1 (\log x)[\log(1-x)]\,dx = 2 - \frac{\pi^2}{6}$

697. $\displaystyle\int_0^1 \frac{\log x}{1-x^2}\,dx = -\frac{\pi^2}{8}$

698. $\displaystyle\int_0^1 \log\left(\frac{1+x}{1-x}\right)\cdot\frac{dx}{x} = \frac{\pi^2}{4}$

699. $\displaystyle\int_0^1 \frac{\log x\,dx}{\sqrt{1-x^2}} = -\frac{\pi}{2}\log 2$

700. $\displaystyle\int_0^1 x^m\left[\log\left(\frac{1}{x}\right)\right]^n dx = \frac{\Gamma(n+1)}{(m+1)^{n+1}}, \qquad \text{if } m+1>0,\, n+1>0$

701. $\displaystyle\int_0^1 \frac{(x^p - x^q)\,dx}{\log x} = \log\left(\frac{p+1}{q+1}\right), \qquad (p+1>0, q+1>0)$

702. $\displaystyle\int_0^1 \frac{dx}{\sqrt{\log\left(\dfrac{1}{x}\right)}} = \sqrt{\pi}$

703. $\displaystyle\int_0^\infty \log\left(\frac{e^x+1}{e^x-1}\right) dx = \frac{\pi^2}{4}$

704. $\displaystyle\int_0^{\pi/2} (\log\sin x)\,dx = \int_0^{\pi/2} \log\cos x\,dx = -\frac{\pi}{2}\log 2$

705. $\displaystyle\int_0^{\pi/2} (\log\sec x)\,dx = \int_0^{\pi/2} \log\csc x\,dx = \frac{\pi}{2}\log 2$

706. $\displaystyle\int_0^\pi x(\log\sin x)\,dx = -\frac{\pi^2}{2}\log 2$

707. $\displaystyle\int_0^{\pi/2} (\sin x)(\log\sin x)\,dx = \log 2 - 1$

708. $\displaystyle\int_0^{\pi/2} (\log \tan x)\, dx = 0$

709. $\displaystyle\int_0^{\pi} \log (a \pm b \cos x)\, dx = \pi \log \left(\frac{a + \sqrt{a^2 - b^2}}{2} \right), \qquad (a \geq b)$

710. $\displaystyle\int_0^{\pi} \log (a^2 - 2ab \cos x + b^2)\, dx = \begin{cases} 2\pi \log a, & a \geq b > 0 \\ 2\pi \log b, & b \geq a > 0 \end{cases}$

711. $\displaystyle\int_0^{\infty} \frac{\sin ax}{\sinh bx}\, dx = \frac{\pi}{2b} \tanh \frac{a\pi}{2b}$

712. $\displaystyle\int_0^{\infty} \frac{\cos ax}{\cosh bx}\, dx = \frac{\pi}{2b} \operatorname{sech} \frac{a\pi}{2b}$

713. $\displaystyle\int_0^{\infty} \frac{dx}{\cosh ax} = \frac{\pi}{2a}$

714. $\displaystyle\int_0^{\infty} \frac{x\, dx}{\sinh ax} = \frac{\pi^2}{4a^2}$

715. $\displaystyle\int_0^{\infty} e^{-ax}(\cosh bx)\, dx = \frac{a}{a^2 - b^2}, \qquad (0 \leq |b| < a)$

716. $\displaystyle\int_0^{\infty} e^{-ax}(\sinh bx)\, dx = \frac{b}{a^2 - b^2}, \qquad (0 \leq |b| < a)$

717. $\displaystyle\int_0^{\infty} \frac{\sinh ax}{e^{bx} + 1}\, dx = \frac{\pi}{2b} \csc \frac{a\pi}{b} - \frac{1}{2a}$

718. $\displaystyle\int_0^{\infty} \frac{\sinh ax}{e^{bx} - 1}\, dx = \frac{1}{2a} - \frac{\pi}{2b} \cot \frac{a\pi}{b}$

719. $\displaystyle\int_0^{\pi/2} \frac{dx}{\sqrt{1 - k^2 \sin^2 x}} = \frac{\pi}{2} \left[1 + \left(\frac{1}{2}\right)^2 k^2 + \left(\frac{1 \cdot 3}{2 \cdot 4}\right)^2 k^4 \right.$

$$\left. + \left(\frac{1 \cdot 3 \cdot 5}{2 \cdot 4 \cdot 6}\right)^2 k^6 + \cdots \right], \text{ if } k^2 < 1$$

720. $\displaystyle\int_0^{\pi/2} \sqrt{1 - k^2 \sin^2 x}\, dx = \frac{\pi}{2} \left[1 - \left(\frac{1}{2}\right)^2 k^2 - \left(\frac{1 \cdot 3}{2 \cdot 4}\right)^2 \frac{k^4}{3} \right.$

$$\left. - \left(\frac{1 \cdot 3 \cdot 5}{2 \cdot 4 \cdot 6}\right)^2 \frac{k^6}{5} - \cdots \right], \text{ if } k^2 < 1$$

721. $\displaystyle\int_0^{\infty} e^{-x} \log x\, dx = -\gamma = -0.5772157\ldots$

722. $\displaystyle\int_0^{\infty} e^{-x^2} \log x\, dx = -\frac{\sqrt{\pi}}{4}(\gamma + 2 \log 2)$

DEFINITE INTEGRALS (Continued)

723. $\displaystyle\int_0^\infty \log\left(\frac{e^x + 1}{e^x - 1}\right) dx = \frac{\pi^2}{4}$

724. $\displaystyle\int_0^\infty \left(\frac{1}{1 - e^{-x}} - \frac{1}{x}\right) e^{-x} dx = \gamma = 0.5772157\ldots$ [Euler's Constant]

725. $\displaystyle\int_0^\infty \frac{1}{x}\left(\frac{1}{1 + x} - e^{-x}\right) dx = \gamma = 0.5772157\ldots$

Index

(Note: Numbers in parenthesis refer to table numbers)

A

B

C